Heritage Traces in the Making

Traces Set

coordinated by
Sylvie Leleu-Merviel

Volume 6

Heritage Traces in the Making

A Communicational Analysis of Modes of Heritagization

Jean Davallon

iSTE

WILEY

First published 2024 in Great Britain and the United States by ISTE Ltd and John Wiley & Sons, Inc.

ISTE Ltd
27-37 St George's Road
London SW19 4EU
UK

www.iste.co.uk

John Wiley & Sons, Inc.
111 River Street
Hoboken, NJ 07030
USA

www.wiley.com

Library of Congress Control Number: 2024933514

British Library Cataloguing-in-Publication Data
A CIP record for this book is available from the British Library
ISBN 978-1-78630-944-0

Contents

Introduction

A Communicational Analysis of Modes of Heritagization

Le don du patrimoine (The Gift of Heritage) (Davallon 2006) followed up the exploration of the symbolic dimension initiated by the study of an exhibition staged as a media device. Observing, with Desvallées (1998, p. 90)[1], that the notion of heritage is a *practical category* (and not a concept), I proposed to move away from the quest for a "good" conceptual definition of this notion and to approach it, in the way that Lenclud (1987) studies tradition, in a "reverse filiation" sort-of way: in fact, it covers a range of "things" (material objects, social processes and complex entities), which are the mediums of a relationship to the past constructed by humans of the present who feel they have inherited from it. The concept of *heritagization* thus refers to the process that transforms these things into heritage through the attribution of a socially regulated status, thus instituting them as heritage objects that we have the obligation to keep, because they have value for the present and the future[2]. In Western society, since the model of social acceptance of knowledge is based on their scientific validation, this reverse affiliation is *scientifically guaranteed*. In order to identify the operations that compose this process, I

1 In the case of certain books, there is a discrepancy between the edition cited and the original edition; thus, when necessary, in order to ensure the historical accuracy of the information, the latter appears in square brackets after the former.

2 This approach to heritagization is clearly distinct from that proposed in the *Dictionnaire de Muséologie* (Mairesse 2022, p. 383), which has no entry for "heritagization" (neither does the 2011 *Dictionnaire Encyclopédique de la Muséologie*, edited by Desvallées and Mairesse), but which states that "heritagization is part of the musealization process". Such a reversal of the relationship between heritagization and musealization, while understandable in the context of a museum approach, is tantamount to reducing every construction of a symbolic relationship to "things" from the past to a technical operation of the physical or conceptual extraction of an "object" from its original context and assimilating all heritage objects to museum objects.

conducted a field analysis and also analyzed texts of a *practical, normative* or *prescriptive nature*, which operate, define and inscribe, in the thing itself, this new status, together with *scientific* texts that frame, describe, report or simply mention this transformation. In a second phase, focusing on an examination the characteristics of the relationship thus established, I formulated, then formalized, the working hypothesis that the recognition of these things as heritage objects amounted to the feeling of being in debt to those who had produced them. This hypothesis was accompanied by two sub-hypotheses: (i) these objects appear to remain the symbolic property of those who created them and who conferred upon them the value that we recognize in them in the present, which gives to the collective[3] that claims this recognition an identity in time and (ii) they must be kept in order to be transmitted, the recognition of a debt forming the basis of the obligation to give[4].

I had postponed the reexamination of the accumulated analytical material to a second phase in order to study the diversity of *production modes* of heritage status according to situations and types of heritage. After the study of the "symbolic" dimension of the process, about which I have just given a brief summary, I had intended to come back to the transformation of "historical practices", if I may borrow this distinction from Augé (1994), who explains that "the symbolism that presents itself as predating any practice, is itself blamed by the historical practices whose evolutions or mutations reflect its effectiveness and extend its influence" (p. 84)[5]. This book is the result of this initial project of a communicational approach to the "historical practices" aspect. However, given the evolution of the heritage category and the development of new modes of heritagization – most especially the

3 In general, unless otherwise specified, I use the generic term "collective" to refer not only to ephemeral groupings of individuals (the audience of an exhibition, for example), but to social groups or perennial communities that regard a thing as their heritage.

4 Back in 1978, Pomian (1987 [1978], p. 42) had noted that collectibles have no usefulness but represent the invisible and are therefore defined by the fact that they are endowed with meanings, that they are "semiophores" (something that is obvious in archaeology). On the other hand, the interpretation he has put forward more recently (Pomian 2003, pp. 13–14), saying that this link, from the 19th-century onwards (only), has been replaced by a sacrifice of the present addressed at future generations, seems to me much more open to debate, since it amounts to denying any transitivity over time: in the case of heritage in general – and intangible heritage or natural heritage now confirm this – it is because we have inherited things from the past that, since we then feel we have in a way received them as a gift which makes us in debt, we think we must spend in order to keep them safe, so that others can have enjoyment of them after us. While undeniably there is expenditure and sacrifice directed toward the future, this could be said to be on account of the value we recognize in these things from the past that makes us in debt.

5 I set out the reasons that made me opt to start with a symbolic approach in Davallon (2010, point 3).

arrival of intangible heritage – it quickly became clear to me that it was no longer possible to limit myself simply to a reexamination of the material gathered for the first phase[6]. I therefore decided to include in my investigation all of the changes and expansions that have occurred in the production of heritage objects since the latter decades of the 20th century and during the early decades of the 21st century. I therefore took a fresh look at this second aspect, which was thus also able to benefit from works that had been published in the meantime.

I.1. An unchanged objective: to understand the heritagization of traces of the past

Describing something as heritage does not mean recognizing it as a transcendent essence, nor does it mean denying it any quality or any operability. This is because the purpose of heritagization is precisely to endow it with specific values and meanings, since it aims to institute a symbolic "recognized" (in reality, attributed) status to objects ("things") of the past from which we feel we have inherited and which we therefore have the responsibility to conserve, safeguard or preserve[7], particularly with a view to transmitting them to future generations. This process, now fairly well established, corresponds to what, in sociological terms, Emmanuel Amougou defines thus:

> Heritagization could be interpreted as a social process through which legitimate social agents (or actors if you prefer) set out, through their reciprocal, i.e. interdependent, actions, to confer upon an object, or an architectural, urban or landscape space or a social practice (language, ritual, myth, etc.) a set of properties or "values" that are recognized and shared first by the legitimized agents and then transmitted to all individuals through the individual or collective mechanisms of institutionalization necessary for their preservation, i.e. for their sustainable legitimation in a specific social configuration. (Amougou 2004, p. 25)

6 It is true that the introduction to *Le don du patrimoine* (The Gift of Heritage), which, being faced with positions defending – or, conversely, stigmatizing – heritage, aimed rather to present a position asserting a benevolent neutrality, focused its attention on the object studied and did not mention (which now seems to me to be an unfortunate omission) the second part that I intended, at the time, to finish very soon after the first.

7 Reference should be made to the distinctions introduced by Micoud (1995 [1994], p. 33) between "conserve", "preserve" and "safeguard" according to the types of assets apprehended as common goods.

This definition puts the emphasis on the social actors and their social practices, whereas on my side I approach it in the meaningful functioning that serves as the basis for a symbolic operativity that we can describe, with André Micoud, as "symbolic work":

> I use the term of symbolic work for the activity of human groups that constantly reconfigure – a condition for the perpetuation of their identity over time – the representations they make of the world, of themselves and others, and of the other human groups. (Micoud 2005, p. 82)

I have had sufficient opportunities to return to the heritagization operations, which I described as "gestures", and have explained their content (see Davallon 2002, 2006, 2018 [2014]). I thus have no need to return to them at any great length. While the selection and arrangement of their elements may have varied somewhat according to the evolution of the situations examined, the definition of the "heritagization ideal type" remains unchanged: on the one hand, the *operations* designated by the gestures remain the same, with a few updates, which I will specify below, and, on the other hand, the *program* for instituting the social status of the object that articulates these operations retains its relevance, until proven to the contrary. However, for clarity of language in this book, I will very briefly summarize the *operations* covered by the five gestures insofar as they will be used as a basis for the analysis of the variations of this program according to the various modes. Having recently effected two updates concerning the last two actions (staging an exhibition and transmission), I will begin by summarizing the first three which are unchanged before presenting the last two.

1) Gesture of discovery: at the root of the feeling of "discovery" is the recognition of the values and meanings of the thing, which come from its origin and its past, by those who discover it for the first time. When a collective feels that this thing belongs to its own past, the members of this collective have recognized a value in it, and they generally show attachment to a thing. I call this recognition of *values*, which can be effected by people outside the collective, and this *attachment* of the members of the collective by the generic expression *mobilization of an interest*. The thing then becomes a *trace*. This interest is at the root of the feeling of "discovery" by the person who "recognizes" these values for the first time. Following publicization, during a visit, watching a documentary, reading a book or an article, etc., outsiders will be able to recognize these values and feel an attachment similar to that experienced by those who "discover" it. It is this process of circulation of representations that is at the root of *interpretive schemes*. This interest occurs in a context where positive representations about this past circulate in this collective and/or in society. It can also mobilize scientists, experts, politicians or audiences.

2) Gesture of knowledge production: while a minimum level of knowledge about the thing is necessary in order to recognize a possible value and to think of it as a trace of the past,

establishing this value and understanding it presupposes the *production of knowledge* to interpret it in order to establish what it really is (its characteristics) and the world that produced it. This operation can be effected by the collective, scientists or experts. It involves at one time or another the use of a reference science (belonging to the historical sciences, social sciences or natural sciences depending on the type of heritage) that guarantees the relevance and validity of the interpretations and therefore the knowledge produced. Attesting to the existence of the world of origin of the thing and the fact that it actually does come from this world, this knowledge has a decisive role in the recognition of the object as an "institutional fact"[8].

3) Gesture of declaration: the status attributed to the thing because of its interpretation, in order to be operative, requires a *public enunciation* of a performative nature, which therefore comes from an actor having social authority, and which is aimed at a generally wider audience than just the members of the original collective. Traditionally, the authority that thus *declares* heritage status is the state (the most common case) or an international instance such as UNESCO; it guarantees both the *status* of the object and the *truthfulness* of its representation, and the declaration is then generally binding, leading to protection. But a more sectoral authority – scientists, the collective, even the audience of various forms of publicization using an interpretive scheme – are able, not to "declare" the status of heritage *object* in the above sense, but to "recognize" and claim the heritage character of the "thing" and enunciate it publicly.

Box I.1. *Summary of the gestures of discovery, knowledge production and declaration*

The processes corresponding to the fourth and fifth gestures (access to the heritage object through exhibition and transmission to future generations) has been rethought in order to take into account both the arrival of intangible heritage and the emergence of new modalities of presentation to the audience on the basis of the extension of the two operations that underpin them (its safeguarding, or preserving in its present state and publicization)[9], insofar as these are better able to cover the variety of situations encountered from yesterday to today and to establish a link in time between the world from which the object comes, the present day and the world of the future, for all of those who recognize the values (and therefore the patrimoniality) of the object.

---

8 Searle (1995, p. 46) posits that "institutional facts exist only within systems of constitutive rules", which have the particularity that they "create the possibility of facts of this type" – in the sense, for example, that the rules of chess create the possibility of the game. It should be noted, however, that this constitutive rule is usually only the crystallization of conventions arising from social practices and the circulation of interpretive schemes.

9 Transmission stems from these two actions under the constitutive rule of heritage.

4) Gesture of maintenance (or preserving in its present state): this operation *aims to keep the heritage object concerned safe, that is, alive and in good condition*, no matter what the risks of disappearance may be. From conservation, it has evolved since the 1970s into *maintenance*, either of the material object or of the (intangible) process. In particular, it involves the state, scientists and the collective, but also the public through their participation in the upkeep of the cultural being, Heritage.

5) Gesture of publicization: producing media traces of the heritage object, in order to make it public, this operation *establishes a link* between the humans of the present, whether they belong to the collective or to the audience[10], and the humans of both the past and the future. However, for material heritages, this operation generally takes the specific, favored form of staging an exhibition; nowadays this uses the various media available, thus usually aiming at a *broadening out* from the interest of the initial collective to the scale of an *audience* that is more open from viewpoint of the society to which it belongs, space and time.

Box I.2. *Gestures of maintaining in its current state and publicization*

To account for the changes occurring in the operationalization of the gestures, I had recently compared two heritagization "regimes" (Davallon 2012b, 2015, 2018 [2014]), corresponding to two types of implementation of the program (respectively, the production of tangible heritage and the production of intangible heritage[11]). But a closer analysis of these implementations, in particular the use they made of knowledge, led me to follow Genette (1994, p. 23, No. 30) in identifying *modes* (possibly with variants) rather than "regimes": for a given situation, several implementations can indeed coexist, interact and combine together in a non-exclusive way, insofar as new modes are added to the old ones.

I.2. The objects studied: media texts as observables

In the introduction to *Le don du patrimoine* (The Gift of Heritage), a *text* was defined, following Umberto Eco (1985 [1979], Chapter 3), as a machine for producing meaning through cooperation between a linguistic *dispositif* (and thus indirectly its producer) and the one interpreting it. Additionally, I had extended this

10 The social ensemble, consisting of institutions and specialists, nonspecialists (audience), together with the various forms of mediation, can be defined as a "heritage public space" as a "place for the development of communicational actions" (Fèvres 2012, p. 34). For official instances, this gesture takes the form of a use aligned to an objective.

11 The program is a typical ideal (in the sense of Weber), resulting from an analytical reduction operated from a basis of situations and texts. What I call "implementation" is in reality the way this typical ideal exists through social practices, seen from this typical ideal.

definition of a text in natural language to complex ensembles that are meaningful, communicational and also social ("socio-semio-pragmatic", if I may use the term), these being exhibitions, the promotion of heritage sites, mediation documents, television programs or video documentaries[12]. This conception of a text, communicational and no longer linguistic, shifted the approach that we could take in regards to its functioning: the enunciation became decisive and, consequently, also the procedures that guided the reader's interpretation of a media *dispositif*, and no longer a simple combination of units of meaning that were shared to some extent because they were socially defined according to a code (or even a language). On the practical level, this *enunciation of a dispositif* took the form of a *writing* effected by means of a material inscription awaiting interpretation during a "reading". Since this point, Yves Jeanneret has extensively developed and set out in detail an approach to a text whose basis is quite similar, but whose wording is more incisive, because it is more complete. Thus, in *The Trace Factory*, he reiterates the three reasons that lead communication researchers to "not regard a text as a simple emanation of verbal language".

> On the one hand, no text exists only as a set of words, it is always deployed upon a perceptible materiality. On the other hand, all texts, even the most literary or mathematical, combine multiple codes; and finally, texts are presented themselves as immediately complex and concrete entities. It is only the eye of the beholder that can choose to divide them into shapes and elements, especially verbal ones. The perceptible, plural, concrete, complex nature of the texts conveyed by the media appears as soon as they are understood within the practical processes of social communication. (Jeanneret 2020, p. 152)

The consequence of this is that the sociosemiotic *dispositif* thus constituted functions as a text, as a document and also as a medium, which he rightly calls a "media text" (or a "media-text complex"[13]). Therefore, fully subscribing to this conception of a text, I adopt it here[14]. My initial classification of media texts involved in heritagization was initially formed from two groups: texts of a *practical*,

---

12 With the semiotician, we can class this as "syncretic semiotics" or "polysemiotics", because they combine multiple languages, sometimes multiple media, in a single enunciation.

13 "Mediatized communication is thus first and foremost a production of representations, embodied in texts, themselves configured in their formal properties by the features of their media" (Jeanneret 2008, p. 166).

14 While the concept of "discourse" introduced by Smith (2006) and Smith and Waterton (2009) presupposes a communicational dimension, it remains, in my opinion, very far short of a consideration of the media text.

normative or *prescriptive* nature and texts of a *scientific scope*. Over the past two decades, the development of documentaries and websites, coming in addition to the staging of exhibitions (and sometimes even replacing it), led me to identify three groups.

The first group brings together *texts that are direct operators of heritagization*. It is possible here to distinguish two subgroups: on the one hand, all of the texts *governing the process*, in particular prescriptive texts of a more or less performative or injunctive scope (regulations, forms, databases, classifications, inscriptions on inventories or lists, the granting of labels, funding, etc.); on the other hand, texts that are produced in the form of *scholarly documents* by specialists, by scientists or by anyone else involved in the heritagization process (scientific, technical, administrative dossiers, but also descriptions, maps, summaries of various kinds, recordings, etc.).

The second group includes all of the texts that make up the *abundant scientific literature* studying, discussing or mentioning the heritagization process, its practices, its characteristics, its actors, its products, its consequences and its evolution, in history, art history, sociology, ethnology, geography, information and communication sciences, philosophy, etc.

The third group is made up of all of the *texts that provide for the publicization* of heritage objects. Firstly, the heritage objects themselves, material or "intangible"[15], textualized by heritagization: the exhibiting of material objects, the publicization of enactments or denotations of intangible heritage, works describing the objects destined to constitute, or constituting, heritage, etc. Secondly, texts based on this first level of media texts: recordings, catalogs, mediation and promotional documents, websites, documentaries, reportages, etc.

Box I.3. *The three groups of media texts*

Even if these three forms of media texts are a priori intended to circulate in different circles (professional, scientific, cultural), the boundaries between these circles being in reality porous, all of these texts more or less form a system: whether by their aim, function, format, circulation, references, transformation and transition from one media to another; in short, by the whole process of writing, interpretation and rewriting that is involved in the circulation and transformation of representations, which Jeanneret (2008) calls "triviality". Also, in addition to their declared function, on which the above classification is based, all of these texts have *two types of operativity* that are variable according to their function.

15 The concept of "heritage object" here covers all heritagized "things", whether physical or processual.

– First of all, is an operativity relating to the heritage object itself. These texts, by offering an interpretation of this object as a *trace of a past social reality* (whether material or processual), present a body of knowledge about this reality that constructs, more or less directly, a *scholarly representation*. This scholarly representation has a crucial role. It provides the interpretation of the characteristics (observable or already known, intrinsic or circumstantial) of any potentially heritagizable reality, insofar as it is the index of what it has been in the past and insofar as it is recognized as a bearer of values. The scholarly representation defines the *heritage value* of the trace object; in other words, its *patrimoniality. The heritage object* is thus a trace and indissociably a scholarly representation. The second is therefore the necessary condition for the first to exist as heritage[16].

– Then, there is an operativity that concerns how, precisely, we should understand this particular cultural form that we call "heritage", the *heritage form*. We should not understand from this that the social world of heritage exists only in the form of the text: it is not a fictional world. It is actually made up of everything that can characterize it as a "social world"[17], and thus made up of everything that can be involved in the *social practice* of heritagization: interactions and collective activities, representations, common perspectives (sense of belonging and awareness of doing things together going in the same direction), but also communicational *dispositifs*, organizational frameworks, institutional regulations, uses, power relationships, etc.[18] This simply means that *media texts are the bearers of indices of this social practice*, either performatively, by contributing to this practice, or, conversely, by being the result of it, or, quite simply, by making reference to it: they contain information about it, they describe it, or they evaluate or discuss it. And this is why an analysis of these indices allows us to grasp the existence of variations in the social practices. But, in practice, how can we grasp these traces and indices?

---

16 On this point, I concur with Fraysse (2006, p. 124) regarding the monument, as he writes in his doctoral thesis: the heritage reality of monuments is constructed by mediation of the knowledge accumulated about it. But it seems to me that this ought to be applied generally and systematized to all heritage objects. The same idea of an indissociable link between object and knowledge, but without the assertion of a consubstantial character, seems to me to be present in Poulot's (2006, p. 16) definition of what he calls "patrimoniality", that is, "the perceptible modality of an experience of the past articulated to an organization of knowledge – identification, attribution – capable of authenticating it".

17 For more details, I refer to the discussion by Hammou (2012) of the notion of "world". Between "social world" and "art world" (Becker 1988 [1984]), he identifies a common origin that gives them their processual specificity, but at the same time their differences: the first corresponding rather to the approach of a "shared culture", the second to that of a "collective action".

18 In short, all of the social processes identified by Amougou in his definition of heritagization.

I.3. The sociosemiotic analysis of media texts as traces of practices

First, it is necessary to clarify a point of method. We are in the presence of two categories of traces. On the one hand, *traces of the past*, and I am seeking to show how these are "things" in the process of acquiring the status of heritage objects (heritagized beings). On the other hand, *traces of the social practices* of heritagization as the bearers of indices of these practices themselves. In practice, making heritagization an object of research[19], instead of heritage, amounts to studying the latter (the *traces of social practices*) in order to understand the former (the emergence of *heritage traces*). This way of proceeding meets two objectives: to grasp the operations that institute "things" thus recognized as heritage traces in heritage objects, and to be able to apprehend the changes in this process of institution. In other words, on the one hand, to understand the heritagization process (which was the subject of *Le don du patrimoine* (The Gift of Heritage)), and, on the other, to identify the various modes of this heritagization process (the subject of this book being *emerging heritage traces*). In both cases, the method is a sociosemiotic analysis of the heritagization operations from the three groups of media texts. This analysis is based, in this case, on an epistemic reduction of observations and texts chiefly by using the *narrative semiotics* model to formalize the operations that are constitutive of the granting of heritage status. It is based on the observation that the way of thinking about heritage is governed by the relationship "object (reality or process) *versus* subject (understood here in the semiotic sense of an actant who knows and who acts)" according to a dualistic logic (according to a naturalistic ontology, if we follow Descola), insofar as the subject isolates, defines and interprets the object that they regard as coming from the past, since it *is already there* when the subject starts to operate, in order to recognize values in the objects (and therefore to endow it with values) and to attribute to it the social status of heritage: in short, to heritagize it. From the point of view of the approach, this analysis therefore does not address media texts (the observables) either as works or even as a simple meaningful ensemble, but in that as they are traces, indices of social practices of heritagization[20]. Narrative semiotics here provides a relatively simple but robust tool, making it possible to formalize both the *operations* and the *program* of transformation that articulates them: to describe the representation present in these texts of the way in which the subject transforms the object. My use of narrative semiotics is therefore fairly close to what Ricœur discusses in *Temps et récit*, even though it is, obviously, for different purposes.

---

19 In the sense in which I defined "the research object" in Davallon (2004). Addressing heritage as an object of communicational research through its operations (gestures) and its social and symbolic operativity (Davallon 2006) has some proximity (although it part of a different approach) to what Heinich (2009), a sociologist, described, a few years later, as "heritage chain" and "heritage function".

20 I therefore subscribe to Jeanneret's project (2019, p. 3) to examine the question of "mediatized traces of social issues: traces produced via media *dispositifs* that claim to be reporting on aspects of society and culture".

– The texts are able to bear traces of the practice, insofar as, as he explained, the practice is "always already *symbolically mediated*" (Ricœur 1983, p. 113), and this is all the more important in the present case, since this is a highly structured process responding to an objective defined in texts.

– But these texts are not merely bearers of indices of practices, and they operate a "configuration" by means of "constructing a plot" (Ricœur 1983, pp. 144–146 and 276–301). What I call "articulation of operations into a program" thus corresponds to a series of operations (the paradigm) that is syntagmized to go from an initial state to an end state in the form *of a transformation program* of the object's status (its definition, its uses, its values)[21]. The first act of the modelization will therefore have to consist of an identification of the components involved in the *operations* (the "doing") of this transformation program, namely, the *objects* (the "actant-objects"), the *operators* (the "actant-subjects")[22], the *values* mobilized (the endowment of values) and the *relationship to time* (the modalities of its representation).

– Then, the media texts operate a "refiguration" of the practice, either directly in the use of operative texts, or through the metamorphosis of the representations resulting from the circulation of these texts. Indeed, if media texts are able contribute to the *heritage form*, they owe it to their variety, their number, their circulation in the form of documents interpreted, used, transformed and "distorted" (Jeanneret 2008, p. 87). Because of the circulation and transformation through the writing-interpretation-rewriting sequence, the heritagization program, in its entirety or piecemeal (operations, object, operators, values, relationship to time), will be able to function as an *interpretive scheme*[23], which will then be mobilizable in whole or in segments by social subjects, in order to interpret situations, objects, parts of media texts, elements of memory or representations whose heritage character is being assessed, with a view to possibly producing new media texts. We should note in

21 This is why I fully agree with Smith (2006) when she considers heritage to be a cultural process. But, it seems to me, Poulot (2006) is basically developing more or less the same argument.

22 In real life, these subject-actants (but the same applies to the receiver-actants) group together the various concrete operators who implement the actions. It is the latter that correspond to what the sociology of the network actor calls "actant" (Akrich et al. 2006).

23 On this concept of "interpretive scheme", I refer to Jeanneret's (2019, pp. 34–35) discussion on this. He reiterates the definition of the concept of "organizing scheme" that he formulated with Émilie Flon: "A structural construction that forms a link between the technical, semiotic and phenomenological dimensions of representation and for this reason defines certain determining conditions for the construction of the meaning from information within a *dispositif*" (Flon and Jeanneret 2010, p. 4). But it introduces a difference: while the organizing scheme finds its relevance in writing, the interpretive scheme finds it "in the reading and the production of knowledge". I use here the concept of the *heritage interpretive scheme* to refer to both forms: inscribed in media texts (in that it assumes an act of interpretation before the act of inscription) and mobilized by the receivers of the text.

passing that this now gives a decisive importance to the gesture of publicization. Then, institutionalized both through its mobilization to interpret and its inscription in new media texts, this interpretive scheme will in a way contribute to the institution of *the cultural being, Heritage*. In this way, all of the media texts will link scholarly representation and heritage form, with the result both that the heritage object is heritage and that the heritage form can cover all heritage objects. In other words, if I adopt the viewpoint of use, by their "triviality" (in the sense of Jeanneret), media texts have the effect that heritage objects and heritage forms form the "cultural being, Heritage", which is in the background of what functions as a generic *heritage interpretive scheme*[24].

In practice, the first part's approach consisted of identifying *regularities* in the indices present within a relatively representative set of the various types of media texts (the observables). The objective was then to assign to each operation, and also to the program as a whole, as many features characterizing them as possible, in order to understand what they taught about the way in which practices transform the status of objects. This explains why the "gesture" that emerged as the pivot marking the change from the status of "thing" to that of venerable object was that of the declaration; the other gestures supported it by preparing for it, and then by operationalizing it: recognition of the object and its study, then conservation and putting on exhibition. Although, in accordance with the traditional conception of heritage, this role of the declaration was expected, the fact of identifying the other gestures (and therefore of identifying an ideal type from the program)[25] revealed the way in which heritagization, as a social process, constructed a relationship to the

24 I call this "heritage interpretative scheme" *generic*, in that it derives directly from the constitutive rule of heritage. I will define this concept in more detail at the end of Chapter 1 (section 1.4). With regard to this point as a whole, it is still in line with what Smith (2006, p. 3) says when she considers the process is self-referential. This self-referentiality is due to a phenomenon of circular causality in which the social practices of the transformation of realities into heritage are the implementation of representations of practices that circulate in media texts related to these practices. This is one of the effects of this process that I addressed with the hypothesis of the debt and the temporal perspective in *Le don du patrimoine* (The Gift of Heritage).

25 Although I have kept the Weberian denomination "ideal type" here to denominate "heritagization", it is actually a production of what Jean-Claude Passeron considers to be a *semi proper noun* – in between common noun and proper noun; in other words, the designation "by typological nouns, of these collective, partially describable 'individuals', constituted by the 'historical individualities' deictically referenced as 'global configurations'" (Passeron 2006, p. 582). For a more developed presentation of this dual status of symbol and index (in the Peircian sense), see his chapter on "historical enunciation" (*ibid.*, pp. 361–383). For the relationship to the ideal type, see also "proposition 3.1" and its four scholia (*ibid.*, pp. 575–591). Veyne (2008, p. 118, n. 1) puts forward a summary presentation of this model.

past – different from that constructed by history and memory – by using certain chosen objects to which it attributes the function of mediators.

But, once the indices representing *what constituted* the process being studied, it remained, in the second part, to examine the *variations occurring in the implementation* of this process.

I.4. Understanding heritagization modes

In contrast to what had been studied previously, this second part turns its attention to the regularities in terms of *changes* affecting the various components: the objects (selection of new candidates for heritage status), operations (e.g. the fact that they can be rolled out in several stages or can overlap), operators (e.g. the arrival of new actors in almost every operation), values (e.g. the introduction of economic or social values), knowledge (introduction of new reference sciences) and the relationship to time (centering on the present of the enunciation rather than on chronology).

The examination of variations does not here take the form of studying particular social situations whose functioning we hear described, or recording the history of an evolution, or understanding the social dynamics involved in the social world of heritage. However, in the light of the effect of the activity of the social world constituted by the world of heritage, new regularities have emerged through media texts. These media texts then become points of singularity that suggest – or even explicitly declare or assert – the emergence of a new way of heritagizing, more or less distant from the previously dominant existing mode, more or less in rupture with it, sometimes even in conflict with it. The new way of heritagizing will either combine with the existing mode(s), influencing them and causing some of them to take a new direction, or diverge from them in the form of a *variant*, sometimes going on to constitute a new autonomous line of force taking the form of a new *mode of heritagization*, or they join together with other lines of force in a recomposition of heritagization modes, resulting in the coexistence of a number of heritagization modes or variants, or combining them. However, this range of lines of force involved social issues that can then be grasped through *certain* media texts which, through the indices they bear, are the *trace* of the implementation of social practices seeking to modify an existing line of force. As such, they constitute *enunciative events*, in that they make legible and visible the presence of what ostensibly presents itself as a new line[26].

26 These events can be regarded as "singular points", to use Deleuze's (1986, p. 29) expression. Indeed, I am borrowing from Deleuze's reading of Foucault's work a notion in terms of the relationship between "regularities", "lines of force", "points of singularity", "force diagram" and "institutions".

I.5. Emerging modes of heritagization

I will start (Chapter 1) by giving an example of how to carry out the sociosemiotic analysis of media texts in order to show how to identify tipping points between heritagization modes. But there is more to it, because the case chosen is also emblematic of the issues currently affecting heritagization, since the starting point is a documentary on the archaeological heritagization of traces of social practices (Aboriginal rock carvings) which leads on to other media texts, either scientific texts, or publicization texts (other documentaries), or operative texts, bearing a different type of traces, those of a confrontation with the social heritagization mode.

Then (Chapter 2), the search for tipping points leads us to first consider two major precursors of the "reversal" of the 1970s, as Nora calls it, which characterize the upsurge of social heritagization. The first of these precursors is the Inventory, which replaces traditional heritagization (in France, an administrative legal declaration for the purposes of protection) with a declaration keeping the traces of the object's characteristics by its inscription in the General Inventory of Cultural Heritage (originally the General Inventory of Monuments and Artistic Wealth of France). The second is the publication of *Les lieux de mémoire*. While, in the conclusion, Pierre Nora defends the traditional conception of a national heritage, stigmatizing the reversal constituted by the emergence of the heritage of groups, he nevertheless develops an approach to it (the "unfolding" of what the place is the memory of) that brings to light the memory dimension of heritage, thus covertly introducing the continuity of memory, as reconstructed by knowledge, in the background behind the rupture with the past that is inherent in heritage.

It then becomes possible to reexamine the tipping point constituted by the social heritagization mode's upsurge in the heritage world during the 1980s (Chapter 3). The analysis of the "emerging heritage" of the traces – in reality, of the remains, be they industrial, rural, urban, memorial, etc. – of a society in the process of transformation, called "transmission in action" by Henri-Pierre Jeudy, makes it possible to identify the major characteristics of the mode of this heritagization in the process of emerging at that time. An analysis of the various types of media texts that contribute to this heritagization mode, or discuss it, reveals two variants that correspond to two lines of force: one centered on the involvement of the population as subject and object of the heritagization; the other focused on the use and treatment of memory as a trace of situations experienced and reported by witnesses.

The sudden advent of the 2003 UNESCO Convention for the Safeguarding of the Immaterial Cultural Heritage (UNESCO 2018) then further disrupted the traditional (including ethnological) conception of a necessarily material heritage by creating the new category of "intangible" heritage. Sociosemiotic analysis shows that by constructing a scholarly representation designed to keep a trace for the future of social processes, this convention invented de facto a heritagization mode that combines the continuation of putting into practice these processes with this scholarly representation. This invention not only led to the creation of a new form of institutional heritage, but more broadly it profoundly changed the way of recognizing how thing did or did not "constitute" heritage, giving new weight to the heritage interpretive schemes circulating in society and more particularly in the media.

Finally, a mode of heritagization, relatively unnoticed until now, is in the process of emerging: complex heritage entities (Chapter 5). The originality of this mode, whose antecedents can in fact be observed very early in archaeology or ethnology, but which was to develop above all with the social heritagization to which it has more or less been assimilated, is particularly well illustrated by urban heritage that integrates heterogeneous elements into a new entity. But it is above all through the publicization of landscapes that we see how heterogeneous elements of culture and nature acquire coherence and value through scientific knowledge in particular. The novelty of this method of heritagization can be said to be the role it gives to the interpretive schemes circulating in society that define what we have an obligation to keep. The premise of a new way of thinking about heritage?

I dedicate this book to the memory of Yves Jeanneret, with whom I had begun to discuss of many of the questions addressed herein.

I would like to thank Sylvie Leleu-Merviel who, by including this book in the "Traces" series, actually led me to look at "heritage traces" in a new way and to deepen my investigation. Also, I would like to thank Maggie Wakefield for the English translation, and especially for her patience in the translation of the concepts.

Finally, my thanks to my editors for their careful reading of all or part of the book and their comments and corrections: Louise Julien, Joëlle Le Marec, Céline Schall and Cécile Tardy.

1

Analyzing Heritage Traces in Media Texts

The French documentary "*Australie: Le grand livre des Aborigènes*" (Australia: The Great Book of the Aborigines), and an episode of Arte's "*Enquêtes archéologiques*" (Archaeological Investigations) series, which presents an archaeological investigation of Aboriginal petroglyphs (rock engravings) in northwestern Australia[1], offer an excellent starting point for illustrating how to conduct a sociosemiotic analysis of the way in which certain media texts are bearers of heritage traces, and identifying how tipping points in heritagization methods can occur.

This documentary illustrates the development of *media publicization* (on television and the Internet), where documentaries are a much less traditional form of promoting material heritage than exhibitions or books. Moreover, since this publicization focuses less on the objects themselves than on the archaeologists' research process in interpreting the engravings as traces of their world of origin, the documentary highlights the importance of the *production of knowledge* which characterizes archaeological heritage: it is in fact the archaeological investigation that serves as a basis for the recognition of the heritage character of the engravings and therefore the need to apply for their inscription on the UNESCO World Heritage List (UNESCO 2021). It is therefore the bearer of the two levels of traces mentioned in the introduction. The primary level is of potential *heritage traces* constituted by the engravings from the moment their interpretation is initiated, and the secondary level concerns *traces of social practices* from this interpretation carried out by

1 This documentary is the 17th episode of the first season (2016–2017) of the *Enquêtes archéologiques* series (a second series of 10 episodes was broadcast in 2018–2019). The list of episodes is available at: https://fr.wikipedia.org/wiki/Enquêtes_archéologiques [Accessed January 24 2022].

archaeologists. Even before being publicized by the documentary, this interpretation fed into various media texts aimed at the scientific community and the bodies involved in World Heritage List inscriptions (the Australian government and UNESCO experts). However, these scientific or administrative texts echo other scientific texts dealing with the study and heritagization of the Aboriginal material culture, which either supports the archaeological approach and its method of heritagization or proposes a different approach; this introduces de facto a bifurcation towards a new method of heritagization.

1.1. The documentary, a media publicization text

The short TV format (26 min) of this documentary is part of the encyclopedia (in Eco's sense) of Arte's audience, since it corresponds to a category of program focusing on civilizations or heritage (monuments, villages, landscapes, gardens, etc.); this is strongly represented on this channel under the headings "Travel and discovery" or "History", or even "Science", which, moreover, remains accessible for a few weeks on the channel's website. I will begin by analyzing it in order to show how it constructs a relationship between audience and engravings through publicizing their interpretation. It has the advantage of being a mediatized mediation, between the archaeologists who worked on the engravings and the audience of the documentary, and is hosted by a presenter who is an expert in the field, archaeologist Peter Eeckhout.

1.1.1. *An original form of publicization: mediatized mediation*

The use of a presenter who is an expert in the field can be found in many other documentary series about heritage, for example: *Villages de France*, with photographer Emmanuel Laborde, *L'Héritage de Rome*, with photographer Alfred Seiland, *Paysages d'ici et d'ailleurs*, with science journalist Raphaël Hitier[2], *Jardins d'ici et d'ailleurs*, with landscape architect Jean-Philippe Teyssier DPLG, or *Habiter le monde*, with philosopher Philippe Simay[3]. The principle of an actor who is both a narrator holding the narrative thread and a representative of the interpretive activity of the audience is thus quite common. This actor plays the *mediator* role between the audience and the object in an in situ visit. The difference is that the producer of knowledge, who, most of the time, is not present in the latter case, is present here; therefore, the mediation involves three actors: the mediator (in this case, Peter Eeckhout), the producers of knowledge (in particular, the archaeologists) and

2 Doctor in Neurogenetics, journalist, presenter, columnist and documentary filmmaker.

3 Doctor of Philosophy, Assistant Professor at the École Nationale Supérieure d'Architecture, Paris–Belleville.

the objects (the engravings and their environment). However, the *Enquêtes Archéologiques* series goes even further in this communicational logic, since Peter Eeckhout is a Professor of Archaeology at the Free University of Brussels, specializing in pre-Columbian civilizations. As stated in the series credits, he goes out to meet colleagues so that they can present the excavation site, their work and their findings[4].

This episode opens with a presentation that establishes the relationship between the narrator (for the narrative)-mediator (for the communicational structure) Peter Eeckhout, and the audience: having identified the region of the excavations (Dampier archipelago in northwest Australia) and swiftly outlined the purpose of the episode[5], we see him arrive onsite, identify some engravings, refer to their importance, meet the protagonists (the archaeologists), and explain how the presence of the petrochemical plants behind him threatens their existence, before turning to face the audience.

Positioned as narrator, Peter Eeckhout explains to the audience what he discovers: engravings, archaeologists' activity, environment, mobilized lateral knowledge, etc. From the beginning of the presentation, the audience is therefore invited to share the view that the narrator is engaged in a *cognitive quest*, whose ultimate ideological and practical stake is an evaluation of the heritage nature of the engravings (should they or should they not be protected). As long as viewers have at least a minimal knowledge of the world of archaeology and heritage, they will recognize the implicit question behind the presentation given by Peter Eeckhout, which serves as a background for interpretation of the documentary by the audience: "Does the nature, and above all the value of the engravings, as brought to light by knowledge, justify protecting them by restrictive measures?". The future action strategy (in other words, the narrative framework of the episode) is thus established in a world of reference which is both that of archaeological research (interpreting relics) and heritage (does the value of the relics imply their conservation?); to put it another way, a cognitive program whose focus is whether or not the status of an object is to be recognized (i.e. a sanction, in the narratological sense).

---

4 This credit sequence is at the beginning of each episode: "My name is Peter Eeckhout. I'm an archaeologist and I'm going to take you out into the field across the planet. We will meet other archaeologists. Together we will follow their investigations and their latest discoveries. A great journey through time and history".

5 "In northwestern Australia, in the Dampier Archipelago, archaeologists are deciphering one of the greatest history books ever written by man. A story carved in stone that may well disappear. Here, millions of years ago, the earth created this strange cluster of red rocks. In this rugged landscape, Australia's first inhabitants, the Aborigines, left their footprint. There were to be a million patterns engraved here over time for 50,000 years. A unique heritage, now under threat".

The documentary begins with a title sequence that is common to the entire series, presented with a voiceover by Peter Eeckhout – against [a series of scrolling images of an excavation site][6]: "My name is Peter Eeckhout, I'm an archaeologist, etc."; then a presentation, still in voiceover, of the site with which the episode is concerned [geographical location and presentation of the topic of the documentary: map of Australia, then an aerial view of the archipelago, then views of engravings][7]; and, finally, the title of the episode ["*Australie : Le grand livre des Aborigènes*"].

Three image sequences (A) follow. From this point on, the documentary alternates between live speech in interviews and voiceover for the commentary.

(A.i) Presentation sequence of the engravings: commentary that develops the topic stated earlier [during Peter Eeckhout's arrival and image of engravings].

(A.ii) Presentation sequence of the archaeologists and their work, with an initial commentary – on [shot of rock clusters with archaeologists working]; then an initial discussion with the archaeologist Jo McDonald; then, again, a commentary that sets out the issues in the documentary: the threat posed to the engravings by industry; the colonists regarding the Aborigines as savages and not having tried to learn their history. However, petroglyphs are indices to the Aborigines' way of life and how it has stood the test of time – with [engraving of arrival of the colonists and photos of the Aboriginal way of life].

(A.iii) Sequence showing meeting with an Aboriginal park warden (Kenny Diamond[8]), who describes the various types of information given by the engravings for Aborigines.

Box 1.1. *The three opening sequences of the documentary*

6 I have put the description of the image accompanying the commentary in square brackets.

7 The commentary sets out the substance of the documentary; so, despite its length, I will quote it in its entirety: "Ah! this is extraordinary: we can see here, engraved on the rock, this big fish, and then right next to it, quadrupeds, we find these pretty much everywhere in this range of great massifs that are literally constellated with what we call petroglyphs. It is an extraordinary place, we find lots of species represented, past, present, kangaroos, emus, etc. It's the largest concentration in the world of this type of relic. It's quite simply extraordinary. But you only have to turn around to realize the problem: chemical and fertilizer factories, which not only gradually eat away at the territory occupied by the petroglyphs, but also, unfortunately, release contaminants that significantly affect the environment. For five years, archaeologists have been working tirelessly to decipher this great book of stone. Because, thanks to these petroglyphs, we have a little-known history that Jo McDonald and her team intend to retrace. That of the Australians from their arrival on the continent, fifty thousand years ago, up to their colonization by Europeans".

8 Hoping not to make a mistake in the transcription of the name... .

Peter Eeckhout's position is both functional and communicational: he is the narrator reporting on the archaeologists' quest for knowledge, and the mediator between archaeologists and audience. For the audience, he appears as a subject engaged in a quest for knowledge. Mediation is not only about knowledge, but about the activity of discovery. Due to this, he appears as a central *figure* who is positioned at the articulation between the syntagmatic dimension (the editing of the documentary) and the pragmatic dimension (the negotiation between text and reality). On the one hand, he is Peter Eeckhout, a human being who has an authoritative status in the scientific field as a university professor of archaeology, and who went to an excavation site in Australia, where he was filmed; he is the one to whom the enunciating instance[9] delegated the task of presenting to viewers the way in which the archaeologists have interpreted the engravings. On the other hand, in the documentary, he is *in representation* in the form of the narrator-mediator-subject who ensures the continuity and coherence of the articulation of the sequences and all the elements that make up the documentary, thus participating in its textual cohesion.

This *figure*, at the articulation between the editing of the documentary and the pragmatic position of mediator, is an interesting case of enunciative construction of a documentary that is focused neither on the relics, directly on the archaeologists, nor on a reconstruction of the past. But with this figure, the documentary plays, in a particularly marked way, upon the boundary between reality (the place, the actors and the activity) and media text (image, discourse, narrativity and enunciation); it thus sets itself apart from a work of fiction, in order to maintain the authenticity of what is shown and the veracity of the knowledge (Schall 2010; Schall and Lemaître 2010; Flon 2005, 2015).

The chief consequence of this interplay between reality and media text is that the documentary is a bearer of traces of social practices, in the same way as a reportage can be. I am leaving aside the fact that it is, by its structure, form and content, the trace of its own production; however, it is above all, at a first level, as I described above, the *inscription* of a singular mediation practice toward the audience by Peter Eeckhout. As proposed by the documentary's opening, any viewer who has an interest (a "predilection", as Yves Jeanneret would have said) for the subject or for this type of program will interpret these traces as the representation of real actions and concrete objects. All these traces are conditions for a successful reading, in particular by identifying the viewer with the Peter Eeckhout figure as the subject of

9 The presupposed recipient of the documentary whose role is taken on by all the participants who have negotiated their cooperation to bring the series into being, and, within it, this episode: television channel, production company, probably Peter Eeckhout, the archaeologists visited, the university, etc.

the quest. By slipping into the position of model viewer, this real viewer will *interpret* this complex media text as a cognitive quest that the mediator embarks upon with the archaeologists. The figure of Peter Eeckhout therefore represents me in what I could do if I were out in the field to undertake the activities that make it possible to answer the big question that the initial presentation presupposes (in other words, "can the nature and value of the object, as brought to light by knowledge, justify protecting it?"). In order to be able to answer, I would therefore have to follow Peter Eeckhout in his cognitive quest that focuses on a second level: the traces of the practice of interpretation of engravings by archaeologists. This is the figure that ensures continuity, based on the pragmatic guarantee of the authenticity of the portrayal, between Peter Eeckhout the "scientific authority" and Peter Eeckhout in representation: it shows, explains and comments on the quest of the Australian archaeologists, which, from the moment I take on this representation, means that I can regard the documentary as a publicization of the production of scientific knowledge and not as a fiction.

However, this quest by the archaeologists ultimately leads to establishing in what way the engravings are traces of the social practices and culture of the Aborigines[10].

1.1.2. *Interpreting the traces of a social world*

The sequence of meeting the Aboriginal park warden, placed at the beginning of the documentary, is used as an exemplification of the archaeologists' survey among the Aborigines to collect the memory of the purpose of the engravings, their function (ritual, utilitarian use, etc.) and the importance the engravings have for them[11]. However, this Aboriginal memory does not help us to gain an understanding of the *raison d'être* of the engravings, of their use through time. This kind of knowledge presupposes an interpretation of the various indices they contain, and this involves making use of lateral knowledge as a rule for interpretation. The quest, in other words the narration of the processes used by the archaeologists and the unfolding of their investigation, gives a summary of this process of interpretation, intended to bring to light its scientific and heritage stakes; this also enables us to read its semiotic nature.

10 This semiopragmatic process brings to mind *mutatis mutandis*, the process analyzed by Soulages (2005) on the topic of television news, following Verón (1983). The difference is that in the documentary, to avoid reconstructing Aboriginal life fictionally, it is the discourse of the archaeologists or the commentary of Peter Eeckhout that describes it.

11 As we will see in section 1.3.1: as a mark of the realization of the meaning these have for them.

The archaeologists' scientific quest provides the opportunity to present a relatively complex example of the interpretation of traces, due in particular to the nature of engravings which are *inscriptions*, and not just objects whose use – and therefore meaning – could be (relatively) easily established. In the documentary, with the exception of the maps, the engravings are almost exclusively figurative, but nevertheless remain enigmatic, because of their location, the presence of sequences, superimpositions, or certain complex compositions. For archaeologists, the interpretation of engravings cannot therefore stop at simply identifying what they depict; such an identification of motifs can only constitute a first descriptive step that is useful in creating an inventory and providing a basis for interpretation.

Each sequence recounts the activities through which each of the two archaeologists, Jo McDonald and Ken Mulvaney, solved an interpretive challenge posed to them by the engravings:

(B.i) Ken Mulvaney, to date the engravings and describe them.

(B.ii) Jo McDonald, to understand the change in the Aborigines' territory that occurred with the rise in water level as a result of the end of the glaciations, 20,000 years ago, and thus to confirm the existence of two ways of life, before and after this change in environment.

(B.iii) Jo McDonald, to find, on Rosemary Island, indices to the way of life of the peoples before the rise in water levels.

(B.iv) Jo McDonald, to interpret the engravings depicting water points during this period.

(B.v) Jo McDonald, to interpret the stone structures indicating a partially semi-sedentary lifestyle after the rise in the water level.

(B.vi) Jo McDonald, to interpret the shell clusters on West Intercourse Island, relics of semi-permanent gatherings.

(B.vii) Ken Mulaney, to interpret the engraving depicting the first English ship to anchor in the bay of Enderby Island in 1818.

Box 1.2. *The seven sequences of interpretation of the engravings*

To understand what they mean, it is necessary to know their *raison d'être*, which involves linking them to the practices in which they occur, to resituate them in their world of origin, in the history of the Aboriginal way of life and culture: the *social life of engravings* is only one element in the life of the people that produced and used them[12]. From that point on, the quest for the meaning of the engravings would take the form of an investigation into the history of the Aborigines. This investigation forms the heart of the documentary (its second part) that unfolds in seven sequences (B).

How can the interpretation of indices help build knowledge about the Aborigines? In order to answer, we must return in detail to each of the interpretation sequences to understand its process.

In the sequence (B.i), the archaeologist Ken Mulvaney cross-references the recognition of world objects, enabling him to categorize the petroglyphs with the material characteristics of the engravings; these enable us to assess their wear and tear as an indication of their age. On the basis of these two indices, he uses the second to move on to a relative temporal classification[13]. This allows him, given the number of petroglyphs available, to identify the existence of a contrast between the species depicted ("old" vs. "recent" species). However, we can immediately see that at this stage, such a conclusion has only inference value (abduction), since he has no rule available to him that can explain the contrast.

The whole question is therefore what hypothesis corresponding to this rule could be formulated? In the second part of the interview, Ken Mulvaney recounts how he hypothesized – from the fact that the oldest of the engravings depicted terrestrial animals, some of which had disappeared, and the most recent depicted marine animals – that the change in the fauna depicted was probably due to a change in the environment (shifting from terrestrial fauna to marine fauna). "The petroglyphs thus suggest that the Aborigines experienced a radical change in environment. So what happened?" asks Peter Eeckhout, following up this hypothesis. This actually corresponds to a new abduction (a "meta-abduction", as (Eco 1992 [1990], pp. 264–265) would say) advancing a rule that explains the first two levels of inference (distribution of species in two periods, the distribution being inferred from the state of wear of the engravings). If this rule is confirmed, the previous two levels of inference will be too.

12 Jo McDonald, answering the question of whether engravings are fundamental to understanding ancient Aboriginal culture: "Yes, rock art shows us how men lived in this part of Australia for 50,000 years, because a recent excavation tells us that they arrived here 50,000 years ago. It is crucial for us to be able to document this period in order to understand how people lived here. Over time, people carved the things that mattered to them in stone".

13 "So, we can date them relatively thanks to their wear and tear. The older they are, the more they have been eroded; the more recent they are, the more contrast they have" (Ken Mulvaney).

The consequence is the start of the quest for an adjuvant giving epistemic competence (the rule), which makes it possible to validate the chain of previous abductions. This second sequence (B.ii) takes place in the excavation house. The audience learns from Peter Eeckhout that scientists have actually established the existence of a climatic change 20,000 years ago[14]. The meta-abduction is therefore not only a valid hypothesis, but it is a verified hypothesis, because it is based on a robust rule, on scientifically established knowledge. The knowledge contributed by this allows Jo McDonald to model the evolution of the environment in the region studied[15], namely that the end of the Ice Age led to a rise in sea level, which itself caused a profound change in the environment[16] to which humans had to adapt. However, the quest that has been initiated does not stop there, because while the result does indeed fulfill the role of adjuvant in terms of the quest for knowledge on climate change, it raises the question of the precise consequences of this change on the way of life of the Aborigines of this region, and on this depends, ultimately, the validation of the interpretation of the difference between the engravings of old species and those of new species. It thus turns into a (new) quest for indices about the Aborigines who lived before the rise in water level, then those who lived after it, in order to reconstruct their history.

Sequence (B.iii) takes place on one of the islands (Rosemary Island), which are the part of the former desert that remained protruding when sea level rose, and is therefore apt to give indices to the way of life before the rise in sea level. The presence of face engravings, identical to those found on the island, located very far away from the former, indicates that the Aborigines were not attached to a territory

---

14 Opening commentary by Peter Eeckhout: "Through an underwater prospecting operation, and an extensive LIDAR campaign, airborne laser detection technology, scientists have established that the Aborigines of this region lived through the last Ice Age, as a consequence of which, 20,000 years ago, sea levels rose". There are two documentaries that detail this change, where it appears that this knowledge results from investigations on several points in Australia (*Australie : L'aventure des premiers hommes*. 1. *Les grands nomades*; 2. *Les grandes inondations)*.

15 "As you can see, the coastline is far away. Then we have hills that turn into islands. You can see how the coast advances over time. When men arrived in this region 50,000 years ago, the coast was 160 km away from where it is today" (Jo McDonald).

16 Peter Eeckhout summarizes the confirmation of the hypothesis as follows: "The territory of the Aborigines of this part of Australia has therefore been transformed. Between 50 BCE and 20,000 BCE, it was a desert. Then the sea rose, and this desert became a string of islands 6,000 years ago. This is the transformation that is illustrated by the petroglyphs".

and traveled great distances[17]. The reason suggested by the archaeologist for these journeys could be the scarcity of drinking water in a desert with a particularly dry and arid climate.

Sequence (B.iv) presents an investigation, conducted at the site of the face engraving, which reveals the existence of an ancient water point, now dried up. The interpretation of two traces, which form a clue, enables us to establish the link between the face engraving and existence of the water point, and leads to new inferences about the Aborigines' way of life. The first of these traces is the presence of a "calcium carbonate crust" in the ancient water hole, that it was possible to date[18]. This knowledge is produced by deduction, which makes the rule of the existence of drinking water very likely (bearing in mind what archaeologists have learned elsewhere from present-day Aborigines and from other archaeologists)[19], raising a new question which is answered by the second trace: the existence of engravings depicting "maps" that locate water points. Peter Eeckhout summarizes the archaeologists' reasoning, which is based on knowledge provided by other researchers:

> These water holes, now vanished, were thus essential to these great nomads in order to survive in the desert. But how could they locate them in this vast territory? It was again the petroglyphs that put the archaeologists on the path. These sets of circles were deciphered with the help of Aborigines in the 1970s. Several have been found on the mainland.

Jo McDonald then shows him, on the sketched reproduction of one of these "sets of circles", how to "read" them as a map[20].

---

17 Peter Eeckhout's synopsis: "Jo has shown that in this desert, before the rise of the water level in 20,000 BCE, the Aborigines were not attached to a territory. They traveled vast distances. She is now able to confirm this, because archaeologists have discovered faces, perhaps the oldest depictions of humans on the planet, several thousand kilometers from here, in the Queensland region. Identical faces have been identified in central Australia; and Jo has found the same depictions carved into the rocks of the region".

18 Jo McDonald gives the following interpretation of this: "There must have been springs and a water hole. We had this crust analyzed. It dates from 38,000 BCE. This means that this water hole was active long before the rise in sea level. That's why people used to travel around in the desert until this point".

19 The other documentary, *Australie : L'aventure des premiers hommes*, gives an idea of this.

20 "You can see here; they have depicted different types of water holes: here there was water permanently; those were less permanent; and these were certainly places where there was water only after the rains. These places had different names. And in traditional songs, the lyrics tell us how the ancestors used to move from one hole to another" (Jo McDonald).

Thus ends the two sequences (B.iii and B.iv), which are at the center of the interpretation, because they have made it possible to reconstruct the way of life of the Aborigines before the rise in sea level. Subject to definitive confirmation of the differences in terms of fauna, the connection between the world of origin of the petroglyphs and the history of the Aborigines is on the way to being established.

Answering the question of how the Aborigines were able to adapt to the new environment after the rise in sea level is the subject of sequences (B.v and B.vi). The investigation, using the same technique as before of interpreting the material traces, reveals a progressive semi-sedentarization of the groups. First of all (sequence B.v), the excavation of temporary habitats from more than 8,500 years ago shows the emergence of the beginnings of collective life. Next (sequence B.vi), a gigantic cluster of shells on another island (West Intercourse), dating from the last 4,000 years, indicates that the local tribes used to meet up periodically. Peter Eeckhout concludes this last stage: "Now attached to their territory, Aborigines will go on to live according to the principles of 'Dreamtime', in harmony with their new environment, for six millennia. An equilibrium that will be swept away by a cataclysm, the traces of which Ken found on the island of Enderby".

Sequence (B.vii) begins with the interpretation of an engraving that depicts a boat similar to the one (the *Mermaid*) on the painting by the Englishman Phillip Parker King, which arrived in Enderby Bay in 1818. At this point, the quest for knowledge on how to interpret the engravings and how to reconstruct the Aborigines' ancient history is complete. Moreover, it is by commenting on archival documents that Peter Eeckhout summarizes the disaster for the Aborigines brought about by colonization: the behavior of the colonists, dispossession of their lands, the Flying Foam massacre (1868), and, finally, the threat that hangs over the petroglyphs today[21]. Then we come to the final sequence (C), which presents the conclusion drawn by the archaeologists from their investigation as a whole.

Semiotically speaking, the interpretation of engravings is what Eco calls a semiotic production through "recognition"[22]: the meaning (the semiotic relationship,

21 On this question of conservation, see a relatively recent point on the destruction caused by the chemical industry (Bednarik 2017).

22 "A recognition process takes place when a given object or event, product of human nature or action (intentionally or unintentionally), made among the facts, is interpreted by a recipient as the expression of given content, either according to a correlation already foreseen by a code, or according to a correlation established directly by the recipient. This act of recognition makes it possible to identify the object as a footprint, a symptom or a clue" (Eco 1978, p. 167).

or semiosis) is operated by the receiver[23]. If these engravings were made by unintentional printing on a malleable medium, they would be footprints; however, they are intentional and moreover interpretable by recognition, and are therefore *inscriptions*. However, the conditions, resources and methods of recognition are by no means the same for the archaeologists and the Aborigines.

Concluding sequence (C) – this sequence takes place on a sacred site of standing stones, which has become the memorial to the Flying Foam Massacre (1868). It opens like the others, with a commentary by Peter Eeckhout – [with a shot of the arrival on the site]; the two archaeologists, Jo McDonald and Ken Mulvaney, explain why they feel they must commit themselves to obtaining recognition of the necessity of applying for inscription of the archipelago site on the UNESCO World Heritage List[24].

Box 1.3. *The concluding sequence*

The latter can compare (at least) the two occurrences, current and past, with a type (Eco 1992 [1990], pp. 249–250), since they possess both the practice of the occurrences and the type through their culture. These occurrences are therefore, for them, a form of *writing*, as shown by the example of the maps in which there is knowledge of the rules of arranging components to enable a representation of space.

---

23 By analogy with the linguistic sign, they could be considered as "formants" (which Jeanneret qualifies semiologically as "*ferment*") for which the interpreter is responsible for creating the content plan. According to another approach, less focused on linguistics and more communicational, Jeanneret (2020, p. 185) explains that the trace has a dual dimension, material and indexical. This conception makes it possible to get out of the difficulty inherent in the approach of the Peircian index, which is understood as a type of relationship to the referent, and *especially* as a type of sign; a difficulty that I had not picked up when combining under this term the materiality of the trace and the semiotic function of the index.

24 I quote in full from their discussion, which sets out the heritage purpose of their work: [Jo McDonald]: "This place really gives meaning to what we do, meaning to archaeology. And I consider that it is my responsibility as an archaeologist to use my learning and my knowledge as a scientist in defense of a place as fantastic as this. I really enjoy working there, you know, but I also feel a sense of responsibility". [Peter Eeckhout]: "But how can we protect this jewel?" [Ken Mulvaney]: "We have here a place of universal importance, because nowhere else in the world is there such a high concentration of petroglyphs as in this archipelago. Petroglyphs that are, moreover, associated with sacred monuments like this. And I really believe that the Australian government has an obligation to present this site to UNESCO in order for it to be classified as a World Heritage Site. This is the only way to preserve it for future generations".

For the archaeologist discovering them, they are essentially an enigma. While recognition of shapes (depictions of animals, for example) leads to the view that the engravings are indeed of intentional human origin, there is no question of interpreting them as writing, since their *raison d'être* has not been established: in this case, the rule that will permit their interpretation is therefore to be constructed. In order to do this, once they recognized them as such, with the significant incompleteness of their figurative dimension, the archaeologists initially called upon the memory of the Aborigines to learn more about the experience, the type and the rules of arrangement[25]. This approach was essential in order to understand the writing and to read the maps, but also to get an idea of the status of the different types of engravings. Then it was necessary for them, as we have just seen above, to implement processes of abduction and analysis of indices by cross-referencing several sources[26]; this allowed them to periodize the production of the engravings (according to their greater or lesser antiquity) and to define a certain number of the lifestyle features of the Aborigines explaining these indices. Therefore, being obliged to decipher inscriptions, they approached these as symptoms of changes in the Aborigines' environment[27]. Proof of this is the final analysis of the second change in the environment, which is more tragic than that of the rise in water level: colonization, of which the risk of the definitive disappearance of the engravings is only one of the secondary consequences; these are social, industrial and political, all of which provide the motivation to apply for inscription on the World Heritage.

On balance, it can be said that the archaeologists' analysis causes the engravings to function according to three complementary forms of semiotic function: (i) as a *depiction* of objects of the world, provided that the discoverer has available in their encyclopedia the features that permit recognition of the engraved form; (ii) as *inscriptions* intentionally left by humans in stone for other humans to "read"; (iii) as *an interpretation* of the indices they bear, such as their presence in a particular place, as noted by the archaeologist.

Through a recognition of an intentional trace used to signify something (as an inscription), of the spatial and temporal context in which they are found (place, medium and time when they are seen), of the mobilization by the model viewer of an encyclopedia, of the importance of the activity and the experience in which the

25 What Schaeffer (1987) calls, with regards to photography, the "knowledge of Archè".

26 This is found in Ginzburg's analysis of the indices, contrary to the reference that is often lazily made to an "index paradigm" that he himself has in any case rediscussed. See on this point Jeanneret (2020, pp. 189–202).

27 For a discussion of the approach to sources as symptoms rather than footprints of the past, see Morsel (2016).

latter is engaged when they apprehend them, the first two modalities of meaning above are indeed part of what Roy Harris (1993) defines as a *writing*.

In the present case, this writing builds on a salience between the object and its medium[28], but also on an interest from an interpreting subject. The object must first set itself apart from its environment, which is the case as soon as it is recognized as an unintentional trace. However, it must also have features that make it a trace that can be interpreted; in other words, it should have potential meaning for an interpretant.

1.2. Specificities of archaeological heritage

While I refer the earlier analyses to the sociosemiotic formalization of the heritagization gestures set out in the introduction, the first two levels (i.e. the first two layers of traces) that make up the documentary refer to two gestures that seem not to follow each other in the heritagization program (even though they have a circular causal relationship): the production of knowledge and its publicization. If in fact we stick to the ideal typical program of heritagization (the narrative schema of the transformation of gestures), publicization focuses on objects that have already been declared part of the heritage. In this case, it would be the promotion of the engravings, or the territory where they are located (e.g. in the form of tourist promotion of the place where they are to be found, after their inscription on the UNESCO World Heritage List)[29], which would drive the publicization.

What is unusual here is that publicization is not focused on the heritage object, but on its archaeological study and what that study tells us, as a consequence, about the production of knowledge. It is no real surprise that the practices are more complex than the schema that operates the ideal typical reduction. Therefore, the process of heritagization does not take the form of a succession of completed gestures but involves interactions between various practices that take place at various times, the gestures being fragmented and incomplete. This phenomenon can be observed especially in archaeology, where research is long, complex and subject to readjustments, depending on discoveries, the development of interpretation tools and the knowledge produced. Nevertheless, publicization of the production of knowledge (and not of the heritage object, as such) draws attention to the heritagization process itself and the specificities that it involves in archaeology.

28 What Leleu-Merviel (2017) calls “diaphoria”. This is because they depict figures (animals, faces, diagrams, etc.), the engravings stand out against a background (the rock) and have features that attract attention and that first pose the question of their presence and thus of their potential significance (Schapiro 1982 [1969]).

29 Like the *Petroglyphs of the Tanbaly Architectural Landscape* in Kazakhstan (inscribed in 2004). Available at: www.whc.unesco.org/fr/list [Accessed 20 February, 2022].

1.2.1. *Knowledge on the life of engravings*

We should remember first of all that the publicization effected by the documentary *Australie : Le grand livre des Aborigènes* is, like all publicization, primarily aimed at creating in the audience an impression of discovering new things. Even if viewers feel that the answer to the question of the conservation of engravings is justified by their value, the narrative structure of the documentary builds up suspense in terms of how to achieve it. Hence, the focus is on the archaeologists' steps to describe the existence and values of engravings and to understand their use by the Aborigines, so that the heart of the documentary is the story of a reconstruction of the past history of the engravings. If we look at the heritagization of the engravings, the knowledge produced by archaeologists is a *scholarly representation* of the engravings that introduces a separation between two moments in time: a reconstructed initial past social life, and their present-day life dominated by the risk of their disappearance. Between these two moments was a period of neglect, during which they were no longer used and began to be damaged or partially destroyed, which ends exactly with their archaeological study.

This scenario of two lives corresponds faithfully with that of archaeological objects[30]. The starting moment of the second life begins with the initiation of heritagization: the simple fact of regarding the object as a trace, if only to save it from possible disappearance, produces a temporal break between the previous life and a new life marked by a potential existence as a heritage object, even if the first life happens to carry on to some extent after the acquisition of this status[31].

From a heritage point of view, nowadays Aboriginal engravings belong definitively to the past, since today's Aborigines no longer produce them, as suggested by the Aboriginal warden interviewed in the documentary, who explains that they were created and used by his ancestors. This first life of *full usage* ended with the colonization that deprived the Aboriginal community of the Dampier

30 I am borrowing the concept of "social life" from the work presented in *The Social Life of Things*, edited by Appadurai (1986). In addition to its introductory chapter ("Introduction: Commodities and the politics of value"), I draw upon three other chapters: naturally, the classic "The cultural biography of things: commoditization as process" (Kopytoff 1986), but also "Varna and the emergence of wealth in prehistoric Europe" (Renfrew 1986) and "Sacred commodities: the circulation of medieval relics" (Geary 1986). On lifecycles in archaeology, despite their differences, three approaches lead to similar conclusions: Pomian (1981), Eco (1993) and Hammad (2015, pp. 172–175).

31 We will see a little later that this division between life according to heritage status and pursuit of the earlier life corresponds to the dividing line between archaeologists and contemporary Aborigines.

Archipelago of the lands that enabled them to live, and which led to the Flying Foam massacre, the engravings having at that point lost both their usage and their meaning.

In this particular case, the new life of the engravings thus begins with their discovery by the archaeologists. In the documentary, the archaeologists clarify that they started working on the engravings five years ago. This means that they more or less "found" them, at least in their form of material reality, because for one reason or another (for instance, the existence of lateral knowledge), they decided to study them, having the intuition that this set of engravings presented a certain scientific and social interest, and therefore deserved a new approach[32]. Whatever the date on which the study of the engravings began, the documentary shows that it is the knowledge of the first life (the production of the scholarly representation) that confirmed the heritage value of the engravings. Hence, its interest in understanding the *archaeological heritagization* mode, since it allows us to share (by a shortcut) the interpretation process leading from the discovery of remains, potentially heritageable traces, to the recognition of an exceptional heritage value composed of *epistemic, social and memorial* values that warrant, according to archaeologists, the status of a *heritage object*.

As all these values have their origins in their first life; the engravings undoubtedly have an autographic mode of existence in their new life, insofar as their value is intrinsically linked to the history of their production, as established by archaeology. We could certainly imagine that, for reasons such as ritual, for example, an Aboriginal group would carve new ones; this would then be a new period in the life of the engravings calling for a specific approach, as is the case for

32 In their 2011 draft report for the Australian Heritage Council, McDonald and Veth (2011) speak of 35 years of research and write that "this part of the Pilbara coast has been subject to decades of intensive archaeological investigation" (p. 22). A swift exploration of the "Australian Rock Art Bibliography extracted from the Rock Art Studies bibliographic Database for the Years 1841 to 2018", by Marymor (2018, 2019), reveals that the first mentions of publications of engravings on Dampier and Pilbara actually date from the 1970s, with authors such as Wright, Virili, Lorbalanchet, Clarke, etc. (the first of these even raising the question of their preservation as early as 1970). Bednarik (2008), for his part, recalls, during a UNESCO symposium in 2008, that he himself began making an inventory of them in the 1960s. Several scientists have denounced the constant opposition of influential politicians, pressured by industrial lobbies and contempt for the Aborigines (facilitated by the elimination of the original group and the greed of certain other groups), to the inscription of these engravings at national and international level (UNESCO World Heritage), while what we have here is (or was) one of greatest concentrations of engravings (amongst the oldest, if not the oldest, in the world) in the country, even in the world, as Peter Eeckhout reminds us at the beginning of the documentary (see, for example, Gregory (2009) or Gonzáles Zarandona (2011, 2013)).

paintings on bark, sand or canvas, even in sacred caves. However, from an archaeological point of view, engravings identified as ancient retain their autographic mode of existence; it is moreover this mode of existence, from the point when it was established as inherent in their status as traces, that has made it possible to regard them as indices to the ancient social practices of the Aborigines. At the same time, without the interest that supported the decision to produce a scholarly representation, the engravings would not even be traces, because they would not have been the subject of scrutiny, either scientific or heritage. Without the production of knowledge, beyond their mode of existence in Aboriginal memory, no redefinition of their status would ever have been considered. They would have remained simple things, subject only to the vagaries of nature and human activity. However, they have an allographic social mode of existence, which is due to the fact that they are a form of Aboriginal writing that renders them "texts bearing knowledge" for those who have learned to interpret them, and also the fact that the scholarly representation that gives them existence as heritage in scientific texts recognizes this allographic social mode of existence. Therefore, they are autographic in their material reality, but (doubly) allographic in their scholarly representation.

Even though they are not officially the subject of an application for inscription on the World Heritage List by the Australian government, recognized as heritage by archaeologists, they are publicized as such. When the archaeologists consider, in the final sequence of the documentary, that the Australian government has an obligation "to present this site to UNESCO for it to be classified as heritage of humanity", and that the reason for this is that "nowhere else in the world is there such a high concentration of petroglyphs as in this archipelago, and moreover associated with sacred monuments such as this one"[33], they are clearly responding to the question of whether the heritage nature of the engravings justifies the need to protect them by applying the generic heritage interpretive scheme.

Mediatization through the documentary then has a leading role, since it makes their value publicly known on an international scale, irrespective of the decision of the Australian government. Through being linked, knowledge, value and mediatization thus constitute a central issue.

1.2.2. *Issues around the heritage value of the engravings*

It is difficult to ask the question of the heritage value of an object without turning to the original text by Riegl (1984 [1903]): *The Modern Cult of Monuments*. This author draws the dividing line between "memory values", conferred by the monument's creator (when intentional) or by us for historical monuments (because it

33 The Aboriginal memorial to the Flying Foam massacre, where the Aboriginal community of the archipelago was eradicated in 1868.

belongs to the past), and "contemporaneity value" when we consider the monument as if it had just been created. From the resulting value system, the heritage world has retained three values as fundamentally "heritage values": the two "memory values", the *historic value* and the *age value*, together with the *artistic value* (former or relative); the use value having been more or less put aside, except when there is repurposing of monuments. More recently, two authors, Xavier Greffe and Nathalie Heinich, proposed, according to two very different but complementary new points of view, a review of these heritage values. Discussing the *Modern Cult of Monuments*, Greffe (1990) proposed an approach to heritage as a cultural property; therefore, from an economic point of view, that essentially takes us far away from the construction of a scholarly representation through archaeological investigation. Greffe introduced, in addition to the three values taken from Riegl[34] and alongside cognitive value and economic value, another value that is less to be expected from an economist: *social value*. The reason for this is that "men need the testimony of other men and that each era draws from those that preceded it the emotions that will allow it to create and produce" (Dupuis and Greffe, cited by Greffe (1990, p. 11)). The characteristic of social value can be said to strengthen *social identity*, either in the form of a national consciousness, as was the case, for example, at the time of the French Revolution; or, as he writes with reference to Lemaire, as "a consciousness of the past makes it possible to create a common identity for the present as well as for the future"[35]. This social value therefore stems from a societal point of view. Moving to Heinich (2009, Chapter 12), at the end of her analysis of the researchers' inventory process, she puts forward a "heritage axiology" consisting of the values of antiquity, authenticity, rarity, significativity (significance attributed by the researcher to an element as representative of a category or testament to a practice) and beauty. It could be highlighted that this axiology partially redistributes Riegl's values in line with the practice of expertise specific to the heritagization mode of the inventory[36]. However, Heinich (2009) presents them as implicit principles common to members of the same culture (p. 236) that allow them to recognize that an object has characteristics that correspond to heritage

34 "For Riegl, a monument has three dimensions: it is a source of history, it is an artistic creation, it represents time gone by" (Greffe 1990, p. 14).

35 But, he continues, "on condition, however, that everyone is very sure about it". However, as the rest of the text appears to express a few reservations toward Lemaire's thesis, it is not really possible to determine whether the above condition is Lemaire's or Greffe's! Especially since it is difficult to go back to Lemaire's text, the reference to this text having been abbreviated to (Lemaire, p. 31), while the bibliography lists two texts by Lemaire co-authored with Ost, but none by Lemaire alone. However, this imprecision does not change the fact that the role of the of strengthening social identity is well attested as being a characteristic of social value.

36 A mode which I will discuss in section 2.1. A wide-ranging discussion of the notion of "value" can be found in Heinich's (2017) work *Des valeurs. Une approche sociologique*. It goes without saying that I am dealing here only with heritage value.

characteristics. The values retained in this axiology are consistent with many circulating values (in particular, the values related to time and the social relationship with which the object is invested) that contribute to constituting the heritage interpretive scheme, thus conferring upon it its generality.

In the case of the documentary on the engravings, these approaches to value led to identifying three ways of implementing their values.

– The first, operational, involves professionals, scientists or the administrative bodies. The *qualities* that the object possesses, based on its characteristics, meet the "criteria" applied by the actors who have these competencies. The axiology developed by Nathalie Heinich, mentioned above, while it finds echoes with the interpretive scheme that allows everyone to recognize what heritage is, is above all a professional axiology. It focuses on a rationalization of evaluation that confirms (or, conversely, may give pause to) the expert's intuition of the object's heritage value. Here, it is scientific, due to the archaeologists who, after their investigation, provide evidence for heritage status for the engravings.

– The second modality touches on what I have elsewhere called the receiver's point of view[37]. In this particular case, this point of view is in the media coverage: it is not developed in a direct relationship with the engravings, but passes through the mediation of the documentary. The documentary, by making use of scholarly knowledge, experiences, representations, the relationship to time, the feeling of belonging and identity, gives an image of the engravings that accords with the axiology of heritage and thus confirms the intuition of the heritage character of the engravings that the viewer may feel thanks to the *heritage interpretive scheme*: the age of the engravings is not in doubt; there is agreement from all the archaeologists who have studied them, and the Aboriginal park warden on their authenticity. The investigation into their use by the Aborigines proves their significativity and they are recognized as belonging to what we call "rock art". As a corollary, it is through its participation in the circulation of the various media texts that the documentary contributes to the development of the features that define what heritage is, in the form of this heritage interpretive scheme, which also includes scientific or professional texts (or texts) publicizing heritage objects and social subjects' knowledge, experiences or representations of them. This second modality is therefore strictly ideological: in the sense that it forms a representational nebula, a background from which the feeling (or rather the intuition) that the engravings constitute heritage arises[38].

– The third modality can be said to be social in that it relates specifically to the relationship of the social subjects to the engravings. Paraphrasing Micoud (1995

37 Linked to the central role given to temporality and the effects of the material presence of the monument (Davallon 2006, pp. 84–88).

38 Here, we could say we have the beginning of a *media heritagization mode*.

[1994], p. 27), it could be summarized as follows: Whose heritage are the engravings? The documentary, like any document publicizing heritage, aims to create an interest in the spectator to make them recognize engravings as *heritage*; both as the heritage of a collective, but also, in a certain way, as heritage "for the spectator"; this is insofar as, being exceptional enough to be regarded as heritage of humanity, they make an impression on the spectator, causing – in representation (including through identification) – an attachment and an appropriation. Actually regarded as the Aborigines' heritage, it is in this respect that their study and recognition by archaeologists (including Peter Eeckhout) assign a social value to them; they also become so through publicization, in a very different way, for the documentary's audience. We thus have an example of the symbolic operativity of heritage, which, through the recognition of the value of objects by means of the heritage interpretive scheme, aims to create a broader collective bringing together various actants: the initial collective and an audience. As a result, this operativity of heritagization raises, in this case, the question of the relationship of the Aborigines to the engravings.

1.3. Aborigines and engravings: another heritagization mode?

The documentary, while explaining the production of the archaeological heritage mode, indirectly leverages two other heritagization modes. First, putting forward the idea of an application by Australia for the engravings to be inscribed on the World Heritage List makes reference to the administrative-legal mode: the government must express its desire to protect them and request their inscription on the UNESCO list as heritage of humanity. Then, focusing on the production of a scholarly representation and publicization draws attention to a major element that is missing: the group that regards itself as heir. Following the archaeologists' investigation presented in the documentary, we would have expected the engravings to be seen as the manifestation of an intangible heritage, since they are a form of writing that is transmitted by memory. However, that is not what happened.

European viewers can only wonder why there has been no reference to intangible heritage. It should be noted that when Kenny Diamond said he became "a national park warden in order to protect this heritage and understand his people's history", he described the use that his ancestors made of engravings, but did not mention heritage in the sense of patrimony. Instinctively, listening to him, we infer that this heritage is part of his people's heritage (in the sense of patrimony): applying our heritage interpretive scheme, we think that he is showing an interest (of the patrimonial type) in the engravings, and, possibly, in what they convey of his people's history. However, in doing that, are we not forcing the interpretation? We should place ourselves in the Australian context and seek a different interpretation in considering that not all Aborigines belong to the same community, and that in any case, they do not (necessarily) think in terms of heritage, *heritage thinking* being a specifically European way of conceiving the relationship to time – such as archaeology or history. Kenny Diamond's absence of recognition of the engravings as heritage (in the sense of patrimony) would then be an indication that, contrary to what we might instinctively expect, there is a divergence between the archaeological heritage mode and the nature of the connection that the warden has with the engravings and those who produced them.

In order to examine these questions, let us return to the heritage interpretive scheme and the three modalities (operational, mediatic and social) mentioned above: recognition of qualities according to criteria established by professionals; recognition of a heritage value that makes the object a being of language; heritage recognition by the social group, which is, in this case, missing. Each of these modalities is produced by a particular instance that can be observed in the media texts: the scientists who produce knowledge about the engravings; the producers or readers of the texts who mobilize representations of heritage value; the heirs of the group that produced the engravings. In the scope of the first two instances, the documentary finds itself caught up in different texts for which it offers a more or less explicit interpretation; and the same media texts, in turn, are used in its interpretation. It is therefore necessary to examine whether there may be texts that, if being linked to the documentary, would provide information on the third instance. Some of these texts could even give it a different status, which would then make them *enunciative events*, which would open up a new heritagization mode.

Two media texts immediately stand out because of their proximity with the documentary. The first is a two-part documentary on the Aborigines at the same time as Arte's documentary on the engravings, and which I mentioned earlier in connection with the two major periods (before and after the rise in water level). Entitled *Australie: L'aventure des premiers hommes* [Australia: The Adventure of Early Man], it provides information on the Aborigine's relationship with landscape features and with memory. The second text is a scientific text with an administrative purpose; it is the first release of a report written by Jo McDonald and Peter Veth for the Australian Heritage Council, entitled *A Study of the Outstanding Universal Values of The Dampier Archipelago Site, Western Australia* (McDonald and Veth 2011[39]). Its argument has a direct focus on the way in which the engravings and the site of the Dampier Archipelago meet UNESCO's "criteria", defined by Icomos

39 This report's title page is dated September 23, 2011. It can be found on the website Academia.edu, but does not appear in Marymor's bibliographic compilation (2019). So, it seems to be a working version (hence its date in square brackets). On the other hand, a 2009 article (entitled "Dampier Archipelago petroglyphs: archaeology, scientific values and National Heritage Listing") had discussed the inscription of the site on the national list with a view to its protection (McDonald and Veth 2009).

(UNESCO 2021), and used to determine the "exceptional universal values" (EUV) of an object[40]. It makes reference not only to numerous scientific texts (archaeological publications), but also to texts governing Australian heritage or the procedure for inscription on the UNESCO list.

1.3.1. *Archaeological interpretation of the Aborigines' relationship to engravings*

The report by Jo McDonald and Peter Veth is a synopsis of the scholarly representation of the archipelago's engravings, of which the documentary clearly publicizes only a small part. The report goes far beyond it on two important points: on the types of engravings, and on the relationship that Aborigines have with them. It thus constitutes the real *enunciative event* characterizing the archaeological heritagization mode, the documentary being only a shortcut.

Reading it shows that the documentary's focus was in fact on the presentation of the engravings under themes that the authors described as "economic" or "functional", as opposed to engravings with "ceremonial" or "abstract" themes:

> During the multiple phases of art production, there are high degrees of stylistic heterogeneity, technological innovation, sophisticated use of perspective, and extensive exploration of themes from the economic and functional (hunting, fishing and prey) through to ceremonial and abstract (dancing figures with accoutrements and large complex revelatory panels such as "The Climbing Men Panel"). (McDonald and Veth 2011, p. 39)

Furthermore, within the first themes, we mainly have engravings depicting animals; they do not show, for example, hunting scenes, fishing scenes or scenes composed of several characters[41].

On the subject of the Aborigines' relationship to the engravings, the report emphasizes the strong spiritual meaning of this site for the heirs of those who created the engravings: sometimes still living in the region, they maintain strong

40 Available at: www.whc.unesco.org/fr/criteres/. As the Dampier site does not appear on the World Heritage List, it is possible that it has never been the subject of an application, the government thus having chosen the industrialists rather than the Aborigines.

41 The spiritual approach and the social approach are thus absent from the documentary. For examples of such scenes, see McDonald and Veth (2011, p. 43).

cultural links with the site[42]. This is indeed the basis of the argument on the site's "exceptional universal values" (EUVs). The authors explain, for example:

> The Dampier Archipelago is part of a living cultural tradition and Aboriginal people hold that the engravings are the work of the Marga – ancestral creator beings – from the Dreaming. They are a permanent reminder of the Law[43] and retain their spiritual power. The petroglyphs are an inherited and ongoing responsibility and the Pilbara groups have songs and mythologies for many of the images depicted in petroglyphs on the Dampier Archipelago. Many of the images have cultural meaning over and above straightforward depictions and would likely have played a role in education and initiation. (McDonald and Veth 2011, p. 44)

Documenting the engravings, which was carried out using, as a basis, the memory of elders belonging to groups that regard the engravings of the archipelago as productions of ancestral creative beings, led the authors to claim that they all had a spiritual meaning:

> We have documented accounts from neighbouring Western Desert senior custodians (Veth since 1986; and McDonald and Veth more recently during the Canning Stock Route Project, e.g. McDonald and Veth 2011) that engravings are from the Dreaming (Jukurrpa) and represent both the travels, markings and embodiments of these ancestral creator beings. (McDonald and Veth 2011, p. 45)[44]

The Aborigines from this region (Western Desert) share with those from Pilbara the principle according to which "Ontologically, all engraved art (petroglyphs) can be seen as significant, tangible expressions of ancestral actions" (McDonald and Veth 2011, p. 45).

---

42 See McDonald and Veth (2011, p. 44). The authors cite, for example, a presentation by the Roeburne Art Group explaining how the Aboriginal cultural practice of petroglyphs serves as a basis for current artistic production: www.roebourneart.com.au/culture [Accessed February 15, 2019].

43 The Law is the knowledge that governs the Aborigines' lives, founded on the cosmology of the Dreaming, which teaches, among other things, everything that must be understood in order to survive in a very harsh environment.

44 This reference (McDonald and Veth 2011) refers to: McDonald, J.J. and Veth, P.M. (2011). The social dynamics of aggregation and dispersal in the Western Desert. In *A Companion to Rock Art*, McDonald, J.J. and Veth, P.M. (eds). Wiley-Blackwell, Malden.

From this, it can be seen that the authors have an approach to the Aboriginal memory of the petroglyphs that is in complete contrast to the approach we might infer from our heritage categories when watching the documentary, in particular those, mentioned above, that characterize intangible heritage. From this point of view, the report provides crucial information on the two universes: archaeology and memory. Its textual and editorial enunciation is unquestionably that of archaeologists and heritage experts, but it nevertheless gives a lot of room to the status that the Aborigines accord to the engravings and the approach they still have to them. Everything suggests that, while the documentary does not mobilize Aboriginal memory to any great extent, there was a deliberate choice; perhaps due to the complexity of the material, the textual volume needed to present the cosmological background, or the need to keep within the operational format of the series, whose topic is archaeological investigations and not heritage or the Aborigines either, unlike the other two-part documentary *Australie: L'aventure des premiers hommes*[45].

The difference in interpretation of the engravings by the archaeologists and by the Aborigines is clearly not concerned with what the engravings show: whether they are animals, humans (in profile, full-face, etc.) or objects such as a boat, the engravings are recognized according to the rules of depiction of world objects by the Aborigines, and with one or two difficulties by the archaeologists (and ultimately by the readers of the report). The Aborigines do in fact have definite advantages over Europeans: on the one hand, having the ability to interpret footprints, they have a skill in semiosis by recognition that goes beyond that of Europeans, as can be seen in other circumstances[46]; on the other hand, with the Law, they have the cultural rule (the code) of interpretation of stylizations and locations. However, the difference lies on another level: to qualify it in semiotic terms, I will say that regarding the engravings as productions of ancestral creators (those who created everything that exists) amounts to positing a continuity that runs from their creation to what they are today. This is not something that is missing, represented by a sign, but a continued presence of the thing that has been created[47]. We could put it this way: putting a durational symbolic continuity (or a numerical identity[48], as

45 So there should be a document that brings together the two documentaries and the report to put forward a complete media heritagization that remains at this stage the product of my interpretation as a viewer and reader.

46 Example of footprint analysis by Aborigines in *Australie : L'aventure des premiers hommes* dans le désert de Nullarbor.

47 This semiotic formulation seems to me to agree with what Descola (2005) writes about the Dreaming: "The Dreaming is thus neither a past remembered nor a retroactive present, but an expression of proven eternity in space, an invisible framework of the cosmos guaranteeing the continuation of its ontological subdivisions" (p. 207).

48 On this form of identity, see Genette (1994, pp. 27–31).

Genette would say) in place of the semiotic break which itself presupposes a difference between past and present, and between absence and presence, and between depicted reality and sign. Europeans are misled by the fact that they instinctively recognize archaeological or heritage traces in the engravings, indices to an event to be identified. If we infer that the same is true for Aborigines, we presuppose that they exercise a reflective activity on these engravings as part of an inherited culture. In other words, we project our *heritage thinking* – even if enriched by an intangible heritage, what is more, separate from us. However, assuming there has been no major misreading of what the authors of the report write, the cosmological meaning of the engravings lies not in the interpretation of indices that can be made by individuals in the present day, but rather in the fact that these individuals belong to the same cosmological group as the "trace" and its "meaning", because they all have the same origin, since their links stem from Dreamtime. This would also mean that the engravings, even if they are the work of ancestral creative beings, could have been produced and used (for example, interpreted as indices to the presence of animals) by humans in the past, just as they still could be in the present (with the production of new engravings, by new members of the group, as is the case for certain "artistic" productions), without their cosmological meaning – or rather their mode of existence, in other words, what they are – being altered. It is in terms of this difference that it seems possible to understand the conclusion of this part of the report:

> Some engraving sites, specifically, and all rock art sites, more generally, have varying degrees of significance in terms of their origin, mythological associations and their possible role as mnemonic devices. (McDonald and Veth 2011, p. 46)

We have seen that archaeologists, for their part, had to deploy a highly complex system of interpretation in order to be able to define the origin of engravings as "things" produced by humans of the past, and thus to be able to reconstruct the use that these humans may have made of them. The engravings are all traces that are meaningful because they are bearers of indices to this origin, which then need to be interpreted using lateral knowledge. This approach, based on a semiotics of the interpretation of objects as traces, will enable the production of knowledge external to the link that may exist between the creator of the object, the object and its interpreter. Contrary to what happens in Aboriginal cosmological continuity, it is a reconstructed (and not a given) continuity which will be found at the foundations of the heritage object, hence the crucial role of the knowledge produced.

We are thus faced with two symbolic regimes that are totally different because they belong to two "ontologies" (this word being understood here in the sense given to it by Descola)[49]: from the archaeologists' side, naturalism; from the Aborigines' side, totemism. It is this difference that confronts the archaeologists, and more generally the world governed by heritage thinking, against the Aboriginal symbolic order that conceives of objects in radically different ways.

1.3.2. *The engravings, "Aboriginal heritage"?*

As the report was written for UNESCO experts, everything suggests that the parallel presentation of the archaeological approach and the meaning that Aborigines attribute to them is in line with the normative rule established by Icomos-Australia (2013 [1979]) in the Burra Charter, which recognizes material heritage objects as having values that are qualified as *immaterial* (in other words, which can be said to be memorial and spiritual for the Aborigines): these values can then take their places alongside other heritage values. A traditional sacred gathering place (like the sites of raised stones) and, much more than that, a place as unique and eminently symbolic as the memorial to the Flying Foam massacre, can be seen as perfect examples. In any case, the affirmation of these "immaterial values" introduces another interpretation mode for these material traces, relics and the engravings. In this sense, it may announce the possible emergence of a heritagization mode different from those already met – or at least, of a consideration the *social values* of these archaeological heritage objects. However, this is a consideration based on what modalities?

A comparison of the report and the documentary *Australie : L'aventure des premiers hommes*, the latter being based not on scholarly documents but on publicization material, provides additional information on the "naturalistic" treatment of the Aborigines' comments. This documentary presents a broad synopsis of the state of archaeological research in order to retrace the adventure of the Aboriginal settlement of Australia, their environment and their way of life, from their arrival on the continent to the present day.

Hence, in a way it deals with the production context of the engravings and, in this regard, exemplifies the results produced by the archaeological investigation of the history of the Aborigines. However, alongside this contribution of knowledge on the Aborigines' way, which I have mentioned before, it is the way in which the Aborigines' comments are presented that I will pick up on here: for each of the sites presented, which are spread across the whole of the continent, the interpretation almost invariably involves anthropologists and Aborigines, with perhaps, in

49 "An ontology, for me", explains Descola (2017 [2014]), "is simply the instituted result of an identification mode, the particular form, identifiable in discourses and images, taken at this or that time, and in this or that region of the world, by one of the four continuity and discontinuity regimes [between nature and culture]" (p. 237).

addition, sequences of archival documents on the life of Aboriginal groups (or tribes?), many of which were recorded in the 1960s. However, it can be noted that the composition of the two media texts, that is, the report and the documentary, is similar: the general framework follows the scientific thinking of a presentation of the results of archaeological investigations, with, in parallel, a presentation of Aboriginal memory, and with the words illustrated by reproductions.

Therefore, a parallel is drawn between the discourse of the archaeologists and that of the Aborigines. However, while the archaeologists' discourse belongs to the scientific register[50], that of the Aborigines belongs to tradition. When they arrive at a site, the Aborigines address the ancestors; they give the name, the meaning of the place and its origin according to ancestral knowledge, perhaps making reference to the Law; they explain the way of life, techniques used in hunting or preparing seeds, the use of engraved maps and water collection, eel farming techniques, and the remains of dwellings; from looking at paintings, they speak about events that marked the life of the tribe, etc. Two methods of approaching and interpreting the same object or the same situation thus emerge, in two registers that derive from universes that are fundamentally different. For example, at the petroglyphs of faces at the Dampier site presented by Ken Mulvaney, who explains the style and composition of a face engraving and begins the process of dating them (25,000 years, perhaps 35,000), the elder with whom he is working, after having addressed the ancestors, comments (in voiceover): "There are many spirit descendants here and they are alive, that is why we must speak to the earth. These stones are very precious to us. They tell the story of creation, like the Bible. Seeing them, one understands the past. Aboriginal Law teaches all this".

Like the report, this documentary seems to follow the recommendations of the Burra Charter of Icomos-Australia, according to which the intangible values of a community or group (in this case, the spiritual value of landscapes, relics, paintings and engravings for Aborigines) are intangible values that form an integral part of heritage[51].

Therefore, this charter specifically aims to reduce the conflict between the archaeologists' approach to the engravings and the Aborigines' approach. According to the charter, all tangible heritage objects have a memorial dimension; therefore, it proposes considering the heterogeneity of what constitutes heritage (or memory) for the diverse social groups that exist in the country. The charter thus avoids any

50 Note that among the archaeologists interviewed, we follow Ken Mulvaney on the face petroglyphs of Dampier, Peter Veth on the search for water and Jo McDonald on the first contacts with the English.

51 This approach was clarified in the revision (1988) of the Burra Charter. The current version is the 2013 version, which includes the documents that were previously appended.

distinction between "heritage" and "places of memory", as does the French model, and brings together their various values (aesthetic, historical, scientific, social or spiritual) under the generic name of "cultural significance":

> *Cultural significance* means aesthetic, historical, scientific, social or spiritual value for past, present or future generations. Cultural significance is embodied by the *place* itself, its *fabric, setting, use, associations, meanings,* records, *related places* and *related objects.* (ICOMOS Australia 2013)

Moreover, for the same object, this cultural value can be different for different groups, or even for different individuals. Among the values that make up this cultural value, a special place is given to "immaterial" values or dimensions, such as symbolic or commemorative qualities, which are presented in the charter as the "meanings" of the place or object.

> *Meanings* denote what a *place* signifies, indicates, evokes or expresses to people. [Explanatory Note: Meanings generally relate to intangible dimensions such as symbolic qualities and memories.] (ICOMOS Australia 2013)

This charter therefore sets out to supplement archaeological thinking – and more generally, heritage thinking – with the Aboriginal symbolic order by taking *social values* into consideration. These values involve the relationship of social subjects to heritage objects, but more than that, they imply giving a particular importance to social subjects that belong to the group of origin of these objects. Hence, there is recognition of a new actant taking part in heritagization, an actor who is neither the producer of knowledge nor the instance declaring heritage status. Although the report gives a key place to Aboriginal knowledge, and with its arguments on exceptional universal values proposes a recognition of Aboriginal memory during the reporting process, its composition and content are within scope of the *authority* (in the double sense of author and institutional validation) of archaeologists. However, if a new heritagization mode is to be created – here, a *social heritagization mode* – the initiative should be passed down to the group; or at least, the group should have participation in the authority. Otherwise, there is indeed a new direction of thrust, but it remains in the shadows: relative and limited. The introduction of the social dimension is the beginning of a line that is in the process of being drawn between the material dimension (objects such as traces) and the anthropological dimension (the way of life of the group at the origin of these objects), but which is not able to detach itself to form a unique and autonomous heritage production mode.

The fact that, in the charter, the archaeological component remains structuring and the social value dimension is only an appendix, being without "authority", brings consequences for the traces, their interpretation and their interpreters. The first (and most visible) consequence is a standardization of the social value of objects under the heading of "meaning", assimilating, for example, Aboriginal cosmology (Law) to "symbolic or commemorative qualities" and positing it as identical to the collective memory of groups of colonists. To look at the scientific aspect alone: by removing the political dimension, such standardization merges a memory attached to places or events with the logic of a culture; it dismisses any difference between the two ontologies in favor of just one. However, even if using the categories of naturalistic ontology to understand a culture (all the more so if it is governed by another ontology, as is the case, here, with totemism, which is indeed the basis of the study of societies), that does not mean it always goes without a hitch, such as when it comes to giving a symbolic status, of a heritage nature (therefore of European origin), to objects belonging to this other culture – all the more so for another ontology. The second consequence is a perfect illustration of this type of problem: the archaeological investigation and the heritagization of artifacts risk de facto introducing a continuity solution between the Aborigines of old and the Aborigines of today; this is because the latter, as bearers of memory, become informers on issues predefined by the archaeologists. It is true that the report, like the documentary *Australie : L'aventure des premiers hommes*, avoids this risk in principle by affirming, on the contrary, the continuity of Aboriginal culture within the group by resituating places, objects and discourses in the thinking of the Aboriginal culture, in a manner distinct from heritage thinking. In a way, they "encapsulate" the Aboriginal universe (how could it be otherwise?). The fact that the choice of these places and objects is made by archaeologists inevitably transforms the memory summoned up and expressed into a contribution of lateral knowledge on the objects being studied; this integration of Aboriginal memory as heritage knowledge thus carries the risk of heritagizing the memory (i.e. the bearers of it), along with the places and objects.

These two consequences, given the contradictions contained within them, show the limits of the archaeological heritagization mode applied to objects belonging to a group, unless the group has the intention of adopting a heritage-related viewpoint, interest or purpose. For the documentaries' viewers, who are familiar with the heritage interpretive scheme, the objects presented "are" heritage. However, this will not necessarily be the case for an Aborigine; or rather, even when this is the case, and they express their culture in heritage thinking, these objects will not so much retain another "meaning" for them (as the Burra Charter suggests) as keep the group's own mode of social existence, the heritage interpretive scheme remaining an expression of this mode of existence for others.

1.3.3. *Challenging the archaeological heritagization mode*

The question of the relationship between culture (in the anthropological sense) and heritage is at the center of a series of articles, published by Laurajane Smith, Emma Waterton and Gary Campbell[52]. The major focus of this is specifically the critique of archaeological discourse as a heritage operator.

In a text written some time ago, dealing with the conflict between archaeologists and the Tasmanian Aboriginal Lands Council (TALC) that resulted from the film *The Last Tasmanian* – which suggested that all Tasmanians had been eradicated and therefore were no longer in existence – Smith attacks both the practices and the scientific imperatives of archaeologists.

These, she explains, are useful to the denial by the government authorities of Aborigines' claims to their cultural identity, insofar as archaeologists consider the elements of Aboriginal material culture as objects belonging to the past, and themselves define their value and meaning:

> Thus, archaeological rationality, emphasized by the logical positivism of processualism[53], became useful in defining populations (be they indigenous peoples or other groups) through both their "archaeological" past and the material culture (or heritage objects) which were defined as representing their past. Further, the application of 'rational' knowledge explicitly renders the social problems it governs as non-"political" and thus more tractable. (Smith 1999, p. 27)

Hence, the question was raised about this power of archaeology by some Aboriginal communities, as was the case during the conflict referred to above, which ultimately led to the cessation of some archaeological research in Tasmania and the restitution of elements of their material culture to certain communities.

This criticism of the archaeological approach can also be found in the discursive analysis of the Burra Charter by Waterton et al. (2006). I would note in passing that this analysis reinforces the idea that this charter has a structuring role in the way of thinking about heritage in Australia. However, more importantly here, the authors

---

52 For example, to name but a few: Smith 1999; Waterton et al. 2006; Waterton and Smith 2009; Smith and Campbell 2017.

53 If I am not mistaken (not being an archaeology specialist), this is an approach (which has its origin in English-speaking countries) that emphasizes the anthropological, functional, ecological dimension and uses measuring instruments (e.g. for dating) that allow improved objectivity. The report seems to agree with many of the characteristics of this approach.

show that the revisions to this charter (especially the 1999 revision) reflect an attempt (unsuccessful, according to them) to integrate societal changes and in particular the growing strength of the demands of the Aboriginal groups to exercise authority over their lands, and also to assert their identity. This confirms that this charter effectively initiates a bifurcation point that does not, however, go so far as a shift toward to a new heritagization mode. A careful discursive analysis of the charter, such as an analysis that looks at the modalization, identifies the reason for this: maintaining the difference between, on the one hand, active subjects (experts) and, on the other hand, subjects who have been consulted or beneficiaries who limit themselves to either recognizing values, providing information, or taking ownership of the representations, but who do not really take any action. Even though a new actant (the aboriginal social group) is actually taken into consideration within the heritagization process, the charter ultimately upholds the privileged position of the experts. This leads the authors to conclude:

> Although laudable and sincere attempts have been made to incorporate a greater sense of social inclusion and participation in the Charter's revision, the discursive construction of the Burra Charter effectively undermines these innovations. (Waterton et al. 2006, p. 351)

We might wonder if this state of affairs is not related to the division between two postures, which, as seen earlier, divides everything: from an ontological point of view, these are two comparable ways of thinking about the relationship between nature and culture, and from a societal point of view, they are two profoundly unequal positions. This is what the authors seem to be suggesting when they set the heritage rational of archaeology against the demands of the Aborigines.

In terms of the question of the bifurcation of the heritagization mode from an archaeological logic to a more social rationale, these criticisms should be placed in the context of a more general and more radical approach that aims to bring a new way of approaching heritage, which would be fully social. In the book edited by Emma Waterton and Laurajane Smith on *Taking Archaeology Out of Heritage*, their own chapter starts with an indictment of the pretention that exists within archaeology, within architecture and art history, to provide the tools for defining, thinking about, interpreting, managing and preserving heritage, as national heritage.

Instead of the knowledge derived from the above disciplines, which are the operators of what Smith (2006) calls the "*authorized heritage discourse*" (AHD), they put forward an approach in which "heritage is a subjective and political negotiation of identity, place and memory" (Waterton and Smith 2009, p. 15). They say that, in doing this, they are drawing upon the *re-theorization of heritage*, developed by Smith in *Use of Heritage*, in which she regards heritage as a cultural process or a performance “that is concerned with the regulation, mediation and negotiation of cultural and historical values and narratives” (Smith 2006), so that:

> Heritage becomes not so much the thing or place identified by the AHD [authorized heritage discourse] as “heritage”, but instead the values and meanings that are constructed at and around them – heritage is what is *done* and not what is conserved, preserved or managed. (Waterton and Smith 2009, pp. 15–16)

It is therefore an alternative to the usual definition of heritage (the famous “authorized heritage discourse”) which is thus asserted.

> Heritage becomes an act of communication [...] and a process of emotional and cultural engagement [...] that is about the assertion and mediation of historical narratives and collective memories, and the cultural and social values that underpin these.
>
> In this definition, all heritage is intangible, as it is redefined as a cultural process in which the values and cultural and social meanings that help us make sense of the present are identified and negotiated. (Waterton and Smith 2009, p. 16)

Moreover, in her 2006 publication, Laurajane Smith gives examples of the emergence of social processes of a heritage or memorial nature, led by groups that challenge the processes based on the authorized heritage discourse, or that, more radically, propose alternatives to these processes. Even if the re-theorization is about heritage and not about heritagization, it operates a series of shifts that de facto begin sketching out a new mode of heritagization: *social heritagization*. With a series of shifts: a shift in the main actor, the experts (archaeologists, in particular) being replaced by the members of the social group; a shift in the major operation, the subjective and political negotiation of identity, memory, values and meanings taking precedence over the conservation and preservation of material traces; a shift in temporality, the present being at the center, and no longer the past; a shift in terms of the object that constitutes heritage, the intangible dimension having more importance than the material object; finally, a shift in the science of reference, ethnography (and, in view of the methods used, probably more broadly the social sciences) replacing the historical sciences (archaeology, art history and history) or architecture.

1.4. To sum up: value of the method and exemplification of the approach

This chapter, whose purpose is chiefly to illustrate how the approach set out in the introduction could be applied to the examination of media texts in view of an understanding of heritagization modes, provides us with a series of lessons that will be examined in detail in future chapters.

A first set of lessons is about the functioning and function of documentaries. First of all, they show how the documentary *Australie : Le grand livre des Aborigènes*, as a media publicization text, reports on the archaeological interpretation process of the engravings: these, bearing in mind the Aborigines' comments, having been recognized as *traces of social practices of the past* by archaeologists who have produced scientific knowledge about the nature of these engravings and the context of their creation. The consequence of this production was a recognition and affirmation of their heritage character calling for one of the highest official pronouncements: their inscription on the UNESCO World Heritage List. Then, we saw how this operation of the documentary effected a *mediation*, especially through the figure of Peter Eeckhout, of the program of production of knowledge, characteristic of archaeological heritagization. The resulting scholarly representation is thus made up of concentric circles that start from a central core, formed of knowledge directly related to the engravings and the Aboriginal way of life, to the most distant and peripheral knowledge, such as knowledge about the rise in sea level or airborne laser detection technology; this is *lateral knowledge* that is mobilized to contribute to the interpretation of this particular type of traces, the engravings.

Second, we observe a shift in the central operation of heritagization from statement to publicization, with an increase in the weight of the latter as mediatization of the production of knowledge. In this case, it is due to the position of the Australian state supporting industry against the results of scientific research. However, we will see that, no matter the reasons, the production of knowledge, which up to now was used for the official statements, is going to be publicized. Its mediatization expands the circle of interpreters of this knowledge and thus arouses an audience's social interest in heritage traces that belong either to their own culture or to other cultures.

Third, we explain the emergence of a challenge to the traditional and institutional heritagization mode that contradicts the "authorized heritage discourse", to use the words of Laurajane Smith; in this case, that of archaeology and the Burra Charter. A conflict is therefore emerging between two heritagization modes, which call upon two different sciences of reference (archaeology and ethno-anthropology). However, the differences between the positions and practices of archaeologists, those mentioned in the Burra charter, those claimed by ethnologists and naturally those of the Aborigines suggest that these two modes are far from being monolithic. Each of these modes assembles several different registers of operations and actants (objects, social actors, sciences of reference, etc.), the combination of which produces *variants*, which define multiple ways of approaching the objects concerned and which should be examined carefully.

The conjunction of these three lessons leads to a dual hypothesis that serves as a common thread in this book. The development of mediatized publicization, especially through documentaries, can be accompanied by the emergence and circulation of new representations on what constitutes heritage, which would be related less to the traditional heritagization mode than to new ways of interpreting and building a relationship with traces of the past.

Second, these representations can be said to be *heritage interpretive schemes*. I argued, in the Introduction, that these schemes make reference to a *generic* form that played on the feeling of conformity to the constitutive rule of heritage, according to which, what "constitutes" heritage is a "thing" (material object or process) derived from a past, during which it was endowed with values, so that it must be kept in order to be passed on. I would like to show, in the following pages, how these schemes are based on a temporality that unites past and future in the presence of a sociality (within the relevant collective) or a complicity (with an audience) around an attachment to this "thing", the result being the emergence of a heritagization mode that mediatizes this attachment.

2

Interpreting Traces, the Principle of Heritagization

Gérard Lenclud's (1987) article on *"La tradition n'est plus ce qu'elle était"* ("Tradition is not what it used to be"), while it leads to a rethinking of the relationship with the past as a process constructed in the present[1], also invites us to consider the fact that in the late 20th century, in France at least, "heritage [was] not what it used to be", which in any case it probably never was. Because at the very moment when the concept of heritage began to come into common usage and to be mobilized professionally (thus, to establish itself as a recognized social *category*), it found itself out of step with "what true heritage was supposed to be" (at a time, let us note in passing, where the concept of heritage was not the reference). This crossover will serve as my starting point to explore the arrival of the "heritage" category, precursor to the extension of the concept with the arrival of "social heritagization".

Desvallées (1998, p. 90) notes that use of the "heritage" category has certainly been quite episodic, and often in connection with the more widely used category of historical monument – thus preparing for a long time its arrival on the cultural and media scene from the 1970s onwards – but that before the end of the 1970s decade, it had never really imposed itself as a general category covering monuments and museums, archaeology, libraries and archives. However, at the very time when it seemed to offer itself as the potentially unifying category of the various types of objects of the past that we feel we must keep in order to transmit them, its relevance, legitimacy and extension, considered unwarranted, were strongly contested. In the conclusion to his study of the notion of heritage, Desvallées (1998, pp. 102–105) offers an explanation for this: that it is not so much a scientifically established

---

1 This was the case when reading it became the starting point of the project for the project to study heritage rather than heritagization. (Davallon 2000).

concept, more an operational category linked to politics, administration and action whose meaning has gradually stabilized through usage. While in international institutions the concept was used from the 1930s[2], at least in France, he clarifies, this concept still remained highly divisive throughout the second half of the 20th century: supported, on the one hand, by "heritage" professionals who wanted to "open up the field, and that a change in language could only help to achieve [this] goal"; ignored, on the other hand, by the group who were "old-fashioned, who had no wish to open up its field of action and thus to change words" (Desvallées 1998, p. 103). There is no doubt that the concept arouses opposition. The reason why it is not neutral is nonetheless worth taking the time to study, since the phenomenon has also been encountered in English-speaking countries (Smith 2006; Harrison 2013). Would it not be a sign of a profound change in the way of interpreting the past and the emergence of a point of singularity in heritagization modes? It appears that in France the arrival of the concept of "heritage", its contestation, its recognition and finally its generalization were spread over a period which, during its intense phase, has extended over 40 years from the 1960s, but whose institutional effects can still sometimes be felt today. There is a growing agreement, however, to consider that it was during the 1960s and 1970s, with a tipping point of 1980, the Year of Heritage in France, that we saw the beginnings of a turning point both in the use[3] of the concept itself and in the way of thinking about historical monuments[4] or more generally heritage policies[5]. A turning point that I am tempted to put in parallel with the mutations affecting the world of museums, such as, for example, the completion of the *Musée national des Arts et Traditions populaires* project (National Museum of Popular Arts and Traditions), the beginning of its interest for the public, or the emergence of the new museology (Davallon 2019a).

This "installation" of the concept of heritage was accompanied by vigorous research activity and reflection; and so it gave rise to a particularly abundant

2 See Desvallées (1998, pp. 97–102). As part of his doctoral research, Leveau (2012) studied the archives of international institutions of the period, a presentation of which can be found in his book published in 2017.

3 Desvallées dates it precisely to 1968 (published in 1969 in the bulletin of the *Association Générale des Conservateurs*): "Heritage is the entirety of all natural or man-made goods without limit of time or place. It constitutes the object of culture. This dynamic, forward-looking concept, acutely manifested in the development of our civilization, is essential to the well-being and survival of Civilization. In addition to the mission of conserving and transmitting, it involves the protection and exploitation of acquired heritage and future heritage" (quoted by Desvallées (1998, p. 91)). Pierre Nora (2011 [2006], p. 110) dates this to 1970.

4 Bercé (2000) chose the 1970s in her book significantly entitled *Des monuments historiques au patrimoine du XVIIIe siècle à nos jours, or "Les égarements du cœur et de l'esprit"*.

5 According to Vadelorge (2003, p. 12), the pivot for heritage policies is found a little later, in the years 1978–1984.

production of media texts, so that there is, over the period, rich material for studying the way in which heritage objects have been redefined as traces of the past. But even before the upsurge of social heritagization, which created real upheaval since it raised so many questions on the traditional way of approaching what came from the past (Davallon 2018 [2014]), there were a quite significant number of early-warning signs that manifested themselves in particular through a set of administrative measures decided upon by the Ministry of Cultural Affairs, then led by André Malraux, to structure the recognition and protection of what, a little later, was to enter into the concept of "heritage": national natural parks (created in 1960); protected urban areas or historic urban areas (created in 1962); or the *Bureau des fouilles et antiquités* (excavations and antiquities office), along with the *Conseil supérieur de la recherche archéologique* (higher council of archaeological research) (created in 1964)[6]. Alongside these measures, one decision may have seemed at first glance a minor event, but was to be of major significance: the creation in 1964 of the National Commission in charge of preparing the establishment of the General Inventory of Monuments and Artistic Wealth of France *(Inventaire général des monuments et des richesses artistiques de la France)* by André Malraux, in accordance with a program designed by the art historian Chastel (1964)[7]. The creation of the Inventory did indeed shake up the French administrative legal tradition of the protection of monuments and national collections, while pushing to its extreme the logic of a systematic inventory of artistic wealth: "For work to be inventoried, it is necessary for it to have become visible", writes Malraux (quoted by Chastel (1964)).

A second, equally significant event occurred toward the end of the changeover period, between 1984 and 1992. This also contributed, in its own way, to a reconfiguration of the traditional conception of "heritage" (in reality, mainly historical monuments), while at the same time stigmatizing this reconfiguration. This enunciative event – the publication of *Les lieux de mémoire* ("Realms of Memory"), edited by the historian Pierre Nora – makes (national) "heritage" a subset of *Les lieux de mémoire* (Nora 1997 [1986], pp. 1430–1643), within the material places of memory of the nation, alongside "territory" and "state". To limit ourselves only to the factual level, unlike the creation of the Inventory, *Les lieux de mémoire* is a priori only a historical undertaking of knowledge production, aside from any operational purpose in the heritage field. On this level, by its scope, it has redefined the social, symbolic and political function of the traces of national memory that had until then had little visibility, either in the social space or in the world of knowledge. Heritage occupies a limited place in it; the section devoted to heritage brings together seven contributions focusing on the concept itself, the birth

6 Law No. 60-708 of July 22, 1960 creating national parks. Law No. 62-903 (known as the Malraux Act) on protected urban areas. Creation of the *Bureau des fouilles et antiquités* by decree on January 29, 1964.

7 Decree No. 64-203 of March 4, 1964.

of museums, the approach to French monuments under the Revolution, learned societies, the policies of Guizot, the creation of the *Inspection des monuments historiques* (inspectorate of historical monuments) and the restoration of monuments by Viollet-le-Duc; in short, with the exception of the first chapter which expands on the concept (Chastel 1997 [1986]) on a history of the turning into "heritage"[8] of French monuments and museums. But this section marks the adoption of the concept of "heritage", and above all makes it into a subset of the "places" of national memory, in this way putting forward a conception of heritage aimed at distinguishing it from other traces of the past.

However, what is most important is the commonality between these two events: they both give the key role to knowledge production, giving primacy to the choice and interpretation of the objects to be protected. They are therefore at the center of a set of enunciative events, introducing a redefinition of the frameworks of thought applicable to the material traces that may constitute heritage, but without directly calling into question the traditional conception which regards the administrative legal declaration instituting protection, conservation and transmission as the act of a "heritage" recognition of the monument or museum object.

2.1. The Inventory, a new heritagization *dispositif*

At the time of its creation in 1964, the Inventory was not yet an Inventory of Cultural Heritage (*Inventaire général du patrimoine culturel*) but a General Inventory of Monuments and Artistic Wealth of France (*Inventaire général des monuments et des richesses artistiques*) (Chastel 1964, 1990, p. 7). It was to become the former in 2004, when, responsibility for it having been assigned to the regions, "heritage" was then mentioned in the title[9]. Whatever the reason for the change of

8 The term "heritage" is used here in the current sense of "cultural (or natural) heritage", although it was not used as such at the time.

9 Article 95 of Law No. 2004-809 of August 13, 2004. "*L'Inventaire général des monuments et des richesses artistiques de la France*" (the general inventory of monuments and artistic wealth of France) takes on the name of "*Inventaire général du patrimoine*" (general inventory of heritage) in the Law of August 13, 2004. A general presentation of the Inventory can be found at: http://www.inventaire.culture.gouv.fr/. On the history of the Inventory, see Chastel (1990), Balsamo (1997 [1994], 2003), Toulier (1997 [1994]), Laurent (2004) and Melot (2005, 2012) and the special issue No. 1 (2004) of the magazine Présence d'André Malraux on "Malraux et l'Inventaire général" (see: www.malraux.org/pamhs1-2/ [Accessed March 18, 2019]). On the practical modalities of heritagization implemented by the Inventory, I naturally refer to Heinich (2009) who provides a particularly detailed analysis. For an overview of recent developments, see Toulier (2016), which also presents a history of major periods of the Inventory up to 2016. It should be noted that the reclassification as heritage of the objects in the Inventory results from the reconfiguration of the domain.

title, we can see that by this last date, the category of "heritage" had become commonplace: having come into common usage through being employed by associations and local authorities, it was no longer really controversial.

From its creation, the Inventory marked a profound change in the way of looking at objects from the past by proposing a new way of dealing with them. It recognizes as worthy of interest objects other than those that were customarily "heritagizable" (a term obviously unknown at the time), and at the same time, it puts in place new procedures. It thus introduces a new way of approaching and categorizing traces that can constitute "heritage", and also a new frame of reference for their interpretation. Balsamo (2003) notes how this change in point of view occurred on the three levels of this new relationship to objects: at the level of practices, standards and representations.

The most obvious change is that of *practices*, which covers ordinary (utilitarian) usage and symbolic usage (the ritual of visits or the transmission of values) of objects, thus bringing consequences for the entire heritagization process. It was reflected in the choice, alongside monuments, of vernacular objects, which at that time were sometimes called "small heritage" (ordinary and common objects and no longer only prestigious and exceptional ones)[10] and which moreover were often still in use. Directly related to these new practices, the second change is that of the *legal and administrative standards* arising from the policy: it is no longer a matter of either declaring a status (for example by "classifying" a building) or applying the "*pouvoir de protéger*" (power to protect) (Lamy 1992, 1993) – this may certainly occur subsequently – but the objective is indeed to establish an inventory. And, for good measure, an inventory that takes into account territorial characteristics, which are part of the history of their development (Lamy 2003, pp. 53–55).

The third change, which gives rise to the two previous ones, is the most important: the Inventory "lists, studies and makes known the elements of heritage that are of cultural, historical or scientific interest"; it thus introduces a heritagization mode that creates a complete rupture with the mode of historical monuments being based on the act of administrative and legal declaration, replacing that with the act of the production of knowledge about the object (and possibly its world of origin): a new modality that constructs a scholarly representation of the object that is *systematized* according to a standardized *documentary* format[11].

10 For example, the Inventory drew attention to "minor architecture" and "industrial architecture" (Chastel 1990, pp. 9–10), *considered* at that time as an index of change.

11 Available at: www/inventaire/culture.gouv.fr, under the tab *"Une entreprise de connaissance"* [Accessed March 16 2019].

These changes contribute to defining what Balsamo rightly calls a "heritage object"[12]. Beyond the broadening of the types of objects that can constitute heritage and their lack of hierarchy that such an expression pronounces, this expression acknowledges that recognition of an object is based on its *heritage value*, that is, as an element of the past in the present. It thus shifts from a purely functional and utilitarian mode of social existence to a symbolic mode of existence. So Isabelle Balsamo has good reason to stress on the importance of the change introduced by the creation of the Inventory: *by positing the principle of heritagization through knowledge production*, this creation does actually confirm that a heritage object is constituted not only of a material object, but also of a scientifically established *scholarly representation* – and even, as we will see, that this is what causes the object to be heritage. We can measure the extent of this change by looking again at the differentiation between the two modes of immanence of works as defined by Genette (1994, p. 23) following Goodman (1990 [1968]): the autographic mode of the material object and the allographic mode of the documents. The Inventory operates a major reversal, no longer assigning the key role to the exclusively autographic mode of immanence that characterizes historical monuments and museum objects, but to the allographic mode of immanence of documents that manifest the knowledge produced about it and its world of origin (in other words, the scholarly representation). Primacy is given to the interpretation, and no longer to the material mode of existence of the object, thus making visible the process by which this object becomes a significant whole that constitutes heritage, very specifically: a "heritage object" understood in the generic sense that I give to it and not only in the "works" in the Inventory.

It is this turnaround that I propose to examine. But before doing so, we must quickly come back to the usefulness and the validity of having recourse to a model developed for works of art for heritage objects and scientific documents.

2.1.1. *The Inventory, between autography and allography*

Material heritage objects being artifacts, both in this respect, have an autographic mode of immanence. The objects studied by the Inventory are also categorized as

12 "The 'historical monument', fruit of a selection by history and of social differentiation, is then replaced, for the first time, by the notion of heritage object" (Balsamo 2003, p. 417). Nora picks up this notion to contrast it with that of (traditional) heritage to identify objects (from the 1970s) that are "constitutive of our group identity and therefore endowed with an affective and emotional charge that transforms them into something completely other than objects of price, admiration or curiosity" (Nora 2011 [2006], p. 111). May I point out that I am using this concept to identify any type of object, tangible or intangible, resulting from heritagization.

"works" in the institution's documentation[13]. This is the case, for example, for architectural buildings, which will be discussed later, but the same applies to movable objects or all of the material objects inventoried. As such, by their presence in the present, they are traces of their world of origin and their characteristics can be interpreted as indices of their production. It is this that makes it possible to establish their authenticity and to extract information from them by treating them as witnesses to their world of origin.

It is the differentiation, but above all the link that the documents relating to the inventory process have with these material objects (these "works") that actually raise questions. In the Inventory, three types of documents are mobilized or produced: (i) documents of type 1, which are technical tools for data collection or which are used to present results (files, notes, listings, images, etc., initially gathered in manual "dossiers" and subsequently computerized[14]); (ii) methodological documents of type 2, which, upstream of the previous ones, define the prescriptions and procedures to be followed (administrative texts, manuals and recommendations); (iii) finally, tools for disseminating the results (type 3 documents), in other words, for publicization (works such as monographs, inventory compilations according to geographic-administrative or thematic areas, the various types of valuation accounts, exhibitions, websites, articles, etc.). All of these documents are of course not works in the sense of literary creation, being frequently of a different nature and generally bringing together several subdocuments; but they are the manifestation of a "work" of the intellect resulting from an original scientific activity and are thus media texts in the sense defined in the introduction. Especially the dossiers (type 1 documents) – and, consequently, the reports (type 3 documents) – which manifest what can be classed under the generic term "description" which is

13 For example, in the manual *Principes, méthode et conduite de l'Inventaire général du patrimoine* (Verdier et al. 2007). For an overall approach, see also Abriat and de Massary (2016). Subsequently, due in particular to the consideration of the production context of the objects, the artistic reference will become secondary to the point where the criterion of beauty will be set aside from the reference values (Heinich 2009, p. 219). I will keep the quotation marks when borrowing this term from the Inventory.

14 The writing is regulated by several systems ("descriptive system" and "national documentary information system"), and the nature of the documents formalized by the "data schemas": "Originally classed as 'DTD' (definition of types of document), the data schemas establish a modeling of the overall structuration of the information present in the dossiers. They make it possible to signpost the various documentary elements according to their function and their semantic content, by arranging them amongst themselves, independently of any form of reporting. Their use ensures that data is stored in the memory in a standardized and structured form that guarantees interoperability and longevity. Finally, they offer the possibility of processing data automatically, in particular to produce reports in various forms and media (paper, screen, disk)" (Verdier et al. 2007, pp. 215–216).

nothing other than a representation of one or more material objects (e.g. one or more buildings) and which evidence their heritage character. But, once it has been recognized that the material objects, on the one hand, and the documents, on the other hand, have an autographic mode of immanence for the former and an allographic mode of immanence for the latter, there remains the thorny question of which of the two modes of "immanence" is the bearer of patrimoniality (of what makes the object heritage, if you like). A question that brings us straight back to the conceptual definition of the "heritage object" or to briefly reformulate it in terms of the generic opposition between "material object" and "knowledge": is it the object or the knowledge that constitutes heritage?

We are touching here upon the change, probably the most conceptually profound, operated by the Inventory. When it introduces a systematization of the production of knowledge about the object, the Inventory seems to make only a minor change, which could moreover be anticipated by specialists who are sympathetic to a more scientific than artistic knowledge of "works"; it certainly does not call into question either the existence, or the importance, or even the primacy of the material object; simply, knowledge production becomes a regulated, standardized obligation, whose practical modalities it defines precisely (in the type 2 documents).

In doing this, it institutionalizes the action of knowledge production as a *sine qua non* condition of the recognition of heritage status of the object. Dissociating *scientific recognition* and the *administrative legal declaration* of status, it may seem to introduce a change with no major institutional issues, as long as the division between the inventory of wealth, on the one hand, and historical monuments, on the other hand, is maintained.

However, we should not be taken in by this minor mode, because it is asserting that scientific recognition is equivalent to an administrative declaration, and that it persists even if the material object is modified or even disappears. In a way, it transfers the numerical identity of the material object to the documents that describe it[15]. The rupture with the administrative legal heritagization of historical monuments is thus very striking. It opens up a real breach in heritage tradition, especially in France: heritage then becomes a matter of knowledge before being a matter of property or protection.

15 As Chastel (1990) himself mentions in passing: "On account of its descriptive and scholarly vocation, the Inventory cannot be confused with the historical monuments department and does not include in its interventions any process of protection or anything else: its first publication, the *Canton de Carhaix-Plouguer*, recorded for perpetuity a good number of typical works: bridges, barns... which have since disappeared" (p. 8).

Examining the heritagization process through the Inventory makes it possible to identify the consequences of basing heritization on knowledge production.

2.1.2. *Knowledge production and the typicity principle*

The central role of knowledge production in heritagization is accompanied by – or even based on – the introduction of a second principle that is likewise a complete rupture with the archetype of historical monuments: *the typicity principle.* Chastel, in his chapter on heritage in *Les lieux de mémoire*, describes the difference between the historical monument approach and that of "disused visual objects" which take on "value as sign of attachment, as indicator(s) of a laborious existence, as human revealer(s)":

> Farms, workshops, shops from times gone by, he explains, are becoming now what churches, sites and castles had been for previous generations. [...] To what extent do they enter into heritage? Through the typicity that is opposed to the uniqueness of the work of art, reply the specialists[16]. (Chastel 1997 [1986], pp. 1459–1460)

The introduction of this principle has a dual consequence. First of all, it considerably broadens the field of what can constitute art or heritage. Melot (2012) explains that "it is no longer a matter of going into the field to identify objects already endowed with their artistic or heritage quality, which we had simply not yet been able to see", but of moving on "to an undefined open research, which grows longer every day and whose configuration can be altered by anyone according to their origins, their tastes, their interests, in short, their culture" (p. 25).

Secondly, and perhaps even more fundamentally from the point of view of the functioning of heritagization, this introduction modifies the concept of what defines the object to be studied: no longer an object that is striking and significant due to its singularity, having the *unicum*'s role of semiophore, but an object that belongs to a collection whose meaning lies in characteristics common to several physical objects that will become as many occurrences of what defines this group. Therefore, it is indeed the relationship that the material object has with this collection that calls for study.

Everything suggests that, in making knowledge production the dominant action in heritization, in other words the action that de facto and de jure commands

16 Let us note the nuance that Chastel introduced at that time between an artistic approach to buildings and an approach to these heritage objects that can be recognized today as "ethnological" ("disused visual objects"). The reader will find a detailed and illuminating discussion of the unique/typical, exceptional/serial report in Heinich (2009, pp. 199–205).

recognition of the heritage character of the object, legitimizing its administrative inscription, in the absence of its administrative legal status, this action subordinates "the work", the object, to what defines the collection, insofar as it is only heritagized through the collection and the comparison that this makes possible[17]. Proof of this is the fact, mentioned above, that heritagization is no longer synonymous with protection and above all that the heritage object can continue to exist in the form of a "monumental document" (Fraysse 2006), even when the material object has been altered, or even disappeared. More than the document itself, it is the production of knowledge that is decisive, because this is what builds the scientific representation that defines the way in which *the collection constitutes heritage*, and is therefore the key to the relationship between the collection and the material objects of which it is composed. Now we grasp the importance given to the organization and the systematicity of procedures, and also the use of predefined grids and criteria, in order to ensure the validity of the result.

The heritagization process operated by the Inventory has been studied, described and analyzed in detail by Heinich (2009). There is thus no need to go back over it. Therefore, my topic will focus on the modalities of the *description* of objects, which lead to the production of documents; or, if you prefer, on how documenting can lead to heritage objects. This description, effected between material object, collection and documents, does in fact make it possible to understand the developing relationship between the autographic and allographic modes of immanence. In order to do this, the analysis will look at manuals (type 2 documents) that explain how to obtain a systematicity of procedures and formats, applied in particular to buildings – which were among the first categories of material objects ("works") to be inventoried. For ease of explanation, I will refer here essentially to one of these manuals – the one that deals with the process as a whole – entitled *Principes, méthode et conduit de l'Inventaire générale du patrimoine* in its 2007 version[18], the date of the transition from "monuments and wealth of France" to "cultural heritage". Defining the objectives, the procedures and also the rules of translating the "work" into the

17 The analysis of a large number of works makes it possible to draw original lessons. Having become an object of study, a current production is worth less in itself than through its inscription in families or series. This approach is rendered necessary by the expansion of the concept of work of art towards that of heritage. In return, it sheds new light on "major" works (Verdier et al. 2007, pp. 4–5).

18 This edition (Verdier et al. 2007), the easiest to consult, obviously postdates the introduction of the concept of heritage and refers to computer techniques alien to practices at the outset, but, although it sometimes specifies them, it nevertheless preserves the basic principles of the Inventory discussed here. The presentation of these principles and the procedure given by Toulier (1997 [1994]) confirms this.

scholarly representation, it allows us to identify the infocommunicational processes at work.

2.1.3. *Translating* realia *into documents*

In this manual, it is posited that each "work" (each building, for example) is obviously inseparable from its physical reality, which means that each "work" is taken into account in its autographic immanence, but it is approached as a source of information[19]. While it is indeed endowed with a numerical identity, linked to the history of its production, which gives it, among other things, its authenticity, the consideration of this history serves above all to describe and evaluate it, not with a view to protecting or collecting it, but to document it in order to learn about it and compare it with others. "Works" are indeed defined as observables, "units of study", which should be identified, listed, selected, classified and described, even before (and with a view to) making them known to various audiences. The central operation will therefore take the form of a *translation* of these *realia*, the "works", into "data ", then into *documents* designed to be preserved and published. Compared to the usual approach of using knowledge for historical monuments or museum objects, the Inventory operates a reversal of the relationship between object and knowledge: here, knowledge results from the study of a material object, to be formatted in documents, and no longer as a means, developed elsewhere, of knowing how to deal with the object, for example, how to restore or promote it. It is no longer an appendix or collateral, it is the very reason for the operation.

So what justifies talking about "translation" in this context? Beyond the data needed to identify, list or localize these study units, the objective is to compare them in order to identify similarities and differences, affiliations and exclusions, consistencies and ruptures in order to classify and "analyze" them, in the primary meaning of the word, in other words to identify their relevant features[20]. Nathalie Heinich gives the name of *"prises"* (captures) to this crucial operation of inputting characters, which at the same time relies on the know-how of experts and mobilizes

19 "The first document on the work is the work itself which thus becomes the central object of preoccupations, without mediation" (Verdier et al. 2007, p. 9). This principle is to be found in the first edition of the manual in 1968.

20 In the following lines, it will be less a question of identification and selection of objects to be studied, which in order to be decisive involves above all an expertise of the actors (these operations were carried out, at the beginning, through a pre-inventory often done by volunteers), than of the phase of the actual study of the selected works involving the process of documentation and "translation". I therefore place more emphasis on the process of interpreting objects than on that of simply identifying them as traces.

a "heritage axiology"[21] that serves as a cognitive framework for interpretation. The manual explains to us that this operation of description opens up the possibility of either collecting the works in "families", or classifying them according to "type". What is the difference between family and type? Families are a grouping based on a similarity of the buildings' function purpose[22]; they are formed "by all individuals of the same type present in a territory" (Verdier et al. 2007, p. 34). The type – type of "works" or thematic type[23] – is an abstract model, which is constructed "a priori and which we compare against the works during the investigation" (Verdier et al. 2007, p. 48); it is defined, according to the "Glossary" at the end of the book, as a "set(s) of characters organized into a whole and making it possible to distinguish categories of objects, of individuals" (Verdier et al. 2007, p. 217). The objective is then (i) to define *families*, possibly to break them up into groups, setting aside the *unica* when there are any; (ii) to establish, if appropriate, monographies (of "singular works" or of "selection of representative works"); or (iii) to produce collective dossiers on "one or more families of works". The principle is therefore to keep going back and forth, on the one hand, between classification and description, and, on the other hand, between family and type, so as to refine the description of the characters both as units and as sets; and, if appropriate, so as to also identify "groups (or classes)" within a family, or so as to determine the "works" to be monographed[24].

This translation operation is a *writing* in the sense understood by Goody (1979 [1977]): we will see how it produces a *scholarly representation* of what constitutes heritage in a given family (or possibly a "work") by generating "writings" that obey a standardized descriptive system of fields, captioned maps,

21 "Observation of the work of the experts made it possible to highlight the "*captures*" used to describe the buildings or objects: shapes, sizes, locations, inscriptions, decorative elements, colors, texts, dates, plans, lists, etc." (Heinich, 2009, pp. 233–234). They are certainly involved in the selection, but especially in the description of the properties. As for the values of the axiology of heritage, they are, according to the author: age, authenticity, rarity, significance and beauty (for the full discussion, see Heinich (2009, pp. 233–246)).

22 By way of example, the granite works or starch factories of the Vosges; the rural houses and farms of Taden in Brittany; or the double-span town houses of Castellane.

23 "The characters may concern all or some of the constituent elements of a work: types of work in the first case, types of elevation, overall composition, etc., in the second" (Verdier et al. 2007, p. 217).

24 The principle of typicity remains there in the background (as a horizon at least) even when there is only one work or when non-architectural "works" are concerned. This possibility of collecting, strongly present in *Principes, méthode et conduite*, allows us to glimpse the importance of the interplay between works and family, with the consequences that we will see shortly.

images (copies, photographs, sketches, etc.) gathered together as "dossiers"[25]. Documentation specialists may point out that, as such, this kind of procedure for writing and producing documents is after all hardly surprising for an inventory; but, in the present case, what matters is the way in which this writing operation contributes to the construction of heritage objects and their mode of immanence, going far beyond the mere fact of inventorying (in the general or technical sense of the term).

2.1.4. *The heritage object as a genre*

An effect of the principle of typicity is that analysis of the "works" that are collected, because they have the same function-purpose (forming a family), allows these "works" to be translated into a *scientific description* revealing in what way this particular collection constitutes heritage. This description consists of the heritage "characters" common to the works collected. Since the writing of these characters gives rise to documents, this description can be considered an "ideal object" manifested by these documents[26]. This swift summary of what defines the collecting of "works" and their scientific description explains how the question arises both of the nature of this collection and of its relationship to "works".

A collection such as this cannot but remind us of what Genette says about literary works constituted from multiple versions, because they have been transcribed from oral literature[27]. His theory, in opposition to Goodman's nominalist position, is that a work like this with multiple versions (e.g. *Roland's Song*) has *plural immanence* because its various versions constitute its immanence and the same applies to the "genre". Because of this it has a *generic identity*, which must be distinguished both from the "numerical identity" of works with autographic

25 Melot (2012, p. 38) also refers to Goody with reference to the tabular writing of the Inventory, which means that it not inscribed in a linear history and opposes any narrative, thus making it "the despair of art historians".

26 The root word "description" here means both the scientific work and its ideal immanence. Note that, according to Genette (1994), in an allographic regime, the ideal object is the immanence of the work, while the document is its manifestation. The operation of the ideal object is explained in Chapters 7 and 8.

27 He places these works in the category of "*Mouvances*" (Movements) (Genette 1994, pp. 224–226) and discusses the relationship between work and version in the point "*Entre œuvre et genre*" (Genette 1994, pp. 228–238). This mode of existence of works is part of what he calls "transcendence", which consists in a modification of the "immanence" (Genette 1994, pp. 185–186): in this case, the very fact of having immanence in versions, and not in an individual ideal object, as a single work would do.

immanence and from the "specific identity" of works with allographic immanence[28], because it is constituted from characteristics present in all versions, including the differences that may exist between these versions. However, an objection can be made to this comparison between the collection of "works" constituted by the family in the Inventory and a work of art with plural immanence: this stems from the fact that, in the Inventory, the description seems to cover only common heritage features, thus introducing a mismatch between the family as a class of material objects (therefore presenting differences), and the description that purports to define it (including differences).

Removing this objection will be an opportunity to identify what defines the heritage object here. The Inventory actually talks about "genre" when discussing the collection of the "works" of a family: these are then said to be "of the same genre in a study area", because they have the same function purpose (Verdier et al. 2007, p. 212, under the entry "*Famille*"). Unfortunately, this kind of purely functional and practical definition of the family would rather tend toward reinforcing the objection. However, when he describes the way the Inventory researchers work, Toulier (1997 [1994], p. 51) explains and completes this definition: while he admits that the genre conceived in this way allows only an "identification" of works, he nevertheless adds a crucial insight: "the *study* by genre," he explains, "tends to reveal the recurrence of the historical and morphological characteristics of the family" (my emphasis). It means, he adds, that, in the practical implementation of the investigation, after a survey of the study area, "only individuals deemed 'representative' of the family, based on purposive sampling, will then be selected". The importance of this clarification lies in its implication that the "genre", as defined in the Inventory, is not limited to a simple collection of objects that are chosen only because they have the same function purpose. A second criterion comes into play, based on the recurrence of historical and morphological characters that are specific to the family: a recurrence that is none other than that established by the "scientific description" of the characters. Two consequences should thus be noted.

– First of all, it should be put on record that it is not usage that, as is the case with versions of a literary work with plural immanence, designates the "works" to be selected here; even though obviously usage has in practice, over time, necessarily shaped their morphology. There is a difference here between a family of works in the Inventory and a literary work in versions. In the former, the selection is indeed made by surveying the family as a collection according to function-destination, but it is tempered by a repeated back and forth through recognition of a typicity according to the historical and morphological characters, which is regarded as

28 On the difference between numerical identity and specific identity, see Genette (1994, pp. 27–31), and on generic identity, see Genette (1994, pp. 233).

defining this family. This confirms, incidentally, the crucial importance of the researchers' work (the "*prise*" that Nathalie Heinich talks about).

– Consequently, this back-and-forth will reduce the gap that may exist between, on the one hand, the characteristics of the "works" that make up the family and, on the other hand, the features identified by the description (the ideal object). If the gap is too wide, the "work" is not selected because it is insufficiently representative, such that the family is ultimately a collection of "works" that are "sufficiently representative".

As a result, even though the "family" refers to two things that are out of step with each other (a class of objects and a description), its definition process is quite similar to that for the work with plural immanence: the sufficient presence of historical and morphological characters makes the family a collection that is not only functionally defined, but which also has common characters that, dominating the variations, thus give it a generic identity. In this sense, talking about it in terms of genre is therefore entirely justified. We only have to look at publications of the Inventory to realize that it is indeed the family of "works" (including differences) that constitutes the *heritage object* as conceived by the Inventory. This is why I discussed, earlier, the "scholarly representation" of the family to designate the ideal object, because it covers not only the description of common heritage features (i.e. the description), but also the characteristics specific to each of the "works" selected (the variations making the variants, if you like). It should also be noted that this slight mismatch has no practical consequences insofar as, as Toulier (1997 [1994]) points out in the article cited above, the Inventory intends to limit its action to an "inventory" process, in other words to establish groups of "works", which are indeed based on functional similarities and historical and morphological characters, but which are not the object of a thorough analysis, either as singular objects (with the exception of singular objects that benefit from a monography) or as sets of objects that correspond either to particular formal characters (according to the principle of the art history institutions), or to restrictive cultural characters as the ethnological survey by the Museum of Popular Arts and Traditions did for rural architecture[29]. In the latter case, it had been necessary to select the items studied according to both material constraints (site, techniques and materials) and practices (type of activity, lifestyle, tradition, skill, etc.) in order to be able to define a "type" of building (Cuisenier 1987).

Ultimately, the prior analysis of the relationship between the heritage object as a genre and the "works" that constitute it, however punctilious it may seem, shows the extent to which the Inventory is indicative of the plurality of modes of immanence of this cultural being, the *heritage object*. Firstly, we have seen that each of the

---

29 An emblematic example of this is given to us, for example, by Raulin's (1977) book on the rural architecture of Savoie. For a commentary on this issue, see Balsamo (2003, p. 420).

material objects, as a "work", had an autographic immanence based on its production history. Secondly, we have discovered that the description of common features on which heritage recognition used to be based was akin to a reductive process resulting in a scientific production (description as an ideal object) manifested through one or more document(s). Finally, we have just learned that the family as a genre, which brings together the above-mentioned "works", "immanated", if I may say so, in these works. This shows the extent to which the production process of the heritage object as a cultural being (in other words, heritagization) makes it a fundamentally hybrid being: it exists both autographically and allographically, and also, it possesses, somewhat paradoxically according to the model borrowed from Genette, an identity that is either numerical, or specific, or generic, depending on how we approach the characteristics of its production: numerical, if we consider its material production; specific, if we pay attention only to the features that define it as a meaningful whole; and generic, if we look at its ability to exist through several variants. This shows the extent of the dominance of the interpretation of "works" as indices of their own production, more or less relative to their world of origin through their functional and morphological dimension, which makes the Inventory an exemplary matrix of the role of knowledge in heritagization.

2.1.5. *The Inventory, matrix of a new relationship to heritage?*

The Inventory has thus contributed to – and continues to contribute to – the establishment of a conception of heritage that is in rupture with what could have been a simple extension of the conception of historical monuments or museum objects to new objects: it makes buildings a subset of heritagized "works". This has probably contributed to the reception, mixed to say the least, that it has always received from institutions that look after monuments and museum objects. A rupture such as this does in fact alter the economics of all heritage actions, leading to them putting forward a redefinition of the approach to artifacts of the past. Even if, with hindsight, this modification may seem like a relatively minor episode in the evolution of what is called "heritage", it has nevertheless contributed to producing heritage as a category, making it clear that *it is the scholarly representation that constitutes the heritage object*.

Sociosemiotic modeling shows the extent to which this process, by rethinking the art history approach, as Chastel wished, builds a heritagization mode quite close to the archaeological heritagization mode that we met in the first chapter, while going into less depth in the interpretation of artifacts as traces of a social world, since the purpose is essentially to establish a purposive inventory of the works, not to produce knowledge of their world of origin.

The central operation is the *standardized production of knowledge constituting a more or less plural object*, and partially on its world of origin, allowing us to authenticate it, interpret its properties as indices of its value and its origin, in order to assign a *status to it* and to *preserve the traces of the knowledge* thus produced in the form of *documents*. This operation clearly presupposes a *prior material production of the various* "works" that constitute this plural object.

This central operation brings into play an initial actant-object (the "works") and produces a second actant-object (the documents). The production of the "works" is, for its part, effected by a proto-actant-subject. The interpretation of these works as heritage traces recorded in the form of documents is effected by the actant-subject producer of knowledge (the investigators) to which may sometimes be added another actant-subject for publicization operations. The counterpart of the latter actant who effects the communication is the actant-receiver, the interpreter of the publicization.

The sender who initiated the program is jointly the political body and André Chastel for the scientific and cognitive organization. This sender, being also the one who takes care of its maintenance by deciding on operations, instigating actants and recruiting actors, is the political authority (government and regional councils).

Box 2.1. *Modeling of the Inventory's heritagization mode*

This modeling immediately reveals the consequences, the issues, but also the paradoxes arising from a focus on the production of knowledge. They mainly relate to, on the one hand, the origin of the object and, on the other, the publicization of knowledge. In terms of the relationship between the "works" and their world of origin, the links with volunteers from learned societies and associations that existed when the Inventory began (at least for the pre-inventory) gave the inventory operation itself a "historical" interest for social groups, in the sense that these links maintained (at least locally) an interest in the relationship between the "works" and their world of origin (Balsamo 1997 [1994], pp. 266–267; Ermisse 1997 [1994], pp. 286–288). However, progressive professionalization has refocused this operation on the description of objects and on a mainly documentary activity that falls within an administrative and technical framework. Looked at from another side, disseminating the knowledge produced is one of the stated objectives of the Inventory; it has therefore effectively set up dissemination tools: publications, exhibitions, conferences and now websites[30]. Nevertheless, Ermisse (1997, p. 287) noted, in 1997, the need to "reinvent a link with the public", dissemination through exhibitions or conferences having been forgotten after a few years. Since regionalization, websites have been created by the regions and the data are used for

30 Research and publicization are moreover linked in the presentation of the Inventory's publications (Gros et al. 2007).

tourism communications, but we cannot help but note the absence, or at least the great weakness, of the Inventory's media coverage. The fact that the latter is in charge of its own "restitution of knowledge" (Toulier 1997 [1994], pp. 154–155) brings the consequence that its publicization is inevitably very limited and essentially documentary. For example, unless I am mistaken, there has never been any mention of recognition of "works" or a family of "works" by the Inventory in documentaries on heritage broadcast on television. And yet, it is the circulation of representations and values that produces a collective large enough to create a social interest that extends beyond just the circle of people interested a priori in what potentially constitutes heritage or the circle of tourists in a given geographical area. Moreover, Ermisse (1997 [1994]), in the conclusion to his presentation, asks a series of questions on the Inventory's input into "heritage awareness", to "the evolution of mindsets and sensitivities", to "public policies in favor of heritage"; summing them up them in the question of "the real impact of the heritage mobilization that the founders were hoping for" (p. 288).

However, there is no doubt that the Inventory, through the use, whether declared openly or not, that has been made of it by heritage management structures (e.g. regional parks), associations, tourist communications, the media or simply (but an important use) by amateurs, has helped to create informed opinions regarding ordinary heritage (as opposed to monumental heritage), and consequently to increase social interest in it. It is even possible to go as far as thinking that the obligations and administrative constraints relating to conservation may, in some cases, have been replaced by a kind of virtuous circle based on the circulation of the representations and values attached to "works", arising from the fact that in scientifically recognizing a family of "works" as heritage, the Inventory was making a public statement of their value. According to this schema, the more publicization about the family is developed, the more obvious the value of each "work" will become, the greater the chances of public interest being boosted, and the more each owner will be invited to recognize and boost the heritage value of the "work" they own. In other words, social interest in objects from the past being one of the effects of "recognition", the likelihood of owners discovering the value of the objects they own would increase proportionately. However, in order to be operative, a virtuous circle such as this presupposes that epistemic value be transformed into aesthetic value and above all into social value, apt to create a positive horizon of expectation on the part of the collective. It is on this condition (necessary, but still not sufficient) that the owner, living their lives as a member of this collective, would perceive themselves as the custodian of an object with socially recognized values, whose symbolic, and possibly economic, value may have increased. If this were the case, the supremacy of scientific recognition over the administrative legal declaration that is characteristic of the Inventory, which may have seemed like a weakening of heritagization – even a questioning the constitutive rule of heritage as things that we have an obligation to keep (in their current state, might I add) in order to transmit them, linked to the fact that a declaration traditionally orders, through protection, the act of conservation – would then become a force in that it would contribute to a communication ecosystem relating to the heritage value formed of several actants implemented by many actors. For example: an epistemic actant (with the scientists who are the producers of knowledge and of administrative recognition); a communication actant (with publicization by media professionals); a political actant (with the promotion of the works of a territory by associations, activists and elected officials); and a cultural actant (with the constitution of a collective in public, consisting at least in part of members of the social group, but also by media and tourist receivers). In the absence of research on the subject, I do not know what the reality is regarding both the existence and the operativity of such a virtuous circle on the conservation of vernacular heritage. This is especially the case since the Inventory fully admits the disappearance of works (Toulier 1997 [1994], p. 154), taking the view that their conservation passes, in some way, through what can be described as a "documentary memory", de facto making its own documents into "monumental documents" (Fraysse 2006)[31].

Whatever the truth of the matter, the fact remains that the primacy given to the production of knowledge gives – in theory, at least – a central role to publicization and social interest, resulting in a kind of dissemination of patrimoniality. From this point of view, its social operativity is aimed at producing media texts that contribute to the definition and strengthening of the cultural heritage being. In terms of the heritage object, first of all: the fact that it is an object with plural immanence that happens to be the repository of patrimoniality has the corollary that the characteristics possessed by each "work" then call upon those of the other "works", involving on the part of the receivers (reader, visitor or viewer) a comparative approach between them. Rather than visiting a "work of art" to be apprehended in all its singularity, the circuit or itinerary model predominates.

From the receivers' point of view, the temporal dimension (belonging to a period) is then relegated to the background. The aesthetic value (the relationship of pleasure linked to discovering the object) and the social value to some extent take precedence over the historical value. Visitors' potential encounter with a set of "works", publicized as such, will result in a transfer of their value to this territory, especially since the Inventory will take its place alongside the action of associations, heritage institutions, tourist action, media communications, etc. This development of a communicational ecosystem of heritage (Greffe 1990, 2014), between the representation of works and the transfer of their symbolic capital to the territory, is perfectly illustrated in programs such as *Des racines et des ailes* (Roots and Wings)

31 It should be noted that the above reasoning is in line with the logic of the Inventory and therefore does not consider what the interested parties (the inhabitants) consider to be possible assets. In the Inventory, heritagization remains above all a matter for specialists.

or *Villages de France*, not to mention all of the other documentaries whose structure is based on a territory's discovery trails. Patrimoniality becomes a priori both unstable over time and unlimited in its extension, since existing heritage objects can be transformed over time, new objects appear on the same territory and new categories of objects enter the Inventory[32].

Although the Inventory thus reveals the role of knowledge production and its publicization in heritage, it still remains, as an institution, dependent on the French Government and its administration.

2.2. Heritage in the face of the interpretation of national memory

Unlike the Inventory, the approach proposed by *Les lieux de mémoire*, a scientific media text published between 1984 and 1992[33], is as far removed as can be from the approach of "heritage" specialists (curators, architects, art historians). Its project aims to make intelligible the production and the development of "places" that crystalize the national memory, with the particularly ambitious objective of beginning a new history of France[34].

It certainly makes us wonder what lessons an undertaking such as this may bring in terms of modes of heritization. Indeed, deducing from the presentation of the section entitled "Heritage", which begins with the statement: "Heritage: in a sense, the word could have covered this entire book...", that heritage and places of memory are intrinsically linked, that, for example, heritage could be the medium or the result of the creation of a place of memory, would be a grave mistake. After listing the (national) heritage institutions (*Archives, Bibliothèque nationale, École des chartes, Comité des travaux historiques, Société de l'histoire de France*, the inspectorate of monuments and sites, learned societies, national museums), the author immediately

32 This virtually endless process of resumption and extension, noted among others by Melot (2012), is very clearly criticized by Jean-Louis Jeannelle (2014).

33 The whole is composed of three major works, spread over seven volumes. The first was published in 1984, the second, consisting of three volumes, in 1986 and the third, also consisting of three volumes in 1992. All were brought together in a "Quarto Gallimard" edition in 1997. I quote from this compilation edition, putting in square brackets the year of the original edition of the referenced article.

34 A perspective asserted from the introduction with the proposal: *"Les lieux de mémoire, une autre histoire"* (Nora 1997 [1984], p. 37), then made further explicit with the relationship between the development of historiography and events (e.g. the Annales school and the crisis of the 1930s), developed in the chapter entitled: *"Comment écrire l'histoire de France ?"* (Nora 1997 [1992], p. 2219); this is referred to again the very end of the conclusion (Nora 1997 [1992], pp. 4710–4715) to say nothing of the entire structure of the work itself, to which I will return.

recalls the point of view according to which these institutions are approached: "All institutions from which this undertaking, to his mind, did not require a retracing the entire evolution, but a clarification of the first appearance and an indication the trace" (Nora 1997 [1986], p. 1431). And the remarks in the general conclusion, stigmatizing "the heritage explosion", presented as specific to "the era of commemoration" (Nora 1997 [1992], pp. 4707–4710), do not legitimize having resorted to this undertaking. And yet... a careful examination, not so much of the chapters, some of which, as I will illustrate later, clearly contribute information on the heritage character of certain places of memory[35], as of the overall arrangement of the work, reveals at least three reasons that warrant discussion.

The first is the shift that the undertaking operates by expanding the investigation of an internal analysis of objects toward the inscription of memory in places. The second is a secondary benefit of this expansion: an invitation to reconsider the traditional conception of "heritage" (in reality, at that time: monuments) by reexamining its establishment and its institution.

The third is, with the apparent paradox between the assertion of an approach to memory and the criticism of the claim to the memorial use of what social groups consider to be their "heritage", the need to ask both the theoretical question of the relationship between memory, history and heritage, and the epistemological question of presentism. On the other hand, discussion of these three reasons requires a detailed reading of the structure of the work and the introduction to unearth its traces.

2.2.1. *Investigation extended to the inscription of memory in places*

As I am not a historian, the following examination *Les lieux de mémoire* certainly makes no claim to discuss the subject from either a historical or a historiographical point of view. I approach this book as a scientific media text in the light of questions posed by the approach of inscription of national memory in "places" to the approach of heritagization modes. Given the complexity of the book's editorial statement, I will begin by examining the two possible ways of accessing it in order to understand its project: firstly through the chapters; secondly starting out from the structure of the book and the introductory texts[36].

---

35 Without actually subscribing on this point to the theory of Rautenberg (2003a, p. 27), according to which we name ""places of memory" objects that belong rather to heritage" (except to include intangible heritage).

36 "Presentation" in Volume 1 (Nora 1997 [1984]) and "Preface to the 'Quarto' edition" (ibid. 1997, pp. 7–8). Insofar as these two texts set out how the book's editorial structure takes into account the epistemic approach, I refer to them extensively here. The terms in quotation marks in this section are those used by the author.

Accessing it through the chapters means choosing a place, a "subject"[37] among all of those presented throughout the book. Alternatively, opting to access it through the editorial structure[38] that consisted of three "parts" ("La République", "La Nation", "Les France"), each composed of sections according to "theme" (sometimes themselves grouped in sub-sections). These sections in turn bring together several "chapters" chosen to correspond to the *project* as set out in the presentations. I am then faced with the presentation of a history of representations (in the operative sense of the word) of "national memory": in other words, a "selective and scholarly exploration of our collective heritage, which draws its truest justification from the emotion still awakened in each of us by the remains of a lived identification with these half-effaced symbols", as mentioned by Nora (1997 [1984], p. 20).

I then have the choice of following the linear layout of the book or opting to go to a particular unit (part, section, chapter, or the presentations). This project's argument is set out in the preface to the American edition: it is about breaking down a synthetic history of France (such as that produced by, for example, Lavisse), to "take the ready-made blocks of our organizational system and of our representations, as they have been conveyed to us by tradition".

Therefore, since the range of objects is potentially infinite, "everything is in the eloquence of the collection, intended to reveal "the figure in the carpet", and in the art of execution, intended to highlight a mirror of identity, a refractive lens, a symbolic fragment of a symbolic whole" (Nora 2011 [1996], pp. 161–162). The point of junction between the "subjects" selected and the structure that supports the collection therefore seems to be the section: bringing together the analysis of emblematic places of memory (the chapters), it constructs a "theme" that contributes to "representing" one of the three regimes of national memory ("The Republic", "The Nation" and "The Frances").

According to the initial "Presentation" of 1984, the undertaking's objective is in fact to analyze the places "where it [national memory] was electively incarnated and which, by the will of man or the work of centuries, have remained as its most brilliant symbols: festivals, emblems, monuments and commemorations, but also eulogies, dictionaries and museums" (Nora 1997 [1984], p. 15). Nora gives us here the four criteria used for the selection of places: national memory, an elective incarnation, the brilliance of the symbols and the ontological diversity of these

37 This is this last term that Nora uses himself (2011 [1996], p. 162) in his preface to the American edition.

38 The full 1997 Quarto edition presents this in its entirety at the beginning of each volume.

symbols. Added to these criteria is the fact that the three "parts" form a "path" relating to three memory regimes[39]. For that reason, when I open the book opting to follow the project, I go from these memory regimes to the "themes", then to the "places" that illustrate them according to the four criteria mentioned above. It is in this way that by highlighting "the secret kinship upheld by true memorials, [...] from objects as apparently different as museums, commemorations, archives, mottos or emblems, [...] from administrative, political or temporal categories", this can give rise to the setting in place and detailed analysis of "the most massive blocks of our representations and our national mythology" (Nora 1997 [1984], p. 7).

Therefore, the way in which a theme participates in the representation of national memory and how, for their part, the places chosen contribute to it means that themes and chapters are linked and their articulation is decisive. They are linked in that the *criteria* for choosing places depend on the way in which the theme represents one of the three organizers of memory (Republic, Nation and the Frances). In terms of the articulation between themes and chapters, that is decisive in that it ensures the relevance of the *project*. In other words, reduced to its simplest expression, "the eloquence of the assemblage", which governs the editorial construction, brings to mind what Nathalie Heinich, after Bessy and Chateauraynaud, calls "*prises*" (captures): that is, "meeting point[s] between the object properties of the elements under judgment, and the resources of the actors who are to pass judgment" (Heinich 2009, p. 234). The properties here are *traces* of political or social functioning, which will be interpreted as so many *indices* of national memory, because they relate to "events", the "will of man" and the "work of the centuries".

I will summarize. According to Nora, producing the history of places of national memory requires an understanding of what has happened between what is at the origin of the place of memory and the present-day of the production of history[40]. An origin that exists in some way before the place of memory, in this relatively indeterminate past, but is in any case lost, before the end of the "memory-societies",

39 Whose organizational criteria are: "From the simplest to the most complicated, from the simplest to date to the least easily identifiable, from the most local to the most general, from the most recent to the most distant, from the most political to the most carnal, from the most unitary to the most diverse, from the most obvious to the most problematic" (Nora 1997 [1984], p. 16). I here refer to the regime of national memory relative to the content of each of these parts (and not of modes or modalities), because despite probable partial overlaps, they are treated in the book as exclusive, the themes that make up each of them not being replaceable by those that make up the others.

40 "Between an integrated memory, dictatorial and unconscious of itself, organizing and all-powerful, spontaneously updating, a memory without a past that eternally refreshes heritage, relegating ancestral past to the undifferentiated time of heroes, origins and myth – and ours which is just history, trace and selection" (Nora 1997 [1984], p. 24).

characterized by "what was lived in the heat of tradition", "the silence of custom" or "repetition of the ancestral". The present, a moment that is ours, certainly characterized by the loss of memory, but also that of a reconstructed history in which the places are "remnants" of the moment of constitution of these places from their origin, both objects of commemorative vigilance and objects of capture by history (Nora 1997 [1984], pp. 28–29). However, with the history of places of memory, this chronological temporality (original moment, moments of places and moment of history) turns into an aspectual temporality that goes from the present of the historian – and especially of the reader – to an original time that seems somewhat mythical and that, in any case, arises from what has been achieved. Belonging to the moment of historical memory, the here and now of the production of *Les lieux de mémoire* identifies the remains (of the places of memory) in order to go back to the moment when these places are constituted as a trace of a lost traditional memory. Therefore, "the eloquence of the assemblage", contrary to what the expression may suggest, is in no way a rhetorical game, but is in accordance with a stated procedure of knowledge production. It does not aim at an inventory of places, or only their description as an object, or even simply a description of their constitution as a place of memory (in the now common meaning of the term), because "what matters to them [that is to say the historians] is not the identification of the place, but the unfolding of what this place is the memory of" (Nora 1997 [1992], p. 2226). *The unfolding of what the place is the memory of*; or, if you prefer: the way in which the interpretation of traces makes it possible to *recreate* the process by which a place established itself (or sometimes "was established") as a place of memory, as a kind of resting place for memory. Setting down a formal narrative of the program of this "production of places of memory" makes it possible to reveal its specificity, and hence its input to an approach to heritage[41].

> The *inscription* of memory in a place, which here is the central operation of the production of places of memory, should be thought of as a result of the major operation carried out by the historian, namely, the unfolding of what the place is a memory of. This means that for a given place of memory, the analysis will interpret some of its characteristics as traces of events, knowledge, values or practices – in short, social processes – which indicate that it falls under one of the three regimes of national memory[42]. In other words, that this analysis will show how it is invested with memory values and becomes what Nora calls a "subject".

41 Note that a program such as this is a literal articulation of operations ("doing" things), objects ("actant-objects"), operators ("actant-subjects"), mobilized values (the endowing of values) and a relationship to time (the modalities of its representation).

42 As an illustration, I refer to two examples where this process is easily identifiable from the very first lines: "Reims, *la ville du sacre"* (Le Goff 1997 [1986], p. 640) and *"Une mémoire-frontière : l'Alsace"* (Mayeur 1997 [1986], p. 1147).

The program presupposes the existence of other operations that happen upstream (*the production of the original memory*) and downstream (*recognition and attachment to the place of memory* by members of the nation), but also an interpretation by the historian of inscribed memory (of the "place of memory" as a media text), then an interpretation of this interpretation by the reader of the work (as a scientific media text), possibly leading to a *revitalization* of the place of memory.

Therefore, these various operations have to do with three actant-objects (*the original thing to be memorized, the memory*, and obviously *the place of memory*) and they are effected by several actant-subjects: three proto-actants (*the producer of the original thing*, the *producer of the memory*, the *producer of the place of memory* who inscribes it and thus endows the place with value and meaning), to whom is sometimes added a fourth (*the user who reactivates the place of memory* by commemoration, for example); two actants in the present (*the historian* who operates the unfolding and produces the book, the *reader* of the book).

Finally, key to the program, it is necessary to add one last actant: the one who, in a fiction, would be the *sender-producer of the program*, putting in place the various actant-subjects of the production process (the producer of the original thing, the producer of the place and the user-interpreter). In social life (and not in fiction), these actants cover various processes and various human or nonhuman instances ("events", the "will of man" and the "work of time", says Nora) that resulted in the subject proto-actants contributing to producing the place of memory.

In terms of the relationship to time, we have said that the temporality was aspectual, because it corresponds to point of view of the historian going back to the past.

Box 2.2. *Modeling the production of places of memory*

This formalization confirms that it is the historian who, by operating the unfolding of the production of the place of memory, gives it its social mode of existence in the present. This does not mean that the historian constructs this production of the place in imaginary form: the chapters, written by historians specializing in the "subject" concerned, are there to prove that the properties and processes are scientifically established. In response to the unfolding of "what this place is the memory of", the analysis of the production of place interprets some of its characteristics as indices of historical processes that have moreover taken place over the long term: the place, in its materiality, acts as a kind of medium of *inscription*. Whether it is a site, a monument, or a tangible or intangible object, this medium can also be recognized as heritage. Among the "places of memory" monuments, I will give two examples of particularly exemplary unfolding in terms of their index of wealth to show their operativity: the Basilica of Vézelay and the Notre-Dame Cathedral in Paris[43]. But what interests us more directly here is how this unfolding reveals that they are the medium of several different modalities of relationships to the past: the modality of places of national memory; heritage objects as classified historical monuments; places of literary memory; fragments of collective memory for associative groups involved in their upkeep; and places of worship continuing a religious tradition of services and pilgrimages (another form of social memory). It is not certain that a study of the strictly heritage character (values and characteristics) of these two monuments would have highlighted with as much clarity the diversity of their relationships to the past as does their study as memorial sites. And what is more, this approach even takes into account the intangible dimension of certain forms of memory, at a time when the knowledge produced on an element of heritage – monuments, museum objects or even "works" of the Inventory – relates first and foremost to the material trace itself.

The "place of memory" historian strives to unfold and decipher the palimpsest of what the place has become the memory of, during the different episodes of its life, how this place has taken on, simultaneously or successively, several modes of relationships to the past. Applied to heritage, this lesson certainly guides us toward the study of heritagization, but above all, it invites us to move away from an overly simplistic approach that would restrict scholarly representation solely to the characteristics of the object. In the two examples above (we can verify the phenomenon for all places whose medium is an element of heritage) the unfolding traces the process by which the place became a *memory of*, and therefore, as a corollary, the operations by which it has also been considered an element of heritage. In other words, this approach encourages us to be particularly attentive to the social environment of the object in a double sense: attentive to its world of origin in the past and attentive to the *representational environment conducive* to its heritagization in the present. Such an environment is in fact one of its conditions of possibility: it will be at the origin of the recognition of the object's value, in that it allows an appropriation that is simultaneously emotive, epistemic and social, both on the part of the members of the group, or possible audiences, and on the part of the instances that, studying this object, are able to inscribe its memory in texts (experts, possibly activists or scientists, the administration) or to endow it with heritage status, through *heritage interpretive schemes*. The existence of this

43 Articles respectively by Lobrichon (1997 [1992]) and Erlande-Brandenburg (1997 [1992]). These two analyses belong to the theme of "Hauts lieux" which includes the analysis of other monuments or sites (Lascaux, Alésia, Châteaux de la Loire, Sacré-Coeur, Eiffel Tower).

environment, which is both institutional and socio-affective, is therefore on the principle of interest in the object as well as the operativity of publicization to an audience, and therefore, ultimately, on the principle of the very production of heritage objects. It is by drawing upon this that the production of the scholarly representation (knowledge about the object enunciated and attested by the scholarly authority) will serve as the basis for the decision (the declaration) to protect, transmit and publicize. All heritagization acts that in return contribute to maintaining the propitious character of this representational environment, at the discretion of the circulation of media texts that are part of the cultural being, Heritage.

However, *Les lieux de mémoire*, in proposing a model of approach to objects of the past that includes the memory of their creation, will create an environment conducive to a renewed way of looking at the different types of traces, including heritage traces, a way of looking that will move away from an “almost notarial acceptance”, to hijack one of Nora’s own expressions.

2.2.2. *Institution of the administrative legal heritagization mode*

We must take into account how the representational environment leads to a rereading of the commonplace on the traditional (national) conception that monuments or museum objects are always-already heritage entities and whose heritage nature can be adequately recognized by an administrative legal declaration to preserve, protect and transmit them. It is this conception that makes the socially recognized mode of existence of monuments – their *instantiation* – that of simple material traces. But, more seriously, as these traces have always been present, since they still are today, it skips over the complex process through which they have acquired heritage status.

It is interesting to note that it is in the article criticizing “the explosion of heritage” – which can be understood here in either the active or the passive sense of the action, or both – that, corroborating the theory he expounded in the conclusion of *Les lieux de mémoire*, making this explosion into a symptom of the “era of commemoration” (Nora 1997 [1992], pp. 4707–4710), Nora (2011 [2006]) begins a rereading of this traditional conception of heritage by opposing the two ways of thinking about the “metamorphosis of the notion of heritage that made it the opposite of what it used to be” (p. 99). A metamorphosis, he continues, that can be understood in terms of either continuity or rupture:

> As the culmination of a continuous movement that has seen the field of heritage enriching and expanding itself since the Revolution or,

> conversely, as a mutation linked to a profound transformation of awareness of the present and relationship to the past. (Nora 2011 [2006], p. 100)

The presentation of the three mutations that took place between the Revolution and the 1970s de facto sketches the story of three moments, three ways of making heritage: "These moments of mutations, of disjunctions in the relationship with time, stand out strongly" (Nora 2011 [2006], p. 100). I will come back to these three mutations to show how these mutations constitute the variants of a heritization mode under construction. In order to do this, I will summarize them by referring to Dominique Poulot's presentation and analysis of them, addressing them in these terms. In his book *Une histoire du patrimoine en Occident* (A History of Heritage in the West), he defines heritage conceptually as "the entry into heritage of objects within a society" (Poulot 2006, p. 21)[44]. This process, thus established in its generality, can be broken down into specific "heritagizations" according to the "patrimoniality" specific to each historical and societal context. "Patrimoniality" is then the "emotive modality of an experience of the past, articulated around an organization of knowledge" (Poulot 2006, p. 16; see also Poulot 2008), which results in a "heritage attitude" combining an assimilation of the past in the present and a relationship of strangeness (Poulot 2006, p. 5). Even if, for my part, I reserve the concept of "patrimoniality" to mean the heritage character of an object and not the relationship that the collective upholds with the past, the fact remains that what he means by this term is close to "the propitious representational, institutional and socio-affective environment", insofar as it translates operationally into the form of both heritage interpretive schemes and the production of "scholarly representation". The difference seems to me to come from a difference in point of view. Poulot emphasizes the socio-historical context which, because of the type of relationship it establishes with the past, results in this object, rather than another, becoming part of our heritage. I emphasize the operativity of sociosemiotic functioning in the transformation of the object's status. But the important thing here is to understand the changes that occur in the ways in which objects are included in heritage (heritagization) and I will therefore content myself with showing how historical research brings knowledge that clarifies my arguments.

Unsurprisingly, Dominique Poulot's analysis overlaps with the three mutations identified by Pierre Nora, but the important thing is that it provides a detailed study of this. In accordance with what he had already shown earlier (Poulot 1992, 1997), the author effectively posits the Revolution as the starting point of the traditional

44 "Conceptually", in the sense that, as for the word "heritage", he does not consider the word "heritagization" as a simple category of use, but as a tool for description and analysis.

conception of heritage, which from the outset is specifically French[45]. It puts into perspective the contradiction between destroying and preserving, with, on the one hand, an absence of relationship and positive attitude toward objects of the past due to their rejection by the Ancien Régime in favor of the creation of a new society, and, on the other hand, the recognition that some of the objects concerned have value that justifies their conservation in the eyes of those who have an artistic viewpoint, interest or practice and those (sometimes the same ones) who have amateur or expert knowledge about these objects. For these artists, amateurs or experts, in spite of the break with past society, such objects can and should be part of the process of building the new society. They must therefore be preserved, because they are of benefit for training artists and for educating citizens. Indeed, they have a social utility based on the qualities they inherited from their creator or on their capacity to represent great men, the memory of the social circumstances of their production and the society that saw their birth being, on the contrary, denied (Poulot 1992, p. 35). I will leave aside the study of the impasses that result from this contradiction, which give rise to complex situations discussed in great detail by the historian[46]. I will confine myself to noting that wanting to reconcile the recognition of the value – artistic, allegorical, "archaeological" (to use a contemporary term) – intrinsic to objects with the negation of an abhorred past will lead to defining a regime of value of heritage objects as "artistic masterpieces"[47], appreciated as the creation of human genius (artistic perspective) and for the knowledge they bring. We therefore see the main acts of heritagization taking shape. In fact, there is undeniably a social interest on the part of artists and experts, who, in view of their political position, will more often than not be able to decide to preserve certain objects by recognizing their value, to declare their status and to make a decision on obtaining national protection. This interest is accompanied by an argument, a kind of sketch of knowledge production, that justifies the choice and the decision. The operations related to interest, knowledge production, declaration, conservation and social use are therefore present, even if the dominant desire for a

45 I say "initially", because this concept has been widely taken up in international institutions (IOM, then UNESCO) to be universalized and globalized as shown by Desvallées (1998, pp. 97–102) and Leveau (2017).

46 This recognition of monuments or works is not made on the basis of the traces of their past usages, but on values defined according to the interest of the present era and their interpretation. Dominique Poulot shows the importance of this when he presents, at the beginning of *Musée, nation, patrimoine 1789–1815*: "La légende du patrimoine" (Poulot 1997, pp. 10–36), which defines the problem of the relationship between vandalism and museum creation.

47 Poulot (2006, pp. 91–117) shows that a similar sharing is operated concerning the exemplarity of great men; one of the examples of this sharing being the discourse that accompanies the presentations in the museum of Alexandre Lenoir. Note that the expressions "artistic masterpieces" or "monuments of the arts and sciences" refer to both fixed and movable realities (that will later be called "historical monuments" and "museum objects").

radical rupture with the past is unfavorable to creating in public opinion a representational, socio-affective environment conducive to conservation, thus putting the defense of objects of the past in a precarious position with the confrontation of contradictory interpretive schemas. Therefore, "nationalization", which transfers ownership, while actually giving a decisive role to the state (Nora 2011 [2006], pp. 100–102) and taking into account the artistic, historical and "archaeological" value, still does not quite equate to heritagization. It lacks a societal recognition, a shared propitious representational environment, a bond of dependence and attachment that turns heritage objects into places of memory that ensure a continuity of belonging to a collective between the past and the present and, therefore, an absence of real knowledge production. In the end, symbolic status comes from what is missing.

It was with the following generations that the environment, both representational and socio-affective, was to be modified. Victor Hugo's famous, emblematic pamphlet "Guerre aux démolisseurs" (War On The Demolishers) is one of the best known indicators. Then came the second mutation noted by Nora, what historians call the "Guizot moment", during which a "theory of heritage" was born (Poulot 1988, p. 44; 1992, pp. 38–45; 2006, p. 133)[48]. The production of the scholarly representation from that point on has focused both on the civilizational context of monuments (the society that produced them, albeit without a direct claim of memory from the groups that produced them) and on their knowledge with the purpose of inventorying them. On the institutional level, it was, in France, the period of the creation of a "heritage" administration: in particular the *Inspection générale des monuments historiques* (General Inspectorate of Historic Monuments), then the *Commission supérieure des monuments historiques* (Higher Commission of Historic Monuments) that was to establish the first list of protected monuments. It was also the period of the implementation of political action toward public opinion, especially proprietors. The principle of a heritagization conceived as a declaration by a modern nation-state of a heritage to be preserved and transmitted on account of its artistic, historical and archaeological value (i.e. the formalization of a scheme mainly representative of the monument), was henceforth initiated, and the administration's actions were in line with the implementation of this principle. Summing up the period "between the late 18th century and the 1830s", Poulot concludes:

> After the Revolution, various processes – from the invention of the museum to the invention of the historical monument, from the reconfiguration of archaeology to the success of the historical novel – invented a *heritage tradition*, which has to do with the new national community, and which was to remain for a very long time the basis of

48 See also Nora (2011 [2006], pp. 102–105). In the words of Poulot (2006, p. 127), this moment opens a new "patrimoniality". For an approach to the societal context of the "Guizot moment", I refer to "Moment Guizot" by Rosanvallon (1985).

> French attitudes towards heritage. (Poulot 2006, p. 21, with author's emphasis shown in italics)

The Third Republic (third mutation, according to Nora), giving a major role to history and national memory, credited heritage with an "ideological importance" heritage, while at the same time practicing a "political indifference toward it", "it is against this contrast that the 'historical monument' system was established over more than a century, an irrefutable testimony of a bygone and threatened past that the national community recognized and designated as representative of its identity" (Nora 2011 [2006], p. 107). Thus, heritage administration, which was to be clarified and consolidated by a number of laws during the 20th century, before and after the Great War, would be able to define what deserves to be preserved, the hierarchy of monuments and the protection rules, and it would begin to extend these rules to sites, urban areas and natural monuments. Historical monuments were therefore inseparable from works and places of national memory: history (especially that of French history) and art history served as the major science of reference for the interpretation and recognition of the value of traces, covering technical disciplines related to restoration (e.g. architecture) or archaeological research. As for public opinion, it was able to subscribe to this national representation of the past, especially through visiting these monuments, textualized both by their restoration and their promotion, but above all through the publicization of their scholarly representation by education. Therefore, this period consolidated and expanded the administrative-legal heritagization model with its artistic, historical and archaeological variants.

Thus, in the 1970s, what I have called "the traditional conception of heritage" was now nothing more than the visible face of a heritagization mode that was the result of successive transformations, in the form, in a way "completed", of the third. Taking these transformations as an object of analysis as heritagization modes thus leads above all to noting the progressive construction of a conception that is hegemonic at the moment when a tipping point occurs. This was still the order of the day when the creation of the Inventory intervened, together with the profound change in perspective that it introduced by considerably expanding what is worthy of interest and by putting a new approach in place. But with no conservation goal and without maintaining a purely scientific national declaration, its institutional impact was to remain small and, in fact, did not lead to any real change. This traditional conception is explicitly the one to which Nora refers and the one he supports[49]. The "new heritages", as Jeudy calls them, threaten the hegemony of that

49 It will be noted that reading Dominique Poulot (2006, p. 21) gives, according to the texts, the impression of a relative ambivalence toward this tipping point, which appears both as a new point of "mutation" (to use Nora's words) – a new form of patrimoniality – and as the moment of an atypical phenomenon: the one where "heritagization seems to merge with patrimoniality", Nora tells us.

conception, because they are laden with a new heritagization mode called for by social groups and based on a new science of reference: ethnology. But the representational, institutional and socio-affective environment had then become propitious for another approach to the traces of the past.

2.2.3. *Heritage, between historical memory and collective memory*

The first explanation that comes to mind when considering Nora's stance toward the rupture of the 1970s is the context of the era when *Les lieux de mémoire* was published. A criticism of the "heritage explosion" was widely shared by many historians and heritage specialists at that time. Faced with the development of "new heritage", stigmatizing the "proliferation", "inflation", "desire" and "obsession" relating to heritage was commonplace[50]. The enunciative event constituted by *Les lieux* was caught up in a shower of media texts developing this theme. So his criticism picks up on the various reasons that can be found in these texts coming from experts in historical monuments or traditional museums – mainly of art – who believe that the emergence of the concept of heritage, accompanied by a massive influx of curiosities, overshadows the recognized works[51]. Not to mention either the criticism from those opposed to a culturalization of heritage, synonymous according to them with enslaving heritage to the pretentions of an "everything is culture" synonymous with commercialization[52]. But when he condenses his criticism of the new relationship to the past with the three words that would "become circular, almost synonymous" under the effect of commemorative practice – namely, identity, memory and heritage (Nora 1997 [1992], p. 4713) – his position also comes closer

50 An overview of this position can be found in Heinich (2009, Introduction) and a critique in Tornatore (2010a, p. 8; 2017b, Chapter 2).

51 See, for example, Pommier (1991) on the subject of the museum. For a reasoned presentation of this position, which is careful not reduced to a simple corporatist defense, even if it sometimes sounds like that (as on the distribution of public funding for the conservation of monuments), refer to Leniaud (1992). Tornatore (2017, Chapter 2) rightly points out that Leniaud subsequently proposed a different approach.

52 The most mediatized author representing this crusade against the "cultural" is certainly Fumaroli (1991). Choay (1992), for his part, put forward (not without reason, even if the charge is sometimes overstated) a virulent criticism of the industrialization of heritage, and also the excesses of "heritage consumption" which transforms monuments into "cultural products". This is a criticism which, even though it points to undeniable trends, excesses and drifts, does rather echo Fumaroli's positions in some of its emphases. For a contextualization of these criticisms relating to heritage policies, it is interesting to consult the synthesis proposed by Poirrier (2003 [2001], p. 583).

to what other historians write about the way in which heritage culminates in its "presentistic use of the past"[53].

However, while these various comparisons show undeniable proximities, they still do not explain either the conflict between the two contradictory acceptations of the term "heritage" or the paradox between continuity and rupture relative to the 1970s. Nora himself put forward the reason for these contradictions as the mutation of society: the fragmentation of the nation and the transition from one general history to the laying claim to memory (and hence places of memory) by social groups. The consequence would be the tipping of the role of heritage from one form of identity to another: "that is to say, that heritage is no longer representative of one general collective identity, of the social body as a whole, but that it is now constitutive of a sectoral identity, of a social category seen only in the cultural dimension" (Nora 1997 [1992], p. 4709)[54]. A few years later, in the introduction to *Science et conscience du patrimoine*, the author is even more trenchant (I therefore quote the entire passage).

> Heritage has become one of the keywords of contemporary historical conscience, moving from the almost notarial acceptance that it still had in the late 60s and 70s to a much more restrictive and pervasive definition: no longer the property that we inherit, but the property constitutive of the collective conscience of a group: a real turnaround. As such, it came to join in the same impassioned constellation the words "memory", "identity", with which it became almost synonymous, words which likewise experienced at the same time, within a very few years, the same devastating reversal of meaning. (Nora 1997, p. 12)

But to understand what serves as the basis for this argument, we must return to the "narrative construction" (Ledoux 2019, p. 12) of the notion of "memory". The

---

53 The last third of the 20th century would thus have given "the greatest extension to the category of the present: a massive, invasive, omnipresent present, which has no horizon other than itself, manufacturing on a daily basis the past and the future it needs day after day. A present already past before it has even fully arrived. [...] We would like to prepare, today, the museum of tomorrow and gather the archives of today as if it were already yesterday; caught as we are between amnesia and the desire not to forget anything" (Hartog 2003, p. 199).

54 Garcia (2000, p. 130) quite rightly pointed out that slipping sideways from the conception of places as traces of memory to a criticism of the hegemony of commemoration and identity was based on "a feeling of loss that makes 'us' 'debtors' towards this 'memory that we no longer inhabit'". This sentiment has as a corollary the fact that "thinking about the movement of contemporary society is equivalent, for Pierre Nora, to a mourning process". This sideways slip had already been noted by Valensi (1975).

starting point is a much earlier text, "*Mémoire collective*", which is one of the chapters of *La nouvelle histoire*, which also constituted an enunciative event by its novel vision of history. Here again, one of the paragraphs, which is also widely cited, draws up a program that appears to be in total contradiction with what he wrote 20 years later:

> One would have to start out from places, in the precise meaning of the word, where any kind of society, nation, family, ethnicity, party, spontaneously deposits its memories or redeems them as a vital part of its personality: topographical places, such as archives, libraries and museums; monumental places, such as cemeteries or architecture; symbolic places, such as commemorations, pilgrimages, anniversaries or emblems; functional places, such as manuals, autobiographies or associations: these memorials have their history. (Nora 2011 [1978], p. 304)

The next paragraph, however, sets out unambiguously how this program is to be implemented, even if it remains schematic at the time.

> But making this history [of memorials], he continues, quickly leads to turning around the meaning of the word to speak out against the memory of places to the real places of memory: States, social and political environments, communities of historical experiences or of generations led to constitute their archives according to the various usages they make of memory. (Nora 2011 [1978], p. 304)

The detailed analysis of the genesis of *Les lieux de mémoire* carried out by Ledoux (2017, 2019) shows that, contrary to what we may think just from the enunciation of the term "memory", this 1978 chapter is built on an opposition of "historical memory" against "collective memory", with the aim of establishing:

> A hierarchization of discourses on the past, restoring the primacy of historical discourse against that of collective memories which is inclined, in his view, toward a certain hegemony: "history is now written under pressure from collective memories", he already asserts. (Ledoux 2017, p. 117)

For this author, all of the ambivalence of Nora's approach lies in the fact that he uses the discourse of memory to reaffirm the legitimacy of (historical) memory against (collective) memory and of the historian against groups that are the bearers of collective memory. This opposition and its ambivalence could thus be said to be at the basis of *Les lieux de mémoire* (Ledoux 2017, p. 120)[55].

In this context, the defense of the traditional conception of heritage must be linked to historical memory (i.e. the memory of historians), and the hegemonic intent of its current metamorphosis – or its "metaphorical extension" in the case of intangible heritage, for example – would be an expression of collective memory. This is followed by an interpretation of the two ways of understanding this metamorphosis of heritage: while it used to accord with the "choice of what was timeless and permanent from the past", nowadays (i.e. from the 1970s) it is the opposite; in other words "the totality of traces of the past as the past" (Nora 2011 [2006], p. 99).

This is why, while this mutation in the relationship that contemporaries maintain with the past is not the first and takes its place as such in the continuity of the previous changes as a way of making heritage, it is nevertheless in rupture with the previous ones. However, a careful reading of the argument reveals an asymmetry between the historical approach of the first three mutations and the criticism of the last one, according to which "the notion of heritage is nowadays entirely metabolized by the transformation of the past into memory" (Nora 2011 [2006], p. 108).

2.2.4. *Reexamining the 1970s mutation*

If we subscribe to Lucette Valensi, the raison d'être of Nora's position vis-à-vis the change in the 1960s and 1970s, namely that "the past becomes autonomous" and "becomes heritage", is

> in relation, with France's departure from the arena of major history after the end of decolonization, with the erasure of the unitary framework of the nation-state, with the 'crumbling of history as a myth that is the bearer of the national destiny' and the dissolution of the national myth which, since the Enlightenment, was supposed to link 'closely the future to the past'. (Valensi 1995, p. 1273)

55 "By setting 'collective memory' against 'historical memory' while proposing to research and establish the historical memory of the French nation and not the collective memories of groups lived as sources of identities, the historian investigates a narrative whose function is the refiguration of a national We and the restoration of a unity assured by historical discourse", specifies Ledoux (2019, p. 18). Nora explains this in: "*Les lieux de mémoire, ou comment ils m'ont échappé*" (2011 [2008]), thus confirming Ledoux's analysis. In this ambiguity of the term 'memory', Rautenberg (2003a, p. 27) sees an appropriation of ethnology by the historian accompanied by carelessness in his methods and his research questions.

Nora lists the symptoms that bear witness to this new form of interest in a past from which we would henceforth be cut off, which range from individual practices, such as an interest in genealogy or terroirs, to more collective or institutional practices such as activism, or more generally investment in the conservation and promotion of "remains" of former lifestyles, of countryside practices and objects of declining industries[56]. It could also be noted that over the past two decades, this interest in the past has further increased with the growth in documentaries that contribute to the representation of a past that is both very close to us and irretrievably lost for us (because literally "imagined"). Hartog (2003) in a way extends the questioning opened by Nora by linking the rupture with the past to the "crisis of presentism" which can be seen as a withdrawal into a present, having turned our backs on the past (pp. 195–206). It is indeed difficult to deny that contemporary French society, even if we stop there, operated a reversal, after the 1960s and 1970s, of the interest that had up to then been directed toward the future (incarnation of ongoing progress), toward a past whose traces are sought, because it is seen as a depositary of "roots", and therefore both the medium of a claim of identity and an object that must be protected. However, we cannot stop there. Thus, Jérôme Baschet calls for a differentiation between the two different processes – the two regimes, one of historicity, the other of temporality – which have become merged under the term "presentism" by François Hartog. Because they actually correspond to different timescales.

> It is proposed to reference the notion of the historicity regime to the long timescale time of history, to the way in which each society thinks of itself in its relationship to its past, its present and its future, while the temporality regime concerns the short timescale applied to the pace of everyday life and the life lived. (Baschet 2018, p. 135)

Hartog does indeed mention these two possible senses of the notion of presentism (he also cautiously speaks of the "crisis" of presentism), which refer both to the mode of temporality of our society and the way in which our society builds a relationship with the past, but he restricts himself to flagging them up (Hartog 2003, p. 18, n. 28). Baschet (2018) notes the asymmetry that exists between the approach of the modern era's historicity regime that is "analyzed firmly as a regime of historicity and presentism which, while also a historicity regime, might appear first and foremost as a temporality regime" (p. 138). Based on this distinction, I will

56 See, for example, (Nora 1997 [1992], pp. 4702–4707). He returns to this in the article on *"L'explosion du patrimoine"* (Nora 2011 [2006], pp. 108–109). We will have an example of this for industry in the next chapter with *"la transmission en acte"*. Another symptom: the Museum of Popular Arts and Traditions which opened in 1976 and which presents the results of work to collect of traces (memory and objects) of this end of the terroirs started in the 1930s. Conversely, Chastel's (1997 [1986]) view is that with the Inventory, the diversity of heritage contributes to national heritage (p. 1455).

point out that alongside the hypertrophy of the present (as a regime of temporality), which probably means that the development of heritage objects and forms of heritage is a reaction to the break specific to our society between us and a past that is finished[57], still the representation of the past (specific to the historicity regime) is necessarily constructed – and a fortiori enunciated – from the present[58]. Also, confining this *aspectual* temporality to group heritage cannot but raise a number of paradoxes, since the unfolding of what the place is a memory of, like any form of knowledge production, also contributes to a construction of the past from the present, as Nora clearly states when he sets out the approach of *Les lieux de memoire*. So if we take at face value, the comment made by Hartog (2003, p. 157), which says that: "*Les lieux de memoire* is intended to be a history of the present, in the present, integrating the fact that the present has 'become the category of our self-understanding'", it seems clear that it is more a position taken vis-à-vis the present, rather than understanding the present, that is the motivation for Nora's approach to the heritages that appeared after the 1970s... While he very rightly notes the change in the instance that is both the holder and the beneficiary (from the national community to the collective), as well as the emergence of new practices, the fact that he reasons in terms of national heritage leads him to make a negative value judgment on these processes, adopting a position that is more normative than properly descriptive or analytical as that applied to places of national memory or to the heritagization of monuments or museum objects in the earlier mutations.

To sum up, I would say that *Les lieux de mémoire* constitutes an enunciative event that has two opposite sides: on the one side, with the unfolding approach, it represents a model for the study of traces of the past; on the other side, it is confined within the limits of the national character that it uses to define memory and heritage (the modern historical regime).

The first side, the unfolding approach as a model, is clearly visible in the difference that exists between it and the operationalization that has been made of the notion. While the unfolding occurs in the present, we have seen that it nevertheless had to do with an operation, the formation of the place of memory, which occurred in the past and of which, to use technical terms, the sender – who is, here, also the operator – consists of events, the will of man and the work of time, as Nora says, so that the inscription takes place according to a layering which is at least in part unintentional. We could say that *memorialization* is inherent in social and political

57 What then changes is the "texture of the present": "The present of presentism is therefore not a pure present, but a present caught up in the moment after" (Baschet 2018, p. 115).

58 It is a temporality built on the basis of an enunciation, therefore aspectual. As Laurent Olivier (2008, p. 153) notes about archaeology: "Not only are archaeological remains discovered in our present, but it is above all in their relationship to our present that the meaning they take on for us is constructed".

functioning: supposed to result from an attachment on the part of the group members that is sufficient to have lasted, the existence of a memory is in a way "recorded" by the historian, even if as a historical memory, it is a memory (re)constructed by him. There we see the difference with the ethnological approach, whose intention is more to record the memory of social groups. But the publicization of the unfolding by definition makes public the process by which memory is inscribed in a place. It is therefore not surprising that *Les lieux de mémoire* should have become a model for collectives who wished to transform their memory into a place of memory, making these places a kind of "monument" (in the primary meaning of the term) responsible for preserving and transmitting it. In a simple manner, the presence of the instituting production of the place of memory, in order to make out of it a media text intended to ensure that in the future we remember past events, will replace the presence of the scientific production of the unfolding[59]. We are indeed faced with two processes that are, semiotically and socially, quite different from each other: the reconstruction by the historian aims to establish an interpretation, as truthful as possible, both of this operation and of what it is about (what the place is the memory of); the production of the place intends, for its part, to preserve the memory of the "thing" (generally an event), in order to be able to revitalize it by telling its story (visits or commemorative reading). On the one hand, there is knowledge production on how a memory trace has been produced; on the other hand, there is the production of a trace through memory inscription of what must be remembered. Either way, there is a common point between the two on which the concept capitalized: the memory inscription – admittedly, not entirely intentionally in the one case and entirely unintentionally in the other case – which gives rise to a media text. Apart the particularly expressive nature of the concept, *Les lieux de mémoire* manifestly endowed it with an epistemic, societal and political value that was all the more effective since a propitious environment was developing in society due to an interest in the memory of significant events of the past, particularly traumatic events. In this sense, the concept has brought heritagization and memory closer together, the knowledge production necessary for the former actually drawing upon the latter. But is this not the same kind of operation as when the historian asserts that the only legitimate form of heritage is that constructed within the framework of national memory?

Because the flipside lies in the limits deriving from the national character of heritage (as a part of national memory, in a way). We may wonder whether the act of reasoning in these terms does not lead Nora to act as if "the whole of the social body", as he calls it, was merely a political body whose construction and maintenance required, and still require, socio-political *dispositifs*, a production of

59 For more details on this replacement, the reader may refer to Ledoux's (2019) comparison between the genesis of *Lieux de mémoire* and the policy of rolling out military memorial sites under Mitterrand.

places of remembrance and national commemorative rituals[60]. He seems to believe that heritagization by groups does not really produce heritage – in any case, not "real" heritage – on the grounds that this heritage is not national. And yet, reducing heritage to national heritage amounts to masking both the nature of what is produced and above all the very existence of the production process itself. Consequently, in the same way that the unfolding of the production of places of memory tends objectively toward establishing that the heritagization mode of national heritage, as it appears in the 1970s, could have been operated by taking into consideration the memory of which the monument (or museum object) is bearer, likewise it should also be accepted that the same can apply to heritagization by groups, in accordance, moreover, with the principle of a generalized operativity of memory as was recognized in the programmatic text of 1978. It brings us back to the conflict between "historical memory" and "collective memory". The truth is that this equates to moving away from the project of a history of France according to historical memory. In other words, as Chastel perceptively remarks, it is then a new project that calls for an "ethnological attention". Having himself introduced a rupture in the traditional heritagization mode with the Inventory, Chastel (1997 [1986], pp. 1457–1458) simply notes a substitution – and not a reversal – in the objects of interest and study: "Farms, workshops, shops from times gone by are becoming now what churches, sites and castles had been for previous generations", but that means that heritage is effectively changing its science of reference.

2.3. In summary: the precursors of a mutation

Of the Inventory and of the theory of places of memory, it is above all the particularities of each of these two practices, and consequently their differences, that we notice first. But despite these differences, these practices have a common point in that they place *knowledge production* at the center of what defines heritage, thus triggering de facto a transformation of how we approach it. The Inventory explicitly begins this transformation by operationally redefining both the heritage object (its typicity becomes the norm instead of its uniqueness) and the way in which it is produced (standardization of the documentary collection). It explodes the range, which all in all used to be limited, of what could legitimately claim to take its place in the heritage category in a sense relegating the monument to one simple heritage category. In terms of the unfolding that is effected by the theory of places of memory, a kind of anthropology of national memory, by making the "place" the medium of inscription of national memory, it leads us to wonder, beyond the

60 On this necessity, I refer to Gauchet (1977, 1981) and Guiomar (1974). Poulot (2006, pp. 93–104) shows the role of the cult of great men in the construction of the nation as well as of heritage in its national version. Going against this oversight, the sum total of *Les lieux de mémoire* highlights the importance of this production and this maintenance.

description of the characteristics of the monument, what memory of what social practices it bears. In any case, this opening up of the monument to (national) memory takes on its full meaning from the new representative, institutional and socio-affective environment, which became propitious to memory (of groups) from the 1970s, during which *Les lieux de mémoire* came along, as evidenced by the operational use that would be made of the concept of "places of memory". When Chastel recognizes the merits of a definition of the patrimoniality of these new objects – the farm, the workshop or the shop of times gone by – he is just taking into account this new environment. Does he not also suggest, moreover, that it is in terms of this new environment that we should henceforth address the absorption of the recent past into the concept of heritage, which has, for example, led to buildings from the 19th century moving from being abominated to being glorified? While the fervor of such an absorption may be surprising, nonetheless it has the effect of having thus "found a new domain to include in the old French assets", he writes (Chastel 1997 [1986], p. 1459). Why would this be? That the first definition of heritage assets – "a historical landscape strewn with ruins and medieval outlines, [which had been] for Romanticism an irreplaceable access to the national consciousness", and which had the effect of "a historical deepening through a selection of remarkable buildings" – has henceforth given way to a new definition that "calls for ethnological attention that can leave nothing out of reach among things and usages", even if this new definition presupposes "a new approach, still imperfectly established" (Chastel 1997 [1986], pp. 1457–1458). We have seen that it was an approach such as this that he had actually begun with the Inventory, foreshadowing what would be the, certainly more radical, path of social heritagization – the very one that is at the origin of the heritage of groups. An "ethnological attention" that would however soon move outside the framework of art history...

Between Nora and Chastel, despite the fact that their analyses and positions may differ, they are complementary. They showed that both in the case of the broadening of the range of heritage traces and in the case of the legitimization of memory, interpretation played a fundamental role in the recognition of traces as heritage. The latter are both traces of themselves in the past (of their characteristics and of their creation) and traces of their social world of origin, that is, of practices and memories of the collective that used to constitute their social environment (as a memory of their function and their usage), with two consequences.

The first was to recognize the importance, for any heritage object, of a dual mode of social existence in the present scenario: (i) a past existence that is necessarily *in representation* because it is interpreted, that is, established through analysis based on the indices of which this object is bearer and through the lateral knowledge produced by historical analysis or borne by the memory; (ii) an existence linked to its actual *presence* in the form of a material trace, since we have not yet come to the question of intangible heritage. Any heritage object is thus, by definition, a hybrid between a *trace* and a *scholarly representation*. There is no "thing" with heritage status that has not been identified, categorized, described and affirmed as an interpreted trace. Or, if you would prefer a more aphoristic enunciation, neither the trace-object without knowledge nor knowledge without the trace-object can constitute heritage. The joint heritage thus conjoins two modes of immanence, autographic and allographic. On the other hand, it will soon become clear that an analysis of heritagization as a production of this heritage object is essential to bring to light this dual mode of immanence. In fact, the administrative-legal institutional history of the monument or museum object, as a prototype of the "object of cultural heritage", has adopted only the material side of the trace and its autographic mode of existence; the allographic mode of existence being related to documents that are bearers of the scholarly representation. To account for the difference between the social categorization and the dual mode of existence, I shall call the former *social instantiation*. Despite this common reduction of hybridization to instantiation, this hybridization nevertheless remains materially visible in two cases: the case, fairly exceptional, of the retention of heritage status in the "monumental documents" of a lost work, for example, in the case of the Inventory or the creation of substitutes (Fraysse 2006); the opposite case, very common, of the documentarization of the material object that is produced by its restoration, its promotion and the addition of integrated mediation devices.

The second consequence, although given little attention, is just as crucial: it is the questioning of the hegemony of the traditional conception of "heritage", which favors the transmission of an already-produced heritage object above its production process, assimilating cultural heritage to an extension of private heritage defined according to the patriarchal heritage model of *patrimonium*. Certainly, neither Chastel nor Nora, who adhere to the traditional conception, go so far as to conceptualize an object's becoming "heritage" in terms of "heritagization". What they have in common is keeping the production of knowledge, a scholarly knowledge, at the service of the administrative and state declaration of heritage. However, by placing it at the center of what gives heritage its meaning, they initiate a shift toward a consideration of the process through which an object becomes heritage (i.e. heritagization): they relate the object to a collective (a nation or a group) and not to an individual or a family. In the Inventory, as Nathalie Heinich has shown, the simple fact of inventorying and documenting objects confers on them, in practice and in law, the status of "cultural heritage". When the unfolding is applied to monuments, it shows that their status is not separable from an inscription of the memory of the national collectivity, even if only through administrative-legal "heritagization"; an analysis of the construction of this mode since the Revolution, showing that this status does not result from a simple transfer through nationalization of an inherited heritage (as the revolutionaries had believed)[61].

In one sense, the conception of heritage in the Inventory, like that of the unfolding of the national memory, helped to give term "cultural heritage" its meaning by establishing a dividing line between *inherited heritage* and *cultural heritage*. This is the same dividing line that Victor Hugo had pointed out. To illustrate this dividing line, the example of the English castles still inhabited by aristocratic families (while sometimes belonging to the National Trust) and open to visitors is, on this point, particularly instructive[62]. In another sense, each castle constitutes a trace of the past accompanied by its memory of lineage, and, on the other hand, it is traced by interpretation of indices of the practices of its own production and social practices related to its usage and its past societal context. In a representational environment characterized by an attachment shared between its occupants, its custodians and the public, this castle will be the object of a desire for conservation based on a recognition of its cultural heritage value, a value which is also shared socially (in the form of interpretive schemes). It then has a dual usage: the traditional usage enjoyed by the occupier (both practical, but also symbolic, especially in the sense mentioned above of the political and lineage "representation"; the essentially symbolic use of a cultural heritage object linking to a past endowed with collective, artistic and societal values (a form of memory). In other words, two forms, complementary because different, of heritage: as an inherited heritage (*patrimonium*) and as a cultural heritage. The dividing line between the two is the decision to open to the public, because this involves a textualization of the object through the establishment of an interpretation designed for a collective of strangers (the establishment of mediation and the public as receiver showing an interest) which ratifies the switch from one form of heritage to another; in other words, its publicization here ratifies its heritagization as a cultural being.

61 Which is not only, let us note in passing, "private" in the current sense, but was moreover derived, at least partially (and even if it seems to have been somewhat "forgotten" by the interested parties themselves), from the logic of the public sphere structured by the representation (Habermas 1978, pp. 17–37).

62 Or in Portugal with, for example, the Palácio Fronteira, in Lisbon.

3

The Social Heritagization Mode

France's declaration of 1980 as the "Year of Heritage" is widely regarded as the institutional event that marked the establishment of the category of heritage in French society and its official recognition. For Dominique Poulot, an interest in buildings less prestigious than monuments, but nonetheless of note locally (for example, churches or what was in the 1980s still called "small heritage"), had already manifested itself in the early 20th century[1]. Likewise, ethnographic museums had also contributed, in their own way, to linking the heritage character of their objects to a study of the lifestyle of groups. In fact, while the objects of the material culture of exotic societies were progressively illuminated by the knowledge produced on their original culture with the development of ethnography, it was with memory studies on the objects of our own past culture, as evidenced by the Museum of Popular Arts and Traditions (Segalen 2010), that the production of memory knowledge on the world of origin of these objects had become the foundation of knowledge on the objects themselves. However, these various preliminary writings remained, for a long time, limited to particular social groups – in particular, to scholars from learned societies, or ethnography specialists – until the emergence of a different relationship with the past linked to the societal shift in the 1970s, in which several phenomena indicated a modification of the representational environment that became conducive to the constitution of *Heritage* as a cultural being.

The most striking of these phenomena was the episode of the demolition, in 1971, of a 19th century architectural iron structure, the Halles de Baltard in Paris, which aroused high-profile controversy and greatly raised public awareness of the issue of conservation of monuments of the past, well beyond the particular case of

---

1 As Poulot (1992, p. 48) writes: "At the most general level, the legacy of mutations from the first half of the century to our current representation of heritage arises from the seeming triumph of two main concerns: a continuous expansion of heritage objects, and their increased socialization".

one building which was largely atypical in relation to historical monuments; this is because it belonged to a "relatively" recent past and moreover had a utilitarian use. This incident introduced a rupture with the conservation of only churches and castles, or even urban areas, especially medieval, which thus went beyond the sole issue of what an inscription in the Inventory was, and which resonated with the increase in the "relics" of a rural but also industrial society, in the process of disappearing or being transformed. There was also the signing of the UNESCO Convention for the Protection of the Cultural and Natural World Heritage in 1972 (UNESCO 2021). While that was not, in itself, an event that made a mark on French public opinion at the time, it would nevertheless have great symbolic and practical consequences during the following decades because of its mediatization and impact on tourism practices. Finally, although it went almost unnoticed, except for professionals, this was the entrance of the notion of heritage into the world of administration, around the edges of the state, with the creation of the departmental service for architecture and heritage in 1979[2].

The declaration of 1980 as "Year of Heritage" was therefore the indicator of a profound change. As Nora has shown, this was not only a "fourth mutation", after the first three that had resulted in constructing the traditional conception of the relationship to the past (Nora 2011 [2006], p. 110), but a real rupture with them, at least in the way of thinking about heritage, operating what he described as an "inversion" or "reversal" (Nora 1997 [1992], p. 4707, 1997, p. 12). This manifested itself more as an upsurge of interest in objects from a past in the process of disappearing. This declaration was in fact only the symptom of a much more profound and widespread process, whose precursors we have seen in the previous chapter and which goes well beyond a simple reclassification of "established heritage" (in reality, bringing together, within the same category, museums, archaeology and historical sites). As historians pointed out, this is a transformation of the relationship with the past due to a transformation of society[3]. The hegemonic – dominant and legitimate – form of "heritage" was to experience a seismic upheaval (Davallon 2018 [2014]), since, in practice, the new denomination would simultaneously: surreptitiously bring in new objects that neither specialists nor public opinion were yet accustomed to regard as heritage (Jeudy 1986; Andrieux 1997, 1998), modify the rules and procedures for selection and introduce new

2 Note that we would have to wait until 1998 (and overcome some resistance...) for the creation of a board of management for architecture and heritage at the Ministry, and 2010 for the creation of a board of management for heritage, with the latter board including the board of management for museums and the Archives de France.

3 For a summary of this point in history, see, for example, Poulot (2006, pp. 154–155). A subtle analysis of this transformation from the traditional French model of the heritage to an open functioning in which the position of the actors involved (*"attachés"*) dominates can be found in Tornatore (2014).

non-specialist promoters and expand the role of public opinion, and also of publicization. To the extent that Guillaume (1980, p. 11) was able to write, in that same year, in a kind of giddiness: "Everything is becoming heritage: architecture, cities, landscape, industrial buildings, ecological balances, the genetic code".[4] Therefore, it is openly about heritagization, the entrance into heritage of objects, henceforth becoming socially and institutionally visible.

Leniaud (1992, pp. 3–4) is the first to have published the name of this process; even if that was, at least initially, in order to lament its existence since it came down to shifting from "the essential reduced to the sacred to the accumulation expanded to the profane" (Leniaud 1992, p. 4), and the distribution of funding would be disrupted by this emergence. To the question "how does man or social group decide that such-and-such item of property should enter into heritage?", he gives this answer, quite novel at the time: an object that has lost its "use value" can acquire, through a process of "appropriation", a "heritage value" – in other words, a process that will justify the obligation to keep it – according to a number of criteria applied to this object by those he calls "mediators". He qualifies this process of appropriation with the word "heritagization". The definition he gives of it is terse, certainly, but it is clear. Not surprisingly, it maintains the dividing line between "real" heritage – that is, "national heritage", the prototype of which remains the classified monument classified, object of protection – and heritage through "heritagization", the heritage of social groups produced through appropriation. It took several years to get away from this opposition and for the category to be really redefined as a unitary concept that covered any heritage category. It would have to be accepted that there is no cultural (or natural) heritage without heritagization, and that the heritage character of an element (its patrimoniality) is nothing more than the set of characteristics that meet implicit (or explicit) criteria defining the value of this element.

The Inventory had expanded the objects that are heritagizable and replaced a binding administrative-legal declaration with the scientific recognition of characteristics; however, at least at first, the origin (the sender) as the operator (the subject) of what was not yet called heritagization was still the nation acting through the administration of the state. However, this is precisely what is changed by the upsurge as a point of singularity: if collectives can posit that certain objects have been preserved and transmitted (thus, heritage objects) and that contribute to their identity, the starting point of heritage action lies with these collectives (even if in the aim of gaining official recognition of the heritage character of these objects by the state), leading to a shift in the end purposes, operations and the actants. Since the relationship with the past is now essentially based on a *social interest*, and no longer only on a national and scholarly interest, a new heritagization model emerges, that

4 An assertion that can be compared to the *Petite "histoire naturelle" du patrimoine* that would be put forward by André Micoud (1996) some 15 years later.

is, of *social* heritagization, as defined by Michel Rautenberg in his book *La Rupture Patrimoniale* (2003a), whose denomination I adopt. This book is also a fundamental scientific enunciative event in apprehending this heritagization mode and remains the one that gives the clearest definition[5] of it.

Without dwelling on the various controversies that this emergence has aroused, and that I have already briefly mentioned in the previous chapter, I shall approach it by examining one of the earliest scientific media texts that was an enunciative event; expressing, in my opinion, in an exemplary way, the questions that this emergence has posed, and that bears a particularly illuminating trace of the manner in which it constituted a point of singularity. This book is *Mémoires du social,* by the sociologist Henri-Pierre Jeudy, which is about "new heritages" – a notion moreover that the author introduces on this occasion – and in which he criticizes what he calls "transmission in action" as a relationship to the past. The criticism, far from focusing on the rupture that these heritages can be said to operate with the traditional conception, is on the contrary applied to the museumization (the museumification) that they operate on the society, and particularly on the memory, of which they were supposed to give an account.

3.1. A criticism of "transmission in action"

It is fairly standard to regard Marc Guillaume's essay, the source of the quote at the beginning of the chapter, according to which everything becomes heritage, as the enunciative event marking the point of irruption of heritage and social heritagization. There is no doubt: this criticism is radical on this point, and it is one of the first texts, if not the first, on the subject. According to the author, the cause of heritage fever should be looked for in a passion for the past which has become a refuge value in the face of the industrial imperative. Although in traditional societies, he explains, conservation of the past has its medium in the burial place that operates the link between visible and invisible, in our societies, the interest in objects as traces of the past and their conservation tends to make this past visible. However, the resulting loss of transcendence makes this quest endless and the accumulation of objects potentially infinite due to never really being operative.

5 We should note at this point that he bases his definition, to which I will return a little later, on the study of a series of situations: the memory of Algerian immigration, the cultural policy of the city of Saint-Martin-d'Hères (compared to Echirolles), the relationship with home (based on the rural Forez, in particular), the project to promote the industrial heritage of the Giers valley, the putting in place of heritage policies (baroque churches, military architecture, rural heritage) by the department of Savoie, and urban policies (especially in the suburbs). As he has written elsewhere (Rautenberg 1997 [1994], 1998, 2003b), his role as an ethnology adviser provided him with many other examples of situations.

"Heterological" conservation can thus be said to have replaced "symbolic" conservation (Guillaume 1980, p. 103). If this criticism, like that of heritage specialists and historians, is aimed at the proliferation of the "heritagizable" due to society's interest in things of the past, its basis is consequently quite different: the essay borrows from philosophy and psychoanalysis to support the theory that the new relationship to the past is a mark of the de-symbolization of our society. This is a theory that is an extension to Baudrillard's criticism, in particular, Baudrillard's *L'Échange symbolique et la mort* (1976).

The theory argued by Henri-Pierre Jeudy in *Mémoires du social* (1986) has many points in common with Marc Guillaume's theory – such as the assertion of a loss of symbolism, and also the role of the industrial imperative in rupture with the traditional conception of heritage – but it nevertheless differentiates itself on two points. First of all, the study is based on examples from the field and the criticism focuses on the way in which objects are handled, while Jeudy obviously does not use terms that did not exist in 1986, on the way in which they are "heritagized". Then, if "transmission in action" – a notion that he was to introduce in *La Machinerie patrimoniale* (2001) – did indeed result in the 1980s from a "collective love of heritage", awakened by "the collapse of the industrial production modes", it is because "instead of being imposed as a writing of history from which people are excluded, as a writing that is written without them and yet of which they are still the living witnesses, this construction of transmission had become at the time [of its development] everybody's business" (Jeudy 2001, pp. 29–30). This is what makes it an enunciative event marking the disruption generated by the emergence of social heritagization. Even if the criticism is aimed at the "collective museophile drive", which was to operate a spectacularization through the production of images and the mobilization of objects, forging a fiction of what the symbolic discussion was in the original social group (Jeudy 1986, p. 42), the study focuses on the modalities of a production that involves the collective[6].

3.1.1. *Industrial heritage, an example of transmission in action*

The fact that "new heritages" are characterized by the desire to preserve the objects as markers of the group's identity and by the principle of museographing these objects, results in a theatricalization that obscures the mutations of the plural memories of social issues, resulting in the production of a "simulacrum". Therefore,

6 This book can be situated in the author's approach which is demonstrated particularly well by two very different works: *La Publicité et son enjeux social* (Jeudy 1977) and *Les Ruses de la communication: L'euthanasia des sages* (Jeudy 1989). On his approach to heritage, we can also look at (Jeudy 2003 [2001]). He formalized his approach in *La Machinerie patrimoniale* (Jeudy 2001).

this "heritage, once constituted, is nothing more than a museum of social issues" (Jeudy 1986, p. 23). Contrary to what they display, these heritages, far from being a "repository of memory", end up being "an obstacle to the movements of memory" (Jeudy 1986). The elements at play are, on the one hand, the population and memory, and, on the other hand, the museography of traces; in other words, the modalities of treating memory and objects. The criticism of the theatricalization resulting from this mode of treatment is picked up and developed – and in a certain way formalized – in *La Machinerie patrimoniale* (2001), a book in which Jeudy explains from the introduction that heritage conservation is based on the twin desire to actualize the past and transmit the present[7]. If the museography of traces does ultimately produce a "simulacrum", this is due to the fact that conservation strategies are characterized by a "process of reflexivity", which is bigger than the simple technical question of staging the exhibition and which has to do with the functioning of society.

> The concept of cultural heritage derives its contemporary meaning from an intensification of global museography. In order for there to be recognizable, manageable heritage, a society must grasp a mirror image of itself, take its places, its objects, its monuments as intelligible reflections of its history. (Jeudy 2001, p. 17)

Actualization and reflexivity can thus be said to be at the very principle of theatricalization, which is inherent in museography, being at the very foundation of heritage. The analysis of industrial heritage helps us to understand this process, because "it is from the constitution of industrial heritage that a craze for the defense of 'new' heritage has spread" (Jeudy 2001, p. 30). Industrial heritage established itself as a novelty, because it:

> Essentially made it possible to legitimize the claims of an identity held as hidden, even guilty. The "world of work" had to show the sovereignty of its identity in the face of that of other social classes that possessed great cultural wealth and enjoyed it in an elitist fashion. (Jeudy 2001, pp. 31–32)

However, this vision of identity was in rupture with the collective memories of the workers, and, by mobilizing "markers, indices of a cultural singularity kept and exhibited", it helped ensure that "the heritage machinery" ultimately followed "only

7 "What prevails is the lure of an actualization of what is kept and transmitted. The rule is clear: if the past is not to be abolished, everything that we live must be actualized" (Jeudy 2001, p. 12).

one path: that of the institutional recognition that heritage confers on any form of claim to identity" (Jeudy 2001, p. 34).

A good illustration of this process of recognition would be the ambivalence of the relationship to work instruments "between the representation of torture and that of human greatness": although we intended to make it objective, as soon as the memory was put in the museum, it "was emptied of all its lived intensity to become the objective summing-up of a concluded history". The reason for this is that it is then divided into two modes of representation: on the one hand, the populist aesthetic of the transmission of know-how; and, on the other hand, the reminder of the sufferings of the working class. With a dual result: on the one hand, "the function of the object sanctions its aesthetic value, because it expresses the beauty of the know-how"; and, on the other hand, "the heritage of 'workers' memories' is a life-saver: it simultaneously enables us to tell ourselves that this kind of physical exploitation is over, and that, after all, at the time of this 'industrial world', life was perhaps not so dark and miserable as we are told" (Jeudy 2001, pp. 36–38). The display thus carries out a "cathartic metamorphosis" by showing another world that will never again be ours.

In what way does this critique of the museographic treatment of an industrial past shed light on social heritagization?

3.1.2. *"Transmission in action" and social heritagization*

In the criticism I have just mentioned, transmission in action is implemented according to two processes: by its functioning and its effect. In a changing society, transmission is the production of a "simulacrum" of what society used to be; this simulacrum will thus actualize this transmission in the form of a "museum of social issues", and the transmission thus operates a "cathartic metamorphosis". However, the production of this simulacrum that is supposed to maintain the symbolic life of the group – the thing that makes it a group – addresses a program that involves subjects and objects through the implementation of a series of operations addressing reflexivity and actualization. Since the discussion is focused on the actual process of heritagization, it is this second process that I propose to model socio-semiotically in order to understand how it characterizes the social heritagization mode.

The major *operation* that forms the center of the whole process is the *production* of the "simulacrum" of social life, which Jeudy sometimes calls "museumization", "museography", or "theatricalization". I will gather up these various denominations (with their connotations) under the generic concept of *representation*, understood both in the sense of "scholarly representation" (the construction of knowledge on the objects to be heritagized), in the sense of "performance" (a theatrical representation intended for an audience); and finally, in the sense of "process of representation", as meant by Louis Marin (i.e. what takes place and what makes present again). Depending on the case, this *representation* relates either to industrial activity that has ceased (partially at least), or more broadly to an aspect of social life that was connected to this activity[8].

This operation presupposes a number of others upstream and downstream. Upstream: the existence (thus, the prior creation) of the *industrial activity* and the connected *social life* that produced *memory*. Downstream: the *preservation* of all or part of the places and objects, as well as a *maintenance* of the memory; then, finally, the making available to the public of the representation thus produced, in particular, by *performances* (use and activation of the objects by group members based on their memory). This manifests an ideal object that results from the description of the characteristics of the activity established by recording the memory of the group members, or by research[9].

First of all, the *industrial activity* that has ceased, necessarily presupposed the existence of at least two groups of actant-subjects: a proto-actant (sender), the *factory owner*, defining the operations that enabled the functioning of the factory, and the repository of the industrial memory; and a proto-actant-subject *performing the activities* that will be the *repository of working memory* (this is the memory that we are mainly focusing on here).

Then, the *production of the representation* itself presupposes an *actant* for each type of operation involved: *memory repositories*, *activists*, *politicians*, *administration*, *specialists* in memory collection or the conservation of places and objects, *producers of scientific knowledge*, *scenographers*, *mediators*, etc.

Finally, the making available to the public (publicization) requires active participation in the functioning of the representation by at least some of the *members of the social group that is the repository of memory*, in one form or another. It also presupposes a *dual receiver* to whom it is addressed: *members of the social group* and *visitors from outside the group*.

The representation produced involves a number of *actant-objects*: *objects* (places, movable objects, possibly documents), *memories* (collective memory, recorded social memory or archives), as well as *arrangements* that make visits possible.

8 Jeudy (1986, pp. 36–39) cites the iconic example of Ironbridge.

9 This allographic functioning of new heritages, while it may not have the very standardized form of the Inventory, nevertheless held decisive importance on the evolution of heritages, foreshadowing (especially with know-how) the functioning of intangible heritage, which I will deal with in Chapter 4.

Finally, it is essential to make a distinction between two *actant-senders*. One proto-sender is at the origin of the *industrial activity and its functioning*[10]; the sender initiating the *representation*, a sign of the rupture introduced not only by the cessation of the activity, but also by the creation of the heritagization process. This sender is implemented by several types of instances or actors: the social group mobilizing reflexivity and actualization on what the activity was before it ceased, and how it can represent that in order to continue it; but also associations, scientists, political or administrative institutions, etc.

Box 3.1. *Modeling transmission in action*

This modeling immediately highlights how the functioning of "transmission in action" differs from the traditional heritage conception. A comparison with the definition of "social heritagization" that Rautenberg (2003a, 2004) gave 20 years later helps us to grasp the way in which it points to the characteristics of this new form of heritagization.

The first striking clarification concerns the central position of *the actant-subject, being the collective* (group or community). In fact, we find it in the same place in social heritagization, which:

> Basing itself on the recognition by the actors of what for them constitutes heritage [...] aims to manufacture "special commons" that are particular first of all to specific social groups who, only then, will want to inscribe them in logics of conversations and exhibitions. (Rautenberg 2003a, p. 127)

The consequence of this observation will be a distinction between *the collective* (the group) and *the audience* (the non-members of the group at whom the exhibition is directed).

The second observation is the equally crucial importance given to *memory* as an element defining the collective: it has a central function in the reflexivity process and is part of the identity. It corresponds to what Rautenberg (2003a, p. 44), here explicitly moving away from Namer's comment about Halbwachs (and partly from the latter too), calls the "collective memory": a "memory borne by a defined social group [which] is constitutive of the image that the group has of itself". In fact, he distinguishes this form of memory from the "social memory", which refers to "the social practices of memory in general", and, conversely, from the "lived memory", which describes "all the memories that each agent can mobilize in response to

10 Not to mention an anti-sender who was at the origin of stopping the activity. It should be pointed out that these senders or anti-senders are, in real life, instances, actors and social processes reported on by scientific and operational media texts.

external solicitation". In addition to memory, three other heritage elements are part of the definition of the collective: the *objects*, *places* and *practices* that come within what I have called representation and which are regarded by Rautenberg (2003a, p. 143) as resources inherited "from generations and an era still very close", which will "be used by a social group seeking to perpetuate itself, to reproduce itself".

These first two points of convergence are fundamental: for Jeudy, as for Rautenberg, the issue the *continuity of the collective* in the transition from past to present. Between the two, there is agreement on the fact that the chief operator of this continuity is the population in the context of mobilization of a recent past that this population has itself lived. As a result, *reflexivity* enables the group to construct an interpretation and representation of the past in the present, both by using memory[11] and by drawing on places and objects that remain almost identical to how they used to be (such as buildings that can of course be modified, but that stay in the same place). On the other hand, to my knowledge, Rautenberg does not use the term "actualization" in relation to social heritagization. The idea of an annihilation of temporal differences between the present, the past and the future, which would make "the past and the future seem to be combined within the present while the present itself becomes the anticipatory tense for the reproduction of the past" (Jeudy 2001, p. 12); this seems foreign to Michel Rautenberg's approach. Social heritagization does not accord with a transmission of objects that is done under orders not to disappear[12] again. The reason for this is that it does not set out to repeat that society: on the one hand, the heritage objects it produces are circumscribed (they are not coextensive with society, or even with an activity); on the other hand, as inherited resources, these objects "seem to have a limited lifespan, at least as long as they have not joined" the category of institutional heritage objects (Rautenberg 2003a, pp. 41–44). The processes and procedures that lead to the transition from memory to social heritage (the "heritage rupture"), then from this to institutional heritage (the "hardening"), show that Jeudy's conception, which reduces the representation to a simulacrum, amounts de facto to passing over these processes and procedures by superimposing (quite understandably, at the time) the hegemonic model of

11 This is even one of the conditions of collective memory for Rautenberg (2003a, p. 43): "The work of memory is a normative social construction that structures the present by reordering the past through selecting historical elements, prioritizing the sequences of history, proposing a symbolic reading of events".

12 Unless I am mistaken, Rautenberg does not therefore subscribe to the relationship, underlying the "actualization" that Jeudy develops between "heritage mindset" ("the logic of heritage conservation") and "transmission". He writes, for example: "Know-how, mills, a fisherman's hut must not disappear again. Between gestures, buildings, language, it is not a matter of choice, everything is to be transmitted thanks to a preliminary conservation operation". Then he concludes a little later, "becoming a stakeholder in the sole logic of reproducing societies, transmission has become a pure artifice" (Jeudy 2001, pp. 12–13).

conservation; this belongs to the administrative legal conception, over the "new heritages", with the consequence of assimilating conservation, continuity and transmission.

In concrete terms, this reduction is translated into the fact that ways of functioning as different with regards to the Creusot ecomuseum, and, for example, the Ironbridge exhibition, are regarded as identical. While the first is in principle centered on the development of the population's capacities to decide its future (de Varine 2013 [1991], pp. 20–26), the second is effectively the reconstruction of a way of life linked to an industrial activity and an era. In terms of the way social heritagization functions, we can say that the ecomuseum puts into parentheses the operation of representation in favor of maintaining the *continuity of the collective* through the use of a reflexivity implemented by this collective, which acts as a community and is applied to the past in order to understand and possibly to have an impact on the future. The traces of the past are resources that characterize this community: there is therefore no production of a simulacrum (unless we regard, as such, the production of a reflexive knowledge or the traces of the past); however, on the other hand, it constitutes an emblematic form of "transmission in action", since it aims to involve the whole community. Still from the same point of view, the case of Ironbridge is, on the contrary, based on an operation of *representation*: on an interpretation, in the present, of traces partly inherited from the past (places and objects), partly reconstructed from knowledge and know-how, with the intention of transmitting these traces, after giving them heritage status, to both a present and a future audience.

We can certainly consider the representation of the past thus produced as a simulacrum, in that there is necessarily a fictional element (I will discuss this point a little later); however, in terms of heritagization, what is more important is the issue of temporality which, according to Jeudy (2001, p. 12), is inherent in the "actualization" in which "the past and the future seem to be combined within the present while the present itself becomes the anticipatory tense for the reproduction of the past". From the moment we postulate that a communicational *dispositif*, such as presenting an exhibition, necessarily produces something other than an imaginary copy of what it presents, "transmission" cannot become, as he posits (Jeudy 2001, p. 13), "a pure artifice". It could perhaps be a partial fiction, but not an identical reproduction. Therefore, an enunciation of the past in the present, even if done with the intention of controlling the meaning that was to be given to it in the future, can only produce a representation (in the form of a media text) whose interpretation will itself remain open[13]. The author's enunciation of this very sentence indicates that this is "false evidence" (Davallon and Jeanneret 2004), since it is saying that the use

13 We should note that the same process occurs in transmission in action, where the interpretation produced by reflexivity and textualized in the form of knowledge and action necessarily remains open in the future.

of the present tense anticipates reproducing; in other words, it introduces – or at least initiates – a representation of the past in the future perfect tense.

On balance, even if the answer that Jeudy advances in terms of simulacra is more critical than descriptive, thus inviting re-examination, the dual question that it poses, both of the *representation* and the *temporality*, highlights the paradoxical nature of social heritagization between "transmission in action", which aims at a continuity of society, and the production in the present, from memory in particular, of *heritage traces of the past*. In other words, this is a representation that makes this past exist in the present for the future. While the two sides of the paradox are indeed inseparable, a *type of variant* of the social heritagization mode nevertheless appears, depending on whether the emphasis is placed institutionally on the one or the other of these aspects. The first type brings together examples of variants of *transmission in action* in which reflexivity dominates, while the representation exists in a weak or non-existent form of actualization. The second type is characterized by a *representation constructed from memory traces*, but also from objects or places in which reflexivity serves rather as a tool, and in which temporality is an issue in the representation. The common point of the two types, which makes them marker posts characterizing the social heritagization mode, is the *mediation role that is here necessarily fulfilled by the institutions* – scientific, administrative and political – that constitutes its condition of possibility.

3.2. Social heritagization as transmission in action

The Creusot ecomuseum, one of the first ecomuseums to have appeared in the 1970s – at any rate, the first urban ecomuseum (the first generally recognized as such being the ecomuseum at Marquèze, in the Landes Natural Regional Park)[14] – is probably the most radical case of focusing on the future of the community. It constitutes the *point of singularity* marking the emergence of the first variant of social heritagization, in that it totally excludes representation in favor of reflexivity, with transmission in action tending to become in some way "total".

As a counterpoint to the Creusot ecomuseum experience, it is worth examining one of its institutional developments, which, if we follow François Mairesse, is linked to the absorption of the new museology movement into the cultural policy of museums, opening up a new way of thinking and addressing the public.

Experiences that stake their claims under various titles, like the successors to the theorization of the ecomuseum experience, can, as I see it, be grouped together

---

14 See the testimony of Gerbaud (2000) who rightly stresses the role of Jean Blanc in the conception of the ecomuseum.

under what can be classed as "community museology". They extend transmission in action, markedly differentiating themselves from traditional museology by affirming the collective and societal character of social heritagization.

Finally, a form of "ethnological intervention", claiming to follow a pragmatic approach descended from Dewey, seems to me today to extend the eco-museum experience by putting the population's experience at the center of the relationship with heritage.

3.2.1. *The Creusot ecomuseum, a project for "total" transmission in action*

Given its primary role and particularity, I will dwell at some length on the Creusot experience, its particularities and theorization. Unlike what used to happen for a large number of industrial sites that have become museums, the creation of the ecomuseum came as part of a development project (de Varine 2013 [1991], pp. 20–26) and was not focused on the heritagization of objects or places that had lost their function. In a way, it did not result from an industrial crisis, but preceded it, hence the absence of representation (production of simulacra, as Jeudy would say, having also noted this particular character)[15].

Its case is particularly instructive in two ways. First of all, the successive occurrence in a relatively short timescale of its creation, then of a crisis in both the museum sector and the industrial sector which proved fatal to it, then of its evolution into a traditional museum, renders the political and institutional issues raised by transmission in action based on the population's involvement eminently readable. Then, and above all, its theorization by Georges Henri Rivière and Hugues de Varine was at the origin of a movement that was to proliferate widely. The texts resulting from this theorization thus constitute a set of enunciative events, giving the ecomuseum an eminently political and museological role that was to be relatively out of step with the functioning and future of the ecomuseum itself[16].

15 "Apart from the case of the Creusot ecomuseum, which, in its anticipation and its prescience, illustrates a model of strengthening cultural identities in the middle of a socioeconomic crisis, the others are mostly based on the ruins of a region torn apart by the enforced closure of large factories" (Jeudy 1986, p. 51).

16 See: Rivière (1992 [1980]) and Varine (1992 [1978]). The fact that it was out of step is clear from the description of this experiment by Debary (2002) in *La Fin du Creusot ou l'art d'accommodation les restes*, and, even more radically, from the analysis put forward by Le Nouenne (1992 [1978]).

I will briefly mention the essential characteristics of the three phases: its creation, crisis and its standardization. In terms of the place given to the population, the originality of the ecomuseum project lies precisely in the aim of transmission in action, such as how Henri-Pierre Jeudy defines it, but without resorting to the representation of a vanished industrial activity, since, at the time of the creation of the ecomuseum, this activity was still in existence.

Therefore, from the actualization that characterizes transmission in action, we have chiefly the principle of a mobilization of all the members of the collective, which consisted of the inhabitants of the urban community's territory, posited as the collective subject of the ecomuseum and as curators of the objects[17]. While exhibitions of objects were "entrusted" to the ecomuseum, these were mainly addressed at the inhabitants themselves, to be used to bolster the culture and share memory among themselves to construct a representation of the group. Therefore, since the ecomuseum was not based on objects, but on a population and its territory, it could be classed as "anti-museum" by Clair (1992 [1974], p. 522). In the context of sustained industrial activity (even if it was starting to face difficulties), the objective of social reflexivity was a raising of awareness that could be used to *construct a new relationship between the community and its future*; community is understood in the sense of a social group that has become aware of its unity. The objective was thus an actualization of the past in the present with a view to the future, without going through an abolition of temporal differences.

This phase of creation came to an end with the dual crisis of the ecomuseum and the factory, the most significant consequence of which was the disappearance of the place and role of institutional reference given to the population in the ecomuseum[18], which was to lead in a third phase to the return of objects, knowledge about these objects and the memory of their owner (the Schneiders). Debary's study[19] shows that the crisis, from the side of the ecomuseum (which is obviously the side that interests us here), was at the point of convergence of a number of types of institutional factors.

---

17 As de Varine (1992 [1978]) explains: "The local population is then not only the object, but also the subject of the institution, not only audience, but also actor in the action and the animation" (pp. 455–456).

18 We can certainly raise the objection that the population's involvement always remained relative. However, this is less a quantitative issue (participation rate) than a founding principle whose operativity lies in its publicization, and which is above all symbolic (in the various senses of the term).

19 Debary's book is arranged in three parts. The first two present the two phases of the development of the ecomuseum: its creation (*"Le temps perdu"*), then its "standardization" (*"Le temps retrouvé"*). The third part is devoted to an ethnological reflection on these two phases and addresses the issue of social heritage (*"Le temps oublié"*).

Internal factors included the progressive deterioration in relations between management and employees (which led to loss of organizational cohesion) and also management transferring money from one account to another to overcome funding difficulties (which was ultimately used as a motive to call for the resignation of the director, Marcel Évrard). External factors consisted of increasingly hostile relations with the factory, which reached breaking point with the ecomuseum's increasing tendency to show the controlling function of paternalism, particularly by initiating a collection of working memory; they also included relations with the management of Museums of France, which did not accept that a museum could be created not from a collection, but from a "social" project, as confirmed by the constraints that would be imposed for its "standardization"[20].

Having been reduced to next to nothing as an ecomuseum, it then began its conversion to a museum (although still retaining its name) at the very moment when the population faced a major economic and social crisis due to the industrial crisis that Creusot–Loire was going through, with the end of a paternalistic industrial order inherited from the 19th century (that of the Schneiders) and its replacement by a system dominated by financial capitalism. The most visible consequence of the end of this paternalistic order was the empowerment of the city in relation to the factory[21]. While the general process of transforming "manufacturing activities" into "practices designated as 'relating to culture and heritage'" illustrates Jeudy's analysis quite well (see Debary 2002, p. 16), they do so in very different ways according to the two phases: the political dimension of reflexivity, which was supposed to turn the population into a community on the basis of its memory, would give way to the development of tourist-related action and a relatively traditional exhibition strategy. A *museological reversal* such as this, just like the contrast between the two phases that it signals, inevitably aroused comments and discussions. In any case, it signaled the disappearance of the population as an operator and as a *raison d'être* of the institution.

In terms of this *raison d'être*, everything suggests that the project, having subordinated its artistic, memorial and heritage role to its function vis-à-vis the population, found itself overtaken by social and political issues, inherent in an aim

20 de Varine (2013 [1991], pp. 20–26), on the other hand, sees the starting point of the crisis more in the mismatch between an initial project aimed at community development and the requirements gradually imposed from the outside (starting with academics). It should also be noted that there may be a gap between a conception of a relatively open museum and heritage (Icom), and their traditional, more conservative French conception.

21 The detailed description of these two crises occupies Chapters 5 and 6 of Debary (2002); the "standardization" of the ecomuseum and the city (in the sense of making it a "standard" museum and a "standard" city, i.e., in conformity with what is usually to be found) is developed in Chapters 7 and 8.

of development, which came into conflict with the traditional role of the heritage institution. I had previously shown, based on a project analysis (texts and interviews), that the axiological foundation (the value system) of this establishment of the community as a collective actor was none other than the democratic model serving as common denominator for the two aims, heritage and political. "The creation of the ecomuseum can be seen as the result of a kind of founding pact in which what were hitherto only groups, individuals with separate interests, become a community defined by a collective interest" (Davallon 1986, pp. 115–116). The establishment of this new actant – a community that exists inseparably in representation (in the mirror produced by the institution) and in reality (as an actor) – is at the center of the ecomuseum[22]. However, the description of the ecomuseum's crisis, as compiled retrospectively by Octave Debary, leads one to question, beyond the model, the effective conditions for success of such a founding pact: as conditions that make the functioning and effectiveness of the institution possible. This then raises the question of how a community can exist as such in the face of the social and political issues inherent in the situation in which it finds itself, given the project that it is developing and the heritagization in which it is engaging. This crisis shows that the fact of laying down this kind of framework for the action, of seeking to bring people together and persuade them through discussion, while using as a basis a knowledge of the past and the current environment, can take on an eminently political character that significantly exceeds the sociopolitical role traditionally devolved to the heritage, insofar as it aims to establish an actant that no one is expecting (and often, not those who think they have the situation in hand). The interest in examining these conditions for success lies in the fact that they are an issue for all population-centered social heritagization experiences.

In practice, the ecomuseum found itself confronted by what we must call the paradox of democratic functioning: to establish an entity endowed with competences (duties, knowledge and power) that enable it to act, while at the same time this entity is physically and socially composed of a diversity of individuals within a restrictive societal environment that has resulted both from the "will of man" and from the "work of the centuries", as Nora would put it. In theory, the democratic model, as outlined above, presupposes at least two conditions for success: first, a legal-political leveling-up of the subjects (individuals or groups) who constitute the population; second, recognition by all concerned that there exists a general interest above private or sectoral interests; in other words, the will and the capacity to accept a common interest. In this case, it was the development of the community as an operator of its future. This assumes implementation of a *dispositif* that guarantees the reconciliation of conflicting social positions. However, what seems to show the practical future of the Creusot experiment is that neither the inventors' strength of

22 de Varine (2013 [1991], pp. 20–26) shows how establishing this was indeed the challenge of the community development project.

conviction on its own, nor the administrative rules or the operativity of the heritage institution were able to deal with the transition from a population of individuals to a "community", in the sense given to it by the ecomuseum project (i.e. as a collective actor conscious of its existence and capable of taking action on its future). Once the ecomuseum was up and running, and the effect of the community work begun before the creation of the ecomuseum itself had disappeared, the groups (or individuals) remained caught up in a past, in a position, in conceptions and interests that were too deeply conflicting. This hypothesis would suggest that, in this specific case, the project to establish a community as the representation of the population appears on the contrary to have exacerbated the conflictual nature of relations between the working population and the factory, as well as the differences of opinion within the working population itself and between the protagonists of the ecomuseum. The various texts that discuss this experiment suggest that the project crystallized these various forms of divisions, the positions of the actants tending to become disjointed, even conflicting, from the point of view of knowledge, values and also powers[23]. We might even wonder whether this situation was not further exacerbated by the factory still being in operation, because had the factory and its management had disappeared earlier, this would perhaps have distanced some of the conflicts, leaving them to exist only in representation. Whatever the fact of the matter, this breaking up of positions left two actants face to face (the ecomuseum's administration on the one side and the factory management on the other, the population now being out of play), who possessed a very unequal and totally unbalanced social power, at the very moment when they found themselves engaged in a conflict of legitimacy relating to conceptions, competences and knowledge about what could, or should, constitute "heritage". To put it simply, on the one hand, the collection of workers' memory, along with the conservation of objects of daily and collective life; on the other hand, the preservation of the official factory archives and industrial know-how (François Bourdon Academy), along with the conservation of industrial objects and power (for example, the pile driver and the Schneiders' possessions). While objectively this conflict resulted in a victory for the power of the factory – and also incidentally in a victory for the administrative legal heritage model – the ecomuseum, in becoming a reference model, nevertheless won a symbolic victory.

Indeed, if the Creusot ecomuseum is a point of singularity opening the series of variants of social heritagization based on the population, it owes that largely to the enunciative event constituted by the theorization about it, carried out by Georges Henri Rivière and Hugues de Varine: on the one hand, this theorization interprets the experiment in a media text that is both programmatic and scientific; and, on the

23 In a completely different register, we will find a particularly instructive example of the existence of differences and conflicts between subjects or between social groups with the preparation of the application dossier of Biou d 'Arbois in Ailhaud and Barbe (2019).

other hand, its analysis helps us, in the same way, to understand how this experiment became emblematic. In the definition of the ecomuseum, formulated by Rivière (1992 [1980], p. 443)[24] for Le Creusot, the population is placed, next to power, as one of the two actants who "design, manufacture and jointly exploit" the ecomuseum. Narratively speaking, it is therefore placed in the position of both sender and subject (in reality, the sender is political). The objective is to turn the ecomuseum into a mirror, where this population can look at itself specifically in order to know themselves, so that they can think about their future. In his definition, de Varine (1992 [1978]) for his part goes well beyond the "role of information and critical analysis" that Rivière (1992 [1980], pp. 443–444) assigns to the ecomuseum, but without going as far as political decision-making. For de Varine, the ecomuseum can be "an instrument of popular participation in territorial planning and community development". Although we certainly fall short of decision-making, there is nevertheless a form of active participation and involvement of the population in the development of the territory that is nevertheless an issue here. In the rest of the text, after having reviewed the characteristics of the first two ecomuseums (Marquèze and Le Creusot), and also their differences, he advances a third form of ecomuseum which he calls the "community ecomuseum", for which he then sets out the theory. First and foremost, this new form shifts the center of gravity from natural ecology to "human ecology", which now includes the first. Above all, it attributes a sociopolitical role to the ecomuseum that goes beyond the basic role of laboratory, conservatory and school, designed to encourage the population "to better understand the problems of its own future", that is given to it by Rivière. Even if its role remains, overall, predominantly heritage-related, in the sense of a consideration and a critical use of knowledge about things of the past (including the environment), the ecomuseum is then given the task of putting the community in a position to play an active part in its development; this is resolutely turning it to face the future and allowing it a real political role that presumes the establishment of a collective subject (a community) capable of taking action on its environment and on its future. We are thus in a radical form of "transmission in action", the purpose of which, far from consisting in producing a representation, is to operate a transformation: to use the knowledge of what makes memory and heritage to turn it into the means to establish the community as a subject capable of playing an active part in its future. Hugues de Varine was conscious of the "usual political tensions" that would inevitably be encountered, and so he specified that the rule must remain precisely that "the eco-museum has the role of changing the forms of the political stakes in the community by including the entire population that has become aware of its power to

24 The full definition can be found in *Vagues* (Rivière 1992 [1980]). For simplicity's sake, I have chosen 1980 as the original date of publication, but this publication is actually the combination of three fragments of which there are a number versions staggered in time with variants set before 1980 for the first two fragments: 1973 for the definition essay and 1978 for the description of the structure of the ecomuseum.

take action on its own development" (de Varine 1992 [1978], p. 485). These two approaches (of Rivière and Varine) sketch out what constitutes the scope of the ecomuseum (its mission), but also all its ambiguity: being a political institution because it is a heritage institution founded on an aim to establish the population as a community, and which will therefore be caught up in political issues as in reality its project is political. An aim and an ambiguity which, as we have seen, were present from the creation of the Creusot ecomuseum and which conditioned its functioning and therefore its future.

Just as, for an understanding of social heritagization, the point that should be taken away from Jeudy's criticism was that, in the background behind the representation, the issue lay in taking population and memory into consideration; in the same way that the Creusot experiment shows that social and political issues are inevitably there behind the declared aim of establishing, based on heritage, a future-facing community and that they determine to a large extent the fate of the institution, and not the other way round.

3.2.2. *A "population" that has become synonymous with "audiences"*

In contrast to this kind of political issue, Mairesse (2000), in his article published in *Publics & Musées* 17-18, *"L'écomusée : rêve ou réalité"*, revisiting the history of the new museology, and with it, the fate of the ecomuseum, argues that this museology has been absorbed by what I call "culturalized museology"; according to which, the publicization of heritage (in particular, the exhibition) is treated in the mode of a cultural activity designed for an audience, and no longer as a tool at the service of a community.

Mairesse begins by looking at the precursors (if not the ancestors) of the new museology: Edmond Groult (cantonal museum of Lisieux, France), Patrick Geddes (Outlook Tower, Index Museum in Edinburgh), John Cotton Dana (Newark museum, United States), and also Jean Capart in Belgium and Theodore Low in the United States, who implemented and thus defined the social role of the museum; it is this role that would before long precisely constitute the challenge of the ecomuseum. After the Second World War, especially from the 1960s and 1970s, new ways of thinking about the participation of the population in museums emerged, based in particular on heritage's role as a factor in the identity of this population, at least for museums that chose to be on the fringes of the traditional museum institution whose the model remains the Museum of Fine Arts. The Creusot ecomuseum then took its place alongside other landmark experiments, such as the Anacostia Neighborhood Museum (in a disadvantaged district of Washington) or the Casa del Museo (in a slum in Mexico City). These various experiments are characterized by the central place they give to the community[25], which would subsequently be at the center of a new conception of museology with the Declaration of Santiago, Chile, in 1972. The project for change in museums that this declaration proposes, which is then inscribed in the project for a change in society, would be supported by Icom, at the instigation of Rivière and Varine. This is the time when many ecomuseums were born; Mairesse mentions two experiments that, according to him, are among the most emblematic: in Belgium, those in Watermael, Gilly and Hamoir (Étienne Bernard), and in Quebec, the one in Haute-Beauce (Pierre Mayrand and Maud Céré). However, he explains, after experiencing a phase of creation and great success, these two experiments went into crisis; then, being no longer funded by governments, they eventually disappeared. As Debary (2002, p. 118) notes, in France, after the period from the late 1970s to the late 1980s, when the ecomuseum model enjoyed great fortune, many ecomuseums created at that time either disappeared or were turned into museums. This was an observation that served as a basis for Mairesse's theory that considers that the desire to put the population at the heart of the new museology (as for the ecomuseum) was carried forward in the attention that museums pay to audiences, thus reconfiguring how they perceive their mission[26]. This desire would ultimately reappear in the commercial twist that turned this population into "visitors" (that we might even call an actual or potential "clientele")[27].

If we follow Mairesse, the ecomuseum and the new museology can thus be said to have served to renew the purpose of the "old" museum, before it generally became subjected to a commercial logic (Tobelem 2005). Undeniably, in France, the evolution of the museum, the beginnings of which could actually be seen emerging in the 1990s with the advent of the category of "audiences" being used as an evaluation criterion (Davallon 1997), today impacts nearly every museum and everywhere that promotes heritage. It is hard to deny, or even to ignore, the fact that the "new museology" has effectively contributed to a modification of the museum. However, the museum, even if actualized in this "cultural" format, is here mainly approached through its mission of publicizing heritage.

This is what makes it so easy for museums to enter into a logic of "cultural offering" whose principle is to address "audiences", conceived as collections of individuals who, while having diverse interests, momentarily converge towards what

25 "Here, the collection and all the technical projects arising from it (conservation, management of objects...) are only of very relative importance in relation to the work on – and by – the community" (Mairesse 2000, p. 44).

26 Perhaps we should this into context with the Icom crisis triggered by the project for a new definition of the museum.

27 As shown by the invention of the category, which makes no sense outside of a marketing point of view – even though museums must realize it – of "non-public".

the museum is offering. For these audiences, the knowledge produced by scientists and specialists is generally used to interpret objects (*realia*) that are more or less foreign to the group to which the members of the audience belong and do not a priori have any relationship with the individual or collective memory of the audience. Whatever the additional social interest of this cultural watershed of the museum or the heritage – its culturalization – which clearly should not be ignored, we are a very long way away from an action by (and for) the population, as conceived by the ecomuseum and its descendants, aimed in particular at establishing this action in a community. Serge Chaumier's (2013) analysis of the future of the Saint-Nazaire ecomuseum in the face of Escal'Atlantic is, from this point of view, particularly instructive on the differences that exist between the two conceptions. At the beginning of his contribution, he takes stock of the reasons why. If we restrict ourselves to the landscape of French museums, few ecomuseums have managed to prolong an "interest in a participative and involving conception of populations" (Chaumier 2013, p. 271). An administrative context and a museum model based on collections of objects from the past have been disadvantageous to a museology that puts the population at the center – and even more, it claims to be at the service of the community (Delarge 2000). From this point of view, we can even interpret the conflict between volunteer activists and professionals as one of the consequences of these disadvantageous conditions. Chaumier (2003, 2005, 2011) has long since shown the recurrent and institutionally structuring effect of this conflict, which has most often ended in the former being replaced by the latter. However, these conditions have not prevented some ecomuseums from keeping themselves going and even growing; and for some, from coming into being[28]. However, perhaps we also have to accept, as both Henri-Pierre Jeudy and Jean-Claude Duclos (1995, p. 296) invite us to do, that it is the destiny of the ecomuseum to be temporary.

3.2.3. *Community museologies and the ecomuseum model*

In the face of culturalized museology, which is most especially dominant in large museums, the ecomuseum model of social heritagization, as transmission in action, has chiefly developed outside France, particularly with, initially at least, the support of Icom (more specifically, under the aegis of Icofom, its committee for museology) and with the support of the International movement for a new Museology (Minom), in experiments that have sought to apply to the letter the principle of a heritage

28 List of ecomuseums not labeled "museums of France", on the website of the Federation of Ecomuseums and Society Museums. Available at: www.ocim.fr/2019/12/ecomusees-et-musees-de-societe-fonctionnement-actions-et-dynamiques/ [Accessed March 22 2022].

institution defined by participation and involvement of the population, even going so far as to sometimes redefine the museology of the ecomuseum[29].

I would like to bring together these various experiments – whether they belong to ecomuseology, socio-museology, community museology *stricto sensu*, social museology, or related forms – under the generic term "community museologies" in reference to the work of Hugues de Varine (2013 [1991]) on "L'Initiative communautaire", which went further along the path he had previously started down with the definition of the community ecomuseum mentioned above (de Varine 1992 [1978]). This seems to me to have drawn a line of force that is common to these museologies. They have been developed and are still being developed, mainly in countries where the heritage and museum model is less fixed and rigid than in France, due to a less centralized state (Portugal and Brazil, in particular), or due to an effective decentralization of decision-making that facilitates the existence of local experiments (for example, Italy and Spain). There is a relatively abundant literature (especially in Portuguese) on the transition from the Creusot ecomuseum to these new experiments[30]; it forms the nebula of enunciative events of community museology in that these experiments sketch out forms of involvement of the population in the extension of the ecomuseum model. Given the abundance of these events, I will confine myself here to referring to a few summary texts that can serve as markers that the reader can consult, but taking care to quickly point out what seems to me to be their specificity. As a starting point, a presentation of the entire journey from Le Creusot to these museologies can be found in two articles by Bruno Brulon for Icofom, one in French (Brulon Soares 2015a) and the other in Portuguese (Brulon Soares 2015b); these can be complemented by testimony from André Desvallées collected by this same author (Brulon Soares and Desvallées 2015 [2014], p. 274). Hugues de Varine had made an assessment of ecomuseums (de Varine 2017) and, with Graça Filipe, written an important contribution, *Que Futuro para os Ecomuseus?"* (2015), which reviews the history of ecomuseums and identifies the risks, responsibilities and trends. This may be supplemented by the interesting semantic analysis of the different conceptions of ecomuseology, community museology and social museology in Brazil contained in Suzy da Silva Santos's Master's dissertation (*Mestrado*) (2017), *Ecomuseus e museus comunitários no Brasil*. For a discussion of the differences between social museology and

29 For an overarching discussion of the principles of ecomuseology, we refer to the first part of Serge Chaumier's book (2003), especially Chapter 3, entitled *"Théorie de l'écomuséologie, point de vue critique"*, which traces the early evolution of this museology and at the same time shows its problems and its limits. This is a very well-documented approach to the political dimension that forms the context of the development of the "new museology" in France and Quebec can be found in Gunter (2019).

30 Summarizing this seems to have almost become an academic topic in museology in some countries, such as Brazil.

socio-museology, from the conceptual, practical and institutional points of view, I refer to the article by Átila Tolentino (2016), *"Museologia social: apontamentos históricos e conceituais"*.

If, as can be seen, these museologies are diverse and plural, they have in common a rupture with culturalized museology and the fact of sharing the very principles that specify the variant of social heritagization mode centered on the population, namely (i) shifting the mission of the institution toward the population through reflexivity; (ii) mobilizing a wide variety of traces that have connections with the population and its living space; (iii) complying with an aspectual temporality where the past is associated with the present looking to the future; and (iv) the communicational split of the receiver between community and audience. I will come back to these various points that specify social heritagization.

First, in accordance with the principles put forward by Rivière and de Varine about the ecomuseum, the population is simultaneously the object and the subject of the institution: it is not only the actant that the institution is talking about, not only the actant first addressed by the institution and thus the chief beneficiary of what the institution offers, but it is established as a collective actant (a community) endowed with the capacity to understand itself, to know its environment and to play an active part in its future. If these museologies are varied and plural, it is because the situations in which they play an active part are very diverse and they must adapt to constraints and expectations. When the population is, for example, to some extent already constituted as a community, the institution (or the intervening body, be it heritage, social or political) can seek to reinforce it by rendering it visible to itself and to others[31]. If, on the other hand, it is a matter of creating a relatively unitary entity from a heterogeneous population, as in the case of new towns, the institution can, on the other hand, use a sharing of memories and experiences as a basis for creating a common culture[32]. Between these two extremes, there is a whole range of intermediate situations, of which the emblematic case is probably the will of a political body set up to involve the population in the territory's development, by seeking to strengthen the sense of belonging by using elements of already existing heritage or elements of the memory of the inhabitants; for example, we find an illustration of this aim in the transformation of museums into eco-museums for example[33]. However, whatever the case may be, the objective is to assign a symbolic

31 A series of particularly illuminating Italian examples can be found in de Varine (2013 [1991]).

32 See, for example, Delarge (2000), Benkass (2012) and Davallon (2012b).

33 By way of example: the transformation of the municipal museum of Seixal, a city near Lisbon, into an ecomuseum, described by Graça Filipe (2000) in her Master's dissertation (*Mestrado*). This is also the case in France for the Viuz-en-Sallaz Paysan Museum, created in the 1980s, which became the PAYSALP ecomuseum in the 1990s.

status not to things, but to a population by conferring upon it a cultural (and political) capacity that enables it to look to the future, and by recognizing the fact that it is located in a space that serves as a base and defines its limits; to recognize it or to endow it with a community existence (perceived as a heritage, as one might say), linked with memory and with objects that have been to an extent heritagized. The corollary of this symbolic status attributed to the population is the existence of societal conditions that favor its establishment, such as the size of the group, its existence as a community, a capacity to self-organize, the presence of a facilitative instance, the existence of political will, etc. This allows the heritage institution to function for long enough to acquire operativity.

Second, "things" are the means used to transform the representation and status of the population. They can be very diverse, not only according to situation, but also within the same situation: spaces (places, arrangements of elements, organization of space, names, etc.); objects (buildings, movable objects, documents, personal objects, etc.); memory traces (recollections, stories, songs, etc.), etc. These things constitute traces as long as they can be interpreted; that is, they can have meaning for this population in this place. The enunciation of this interpretation will therefore produce heritage objects that are both complex and situated, made from interpreted traces linked to a population perceived as an entity located in a space and facing toward action.

Third, these heritage objects are governed by a temporality that is not chronological, but aspectual. Their enunciation defines the present, starting from which, as traces, they "have been"[34] (in fact, in relation to the enunciation's present, they necessarily belong to the past); but as symbolic resources for the population, they enter into one or more media texts that endow them with meaning – a scholarly representation – in a present that is inclined toward "what must be made to happen".

As a result of this, the traces take on meaning not only in relation to the past (due to the fact that they have been), not only in relation to the present of their enunciation (the meaning that these traces have here and now), but because they are caught up in the population's impulsion toward future action that articulates a relationship to the past and present. From the point of view of temporality, community museologies are thus to an extent an ideal for all museologies governed by the social heritagization mode.

Finally, by giving the population the leading role, this ecomuseum model of heritagization formats in its own way the split in the receiver of the enunciation of the heritage object between, on the one hand, the population (the collective, group or

34 I am picking up the formula ("it has been") that Yves Jeanneret (2020) himself borrows from Barthes, adapting it to social traces.

community); and, on the other hand, the audience. We know that Rivière had given the ecomuseum's productions (conversations, collections and media texts) a mirror function for the population, which then occupies the position of audience.

The communicational circuit was then essentially participative and reflective. While this is a theoretical schema (or an ideal), because communication, even if participative and reflective, is necessarily addressed to another party using mediation, the fact remains that community museologies invert the priority between communication towards an audience (specific to culturalized museology) and communication aimed at the community: they seek to produce a reflexivity on the part of the population, and, when there is in fact an audience, to create empathy based on a partial identification of that audience with the population.

3.2.4. *The ethnological intervention and heritage as an experience*

While few of the texts, which make up what is after all a substantial field of ethno-anthropological literature dealing with heritagization and ethnological intervention in heritage matters, refer, in one way or another, to the ecomuseum model; nevertheless, many of them do share with it the particular attention they pay to the population. This attention then takes very varied forms, ranging from a theoretical (or even epistemological) reflection on the ethnology or anthropology of heritage, to a report on a study or a project. Although such attention does not imply a reference (let alone claim any affiliation) to the field of museology, it does, however, presuppose a recognition of the specificities, capacities and future development of the population; therefore, its situation, culture and also its action with regard to the heritage that makes these texts media texts in the field of heritage.

This level of proximity is particularly clear in the works of Jean-Louis Tornatore, but also in those of Noël Barbe, as well as in the research or studies of those who collaborate with them. Therefore, for example, in an article co-authored by Tornatore and Paul (2003)[35], which looked at four industrial wastelands, the two authors explicitly established a similarity and a comparison with the Creusot ecomuseum. This similarity, which also came through de Varine and his criticism of museology, went hand in hand with the rejection of the name of "museum" in favor of "intermediate spaces"; in other words, heritage spaces with a cultural vocation appropriated by the population. However, it is in a recent synthesis of an initiative

35 The article is based on research, ongoing at the time of the article, on the repurposing of four industrial wastelands: the glass site of Meisenthal (Moselle), the former forges at Écurey in Montiers-sur-Saulx (Meuse), the former Japy spinning mill in Audincourt (Doubs) and the textile site at Lepuix-Gy (Belfort Territory). The challenges are described in the article by Tornatore (2010a) on *"Les banquets de la mémoire à l'auberge du patrimoine"*.

that Tornatore (2019b) has been following for a long time[36], looking at the relationship that the population has with heritage (but also building on this relationship, which seeks a redefinition of heritage itself), that we can best see the way in which the ethno-anthropology of heritage renews the questioning opened up by the ecomuseum variant of social heritage. As is clearly suggested by the title of the book, *Le patrimoine comme expérience* (2019a), this initiative is based on Dewey's pragmatist philosophy, from which he picks up one of the basic principles, which Jean-Pierre Cometti had pointed out in his time: to prioritize not the objects; but the *use* – in this case, the subjects' *attachment*. He writes, for example:

> An anthropology of heritage necessarily goes beyond the institutional and administrative construction of heritage, which as we know from the outset has historically been in league with art history, to question heritage as a diffuse social operator centered on the assertion of an attachment to things or beings in the name of variable conceptions of the temporal continuity of human societies. (Tornatore 2019a, p. 47)

This position, he explains, involves three paths of analysis – in reality, in my opinion, it rather operates three shifts. The first shift is from strangeness of the object (for example, of the building) to a relationship of familiarity, and also from distance to proximity. The second goes from transmission to an interrogation of experiences "in the density of the present as gesture both of problematization and of actualization of interactions with an environment" (Tornatore 2019a, p. 47). The third encounters a gulf between the nation's experience, and, for example, the life experience in which it is caught up, such that one is led to wonder what, ultimately, a heritage experience can be made of and consequently to imagine its historicity and the possibility of it being exceeded. These shifts have two notable effects on the population approach: the giving of visibility to those who are usually invisible and voiceless, and also the importance assigned to culture – "namely, what creates commonalities while circulating" (Tornatore 2019a, p. 51).

It goes without saying that this mode of social heritagization based on the population's attachment relegates the gesture of declaring heritage status to the background, when not sidelining it completely[37]. To use the words of Rautenberg (2004), this mode could be said to operate by "appropriation". In the same way as the Inventory makes knowledge production the operation that gives a thing its heritage status, basing the social heritage mode on attachment – which Tornatore

36 See Tornatore (2007, 2010a, 2010b, 2017a, 2017b).

37 As the author explicitly argues in the conclusion of the article entitled *"Pour une anthropologie pragmatiste et plébéienne du patrimoine : Un scénario contre-hégémonique"* (Tornatore 2019b). An article that in a way continues and completes the introduction of the collective work (2019a).

(2019b, n. 18) defines as "what we care about and depend on" – gives pole position to the population's interest in the "thing" to be heritagized (the gesture of rediscovery) or the relationship that this population has with the heritage object (the gesture of publicization), but as part of an approach that is nonetheless very different from the ecomuseum model. The examples studied show the extreme diversity of forms that the population's attachment can take. On the one hand, it could be the primarily aesthetic and recreational attachment of amateur practices of exploring industrial wastelands; on the other hand, the emotions stirred up by the fire of the castle of Lunéville. While the former occurs ahead even of any recognition of these things as traces, apt to cause them to be considered as heritage, thanks either to an interest shown by a population wider than just amateurs, or to an inventory by specialists, or to the production of knowledge on these traces (Rojon 2019)[38], the latter reflecting a use of established heritage that brings into play a competence of the population (in the semiotic sense of a capacity to interpret and act), based chiefly on the memory of past experiences accompanied to a greater or lesser extent by circulating knowledge (Tornatore 2010b). This is a competence that is usually left entirely in the shadows by the enhancements operated by specialists[39]. These two examples lay down the two marker posts between which there unfolds a wide range of forms of attachment, among which we can identify some that relate to traces that have already acquired heritage status and others that, on the contrary, contribute to recognition of this status, or may even, as in the case of the wastelands studied by Sarah Rojon, precede it or escape them.

When the attachment relates to objects that have already been heritagized, it appears as the memory of various practices or experiences, whether in the form of recollections or through the mobilization of a transmitted memory. Today's experience of the object, by activating this memory, makes it a *trace* of these practices and experiences of the past, thus weaving a strong link between the interpreter and the object that is the medium of the attachment. This link will give the object a value for the interpreter, as well as the specific status that the interpreter attributes to it. This bond of attachment fundamentally differentiates this mode of social heritagization through attachment from the mode of heritagization through administrative legal declaration (designation and protection by the State), or through the production of scientific knowledge (for example, scholarly recognition of the Inventory); even if, in practice, it also effectively has a role in the latter. Indeed, there is no shortage of examples where the identification of "things" worth

38 A recognition process that in a certain way triggered the author's ethnological study, as has been shown by the existence of the Urbex exhibition at the Mining Museum, Saint-Étienne.

39 A competence that Tornatore (2010b, p. 112) calls "expertise", defining it as "proximity to an object of the past or to what it represents (the values of value) [which] gives, without overestimating its expression, a capacity to act as its spokesperson, to initiate action aimed at promoting it".

protecting, or inscribing in the Inventory, arises from an experience often classed in this case as aesthetics kindling an attachment. For example, the way Sagne (2019) portrays the relationship of attachment of mediators to the heritage object they present to the public is, from this point of view, particularly revealing of the importance of the attachment that exists among professionals.

When the attachment contributes to recognition of the object's heritage status (heritage in the process of being made), in substance, the process is more or less the same, but the seeking of recognition of an institutional status changes the game profoundly. The diversity of experiences at the origin of the variety of forms of attachment cannot be translated "naturally" in the sharing of the recognition of heritage status. The first condition of this sharing is that the result of the experiences of the social subjects (value conferred and status recognized) must be enunciated in order to become an institutional fact, not in the sense of an "institutional and administrative heritage construction" coming from an administrative and political instance, to which Tornatore refers above (2019b, p. 47), but in the sense of the *establishment* of the constitutive rule of heritage (the obligation to protect), of which the administrative legal procedure is only one of the modes.

However, an example of procedures specific to the mode of social heritagization through ethnological intervention linked to attachment is precisely what we find in the contribution of Ailhaud and Barbe (2019). This focuses on the manner of collecting and enunciating the variety of practices, experiences, interpretations and attachments related to Le Biou d'Arbois according to the "audience" in order to produce a media text of the scholarly document type: the application dossier seeking inscription on the list of intangible cultural heritage (UNESCO 2021); or, as another example, the contribution of Berthelleu and Dassié (2019) on the analysis of the political dimension of heritage and the effects of this analysis on how to investigate and implement a collaborative project in order to produce knowledge shared between individual, or collective memory and specialist knowledge in the context of creating an exhibition. In this project, the researcher is "the one who leads the initiative, takes part in its realization and [who is] the observer of the scene in which he is involved" (Berthelleu and Dassié 2019, p. 264).

In a completely different situation – the case of the destroyed black village (Losonczi 2019) – the function often performed by the ethno-anthropologist, which consists of contributing to creating communalities from the variety of attachments, is performed in this case by the very culture of the inhabitants (especially through the importance of the relationship to death and spaces); however, the fact remains that the researcher's work, as produced in the scientific media text, makes the inhabitants' attachment to the place exist in the public space; whereas, "media" and political practices tended on the contrary to deny it.

Therefore, while ethno-anthropological intervention certainly does not lead directly to declaring heritage status, it nevertheless plays a part either in the preparation of such a declaration, or at the very least in establishing recognition of the status. This is by contributing to making certain things a communality that constitutes heritage or by making it public[40]. Once the manifestation of this recognition is circulating (in presence or mediated), it actually produces a dual representation: representation of the value, generally social, attributed to the heritage object and representation of the collective of those who are attached to it. This intervention then appears as part of the heritage institution, and it becomes in some way a replacement of the reason for making heritage that art historians used to base on the nature of the object[41].

3.3. Representing practices through memory and traces

A careful examination of the ecomuseum made it possible to understand how the use of reflexivity was a means of constituting and involving a collective; similarly, use of the collection and processing of memory helps us to understand the functioning of the representation in social heritagization. Memory is necessarily present in transmission in action, but it remains at its service. In the second type of variant of the social heritagization mode, it is, however, at the center of the process. To define the place of memory in the representation and the way in which memory has been gradually heritagized, in a first point, section 3.3.1, I will focus on Florence Descamps' book, *Archiver la mémoire : de l'histoire orale au patrimoine immatériel* (2019), which shows how memory has been gradually heritagized. In a second point (section 3.3.2), I will come back to the way in which the various types of museography mentioned by Henri-Pierre Jeudy mobilize memory, as well as those who bear it, especially in order to interpret objects and places. Finally, in a third and final point (section 3.3.3), I will examine the experiment conducted by the Musée Dauphinois with the exhibition of the testimony.

40 The article by Tornatore (2019b) concludes with an interrogation of the relationship between commonality and heritage.

41 This question of the position of ethno-anthropological intervention as part of the heritage institution (in particular, through the intermediary of advisers in ethnology and of the Ethnological Heritage Mission) has been the subject of several publications (see, for example, Rautenberg 1997, 1998, 2003b; Barbe 2008). Barbe and Tornatore jointly organized, between December 7 and 9, 2016, a symposium at the University of Burgundy entitled *Du moment du patrimoine ethnologique*, when the call for papers focused on the institutional particularities of the Mission du patrimoine ethnologique in the emergence of social heritage in France (for which, as far as I know, no statutes exist).

3.3.1. *The enunciation of memory and the heritagization of testimony*

Compared to the complexity of the process of identifying and interpreting traces, memory collection may seem an extremely easy operation: all that is required is in some way to "pick" it. However, in reality, that is not at all how it works. It is true that the world of museums, confronted by the silence of objects from other contemporary cultures, had found a lifeline in collecting information from members of the community in order to *document* the elements of material culture that ought to enter collections or be exhibited. This information collection may have given the impression that memory collection was like this. However, the enunciation of memory constitutes a unit of discourse in itself, which needs to be interpreted. To appreciate the difference, we simply have to place the collection of information in order to document objects, side by side with the gathering of life stories of migrants from Aubrac, living in Paris, who were involved as part of the major multidisciplinary study on Aubrac, which was conducted in the 1960s by the Museum of Popular Arts and Traditions, under the direction of Georges Henri Rivière. First of all, the status attributed to the person enunciating the memory is not that of a simple informant: in an articulation of memory and collective, the collection of memory, which was then a methodological novelty (Segalen 2010, paragraph 33), actually started the sideways shift from informant to *witness*. However, this first difference concealed a second: the underlying logic of the discourse (a position of the person facing a situation and events) which de facto makes the enunciation of memory an autonomous unit of discourse and not a simple compilation of information about other things. The proof of this is that many decades later, when Florence Descamps discusses this novelty in *Archiver la mémoire*, she notes that "at no time is any thought or planning given to the conservation and archiving of the materials produced, nor a fortiori their heritagization" (Descamps 2019, p. 48).

We have to read Florence Descamps' book to understand the road that had to be traveled – and the barriers that had to be lifted – for these units of discourse to acquire the status of archives. Here, she is in fact proposing a real archaeology of the constitution of "oral archives", as she calls them, in a methodological and epistemological perspective that gives substance to memory by showing the operations that, constituting as many mediations, make it exist socially. In consequence, this archaeology sheds light on the place and use of memory in social heritagization. After the Aubrac experiment, the first stage in the development of the memory study was the division of the process into two phases – *collection* proper, and *archiving* – with each being institutionally regulated. Even so, arriving at this dual process, as Descamps points out, was not accomplished without the removal of a number of obstacles, difficulties and uncertainties by defining precise methodological rules.

Discussing the reluctance that existed in France for the use of memory as a historical source in the context of an oral history, she highlights how, in the 1980s, in parallel with the development of an interest in heritage, there came a "heritage turning point" in the collection of memory. This was with, importantly, the introduction of the concept of "oral archives", which from that point on were perceived as material for research: the collection could then be regarded from the archival and documentary angle. Indeed, with the creation of places of memory and various types of media texts, the collection and archiving, especially those dealing with the traumas of the 20th century (wars, the Holocaust, and other genocides), became very institutionalized. As a result, they have given rise to a double phenomenon: on the one hand, the installation of the *figure of the witness bearer*, then of the *witness*[42]; and, on the other hand, the *heritagization of the oral archives*: the testimony collected being classified, catalogued, documented and preserved – in short, archived – it will tend to constitute a "memorial heritage"[43].

However, this approach to the production and heritagization of testimonies as an expression of memory is absolutely crucial in understanding social heritagization, because it raises the question of the transition from the production of testimonies to their heritagization; and, more generally, the question of the development of memory in social heritagization (Rautenberg 2003a). It is therefore essential to carefully examine the socio-semiotic operations that articulate the three phases according to three processes: collection, archiving and heritagization.

The first phase is preceded by a preliminary operation that remains invisible and imperceptible to the investigator, even though it is essential. This is the transformation of the experience of the witness bearer (their experience of a social reality at a given place and at a given time) into a memory trace; in other words, the operation of the actual production of the memory. If this transformation is not a direct part of the collecting, it is nonetheless the condition of possibility for all the operations that follow: "Making reference to a personal experience gives a special quality to what is said, the narrative is grafted onto a past that is in some way presentified by the words of one who participated in it", writes Dulong (1998, p. 44). The second operation is the act of speaking of the witness bearer, the actual enunciation of the memory in the form of a narrative, that is, the testimony that establishes the witness bearer as such. During a memory study, memory is made both possible, and at the same time defined by the *communicational situation*

42 The witness bearer enunciates a testimony; they become a witness when their connection with the social fact to which the testimony relates is recognized in reception. This study is the subject of Chapter 2, *"Les archives orales ou le tournant patrimonial"* of Descamps' (2019, pp. 63–90) work. On this type of "sensitive memories", see Cendoya et al. (2015).

43 This is the subject of Chapter 3 *"Vers un patrimoine mémoriel"* (Descamps 2019, pp. 91–121).

created by the research institution and/or heritage institution. It is in fact this institution that chooses the witness bearer according to their characteristics, including their participation in a type of reality; it is what defines the place and the modalities of collection, the procedures and the tools used, etc.[44] This situation will give the testimony its narrative form addressed to a receiver institutionally mandated to collect it. However, still in the case of the study, this second operation is inseparable from a third: the collection of the utterance (the memory) which formerly used to be carried out by taking notes, then as soon as the technique allowed it, by means of first audio and now frequently video recording, because the latter makes it possible to keep a trace of the enunciation (the way of enunciating the memory). The testimony will thus have been inscribed in a *document* that fixes the story as is, on a medium at the very moment it is enunciated. Therefore, as it is documented, memory becomes *media text*, and its material medium of inscription, even if it is digital, in fact an *artifact.*

Even though, in the case of a recording, the second and third operations are, as I have just mentioned, paired and hence inseparable from each other, since it is the recitation of the testimony that is recorded; from a socio-semiotic point of view, there is nevertheless a tipping point between the two. This is usually visible when the recitation is addressed to a receiver *in praesentia* or, obviously, when there is no recording, as in the case of on-site mediation by the memory bearers[45]. The witness, as Dulong explains (1998, p. 73), is the trace of the social reality being reported: "He offers his person as the 'speaking trace' of the event". The symbolic power of the testimony lies in this enunciating physical presence that makes traces, such that the words being uttered presentify the past, while in turn giving this trace its meaning. Comparing this process to what happens for material heritage helps us to grasp its logic: the speaking body would be comparable to a heritage object from the past that would at the same time be its scholarly representation (which would no longer be silent). It thus gives the testimony an autographic social mode of existence: that of a piece of the past present in the present of the reception of the testimony[46]. This link between testimony and witness is preserved when the witness bearer's speech is textualized in the form of a mediation, such as, for example, a guided tour given by former miners. However, with the inscription of the utterance (and a fortiori of the recitation) in a document, the tipping point has disappeared: taken down in note forum or recorded, that is, documentarized, the testimony is now *separated from the body*. Its mode of social existence has thus become allographic.

44 What Descamps (2019, p. 121) calls a "provoked" situation. Details of the rules of its construction can be found in the methodological guide entitled *"Ingénierie des archives orales"* given as an appendix to this work (Descamps 2019, pp. 159–186).

45 This is the situation to which Henri-Pierre Jeudy usually refers.

46 For Dulong (1998), it is this mode of existence that makes testimony a "natural institution" (p. 56), reflected in the fact that a priori, we believe the witness.

In the same way as a book manifests, for example, the "text"[47] of a work (e.g. a novel), the document manifests the "text" of the testimony; the words uttered give the memory its social existence according to the rules and procedures defined by the communicational situation. Even if the act of speaking is partially (i.e. depending on technical constraints) represented, in a video recording, for example, the testimony (as an utterance of memory) no longer belongs to the body of the witness bearer with which it was hitherto consubstantial: it exists henceforth as made manifest by the document[48].

The phase of archiving is from this point on the necessary extension of the recording. *Documentarized testimony* has every likelihood, in fact, of being subjected to a fourth operation: its archiving and preservation. From a socio-semiotic point of view, it must be remembered that the testimony has become an autonomous *media text*: it can now circulate and be transported, copied, repeated, fragmented, quoted, commented on, studied, compared to others, etc.

The question of heritagization arose, according to Descamps (2019, pp. 91–110), during the last 30 years of the 20th century and the very early years of the 21st centuries. With this fifth operation, the corpuses of documentarized testimonies reflect the gestures characteristic of social heritagization: the manifestation of society's *interest* in the past, the production of knowledge enabling the *attribution of values* and the implementation of processes, making it possible *to preserve* and *disseminate* them. The gestures of rediscovery and publicization are therefore front of stage, with those of knowledge production and conservation forming the backdrop. As for the gesture of declaration, it is present through the institutional recognition of things to be kept and transmitted by a cast of actors, belonging to the scholarly world (the "practitioners of the oral source", as Descamps said), to the political world (with the development of regional and particularly local authorities), and to the associative world. However, the 2003 UNESCO Convention for the Safeguarding of the Intangible Cultural Heritage (UNESCO 2018), as Descamps notes (2019, pp. 108–110), has bolstered the expansion and promotion of oral archives, because it has helped to get them regarded by certain groups as "memorial heritage", opening up a discussion on its possible status as intangible heritage. However, an archive cannot be recognized as intangible cultural heritage within the meaning of this convention, since the production of the testimony was prompted. Also, once it had been documented, it became a fixed "memory prosthesis" that can thus no longer evolve over time (Descamps 2019, p. 115). However, she concludes (Descamps 2019, p. 121) that the oral archive can in spite of everything be regarded as an intangible heritage – but not within the meaning of the Convention – when

47 "Text" is understood here in the sense given to it by Genette (1994, p. 105) of an ideal object manifested by a book, and not in the sense of the *dispositif* constituted by media text.

48 Corresponding to a "heritage rupture" (Rautenberg 2003a, p. 152).

"the producers decide a priori that the traces they are going to manufacture will be instantly and deliberately endowed with a symbolic, cultural conservation value that makes them worthy of being heritagized and transmitted to future generations" (when they are "actualized", in Jeudy's sense). Oral archives, Descamps explains, are "'wanted monuments', with commemorative and historical value". They therefore constitute heritage because of the value of what they represent as traces: trace of testimony, which is itself a trace of memory, which in its turn is a trace of the relationship of the witness to the social reality. Faced with this *mise en abyme* of traces, the author is thus correct in pointing to the decisive role played by the institution that guarantees and certifies it – whether this institution be, from an organizational point of view, unique or plural, producer or user of the archives. It is thus, as the narrative semiotics would say, in the position of a "sender" who controls the whole *dispositif* by programming the five operations.

I will conclude by coming back to the tipping point between the second and the third operation. Contrary to what might be suggested by an overly rigid interpretation of the conflict between *in præsentia* enunciation and documentary enunciation, which puts the emphasis on the rupture of the link between documentarized testimony and the past, documentarized testimony is in fact also reconstructed in representation: its being hit sideways by the institution maintains the presence of indices from the past, which turn documentarized testimony into a monument, a sort of place of memory[49]. Certainly, the testimony is effectively separated from the body of the witness bearer, but the fact that at the time of discovering and reading the document, the interpreter knows (by the knowledge of Arche) that the interlinking of traces is institutionally guaranteed, the testimony, although documentarized, remains in a way grafted onto the lived experience of the witness bearer; in fact, the traces of production and enunciation contained in the document presentify as indices the enunciation of the narrative by the witness bearer. A little like when Arlette Farge discovers objects in the middle of the archives, everything happens at the very moment when we begin to discover and read a documentarized testimony – before even approaching it from the angle of the knowledge contained in the document.

This hybrid character of documentarized testimony, which preserves something of the autographic mode of existence of the "speaking trace", while at the same time having the allographic mode of existence of a media text, is one of the keys to understanding social heritagization. It is because of this character that this mode of heritagization can cover a range that extends from the memorial relationship to the

49 Given this role of accreditation of the institution, insofar as it requires a distribution of the rights of the various parties (witness, oral archivist and institution), in terms of the testimony thus processed, Linda Idjéraoui-Ravez (2012, p. 101) was able to speak of "trusted testimony".

past to the patrimonial relationship. It is therefore at the root of the possibility of operating a transformation of a past that is still "hot", "living in consciousnesses and memories", into a "cooled" past characterizing heritage creation (Rautenberg 2003a, p. 152); and yet without the heritage object totally losing this vivid relationship with memory that opens up the anticipation of a revitalization and an appropriation in reception. To summarize, we could say that in spite of the "heritage rupture", these heritage objects preserve a dimension of places of memory[50].

3.3.2. *Mediation by memory bearers and reconstructed "memory"*

Among the wide array of examples mentioned by Henri-Pierre Jeudy, it is possible to distinguish three different uses of memory. An examination of these three uses makes it possible to reconstruct the representation process.

The first is the most obvious, since it is the most discussed by the author from the point of view of "transmission in action": the reactivation of practices or their interpretation for mediation purposes as part of setting up an exhibition of former industrial or artisanal activities. There is a whole gradation of this "memorial" mediation, from the simple reactivation, which, while inevitably always having an element of interpretative mediation, is based on a resumption of the performing of an activity[51], to the guided visit conducted by former workers; for example, mine museums in which, since reactivation is impossible for technical or conservation reasons, former miners conduct a mediation of the guided visit type[52] or in galleries that have been laid out museographically to allow people to discover, over a relatively small area, the extraction activity (or that are sometimes merely a reconstruction, as was the case with the Couriot shaft, in Saint-Étienne, at the time of the creation of the mine museum established on the site, but nevertheless using machines and tools that were heritage objects). However, in every case, whether it is the reconstruction of an activity or an interpretation, memory is mobilized and

50 The entire work of Rautenberg (2003a) can be regarded as a discussion of this hybridity.

51 As was the case for a while for the cutlery industry of Thiers or the felt hat production in Chazelles.

52 This practice of "memory" interpretative mediation is still present in the Nord-Pas-de-Calais mining basin, inscribed on the World Heritage List (2012), where miners still lead visits or take part in encounters, in particular, at the Historical Mining Center of Lewarde and the Wallers-Arenberg mining site. With, in addition, a reference to the latter site, in the shooting of the film *Germinal* (by Claude Berri based on Zola), some former miner- "mediators" having featured in the shooting (inscription on the UNESCO list at the following website: www.whc.unesco.org/fr/list/1360. For a presentation of the site, see: www.chm-lewarde.com/wp-content/uploads/2022/02/Dossier-de-presse-francais-2022.pdf or www.fr.Wikipédia.org/wiki/Bassin_minier_du_Nord-Pas-de-Calais).

enunciated by the memory bearers themselves, by former workers, in the most direct form possible, that is, *in praesentia*. It is therefore a memory (generally integrated) of know-how specific to the profession, accompanied by some degree of ancillary knowledge[53]. Through the memory that the witness enunciates, the audience is supposed to enter into a relationship with social practices that belong to the original world of this memory. Therefore, something happens that could be likened to the impression of "touching reality" that Arlette Farge says she feels when she unfolds her archives and that Dulong describes as "the sublime of the past" (Farge 1989, p. 18; Dulong 1998, pp. 180–184)[54]. This is probably one of the reasons why this use has, as we have seen, an emblematic status in the criticism of Henri-Pierre Jeudy who decries the distortion, the museumification of (workers') memory produced by this kind of representation of an activity or a social life. This use of memory is not aimed only at memory transmission, in the sense that it is only the encounter with the witness that is important in the objective that those to whom it is addressed will keep alive this memory of a situation or an event, as is the case with the testimonies of Holocaust survivors. This use primarily constitutes a form of mediation, in the present, of the site and the heritagized objects; a mediation that mixes to varying degrees, on the one hand, certainly, the logic of testimony, in terms of reporting the memory of their practices and their experiences – possibly the production of a sample of the original world in the form, for example, of executing it – but also, on the other hand, an interpretative mediation based on knowledge about broader phenomena than a simple reproduction of the former practices of the memory bearers. The mix of these various types of mediation is also found today in a completely different context: that of visits to companies labeled *"entreprises du patrimoine vivant"* or *EPV* (living heritage companies)[55]. Such reactivations or interpretations, as representations, result from a semiotic media construction that focuses on what has been deemed relevant to give an account of what the former activity used to consist of. Consequently, this representation is a textualized practice (a media text) which has, we suspect, a dual mode of existence, autographic in the objects it mobilizes and allographic as a text structured by the cognitive construction of memory (and ancillary knowledge). This is why I speak of "representation" to refer to what Jeudy, for his part, calls "simulacra": the objects and practices of which it is composed presentify the objects and practices of the past, which become traces of which the memory delivers a certain interpretation by revitalizing them (or reactivating them). Indeed, no matter what the impact on visitors that may be produced by a relationship with members of the social group that performed the activity, it is in reality an exhibition that is being staged in a heritage context. This use of memory, in addition to making it technically possible to "exhibit" processes

53 We are not very far away from the approach to heritage introduced by intangible heritage.

54 Descamps (2019, p. 116), for his part, talks about implementing a reappropriation and revitalization.

55 Like the companies united under the *Fédération française de la ganterie*.

and not only inert objects, is also an opportunity for the institution to state and attest to the heritage status of the site and the objects by means of a form of mediation that has the power of the memory of lived experience specific to testimony. This "naturalization" through mobilization of memory constitutes an essential tool of museography, and I do not think anyone will deny: someone who uses the memory of their practice transmits by means of their whole person, thus differently from a presenter who explains the practice. However, this use often takes place in an officially recognized heritage organization that not only implements "social" heritagization, but is also one of the "institutional" heritage institutions, to pick up on the objection of Rautenberg (2003a, p. 143).

The second use is that of recorded memory. When the memory bearers belonging to the population disappear, there is no one left who can assure continuity between past and present. The institution must then find an alternative solution to replace "memorial" mediation and its role as a third-party mediator between memory and audience becomes clear. Envisaging the theoretical hypothesis of an alternative solution in which actor-mediators come in to "play the role of the memory bearers" actually helps us to understand what is at play. While this manner of proceeding does not a priori lead to an alteration in the heritage dimension of the place or the content, it may, however, undermine the constitutive rule of the testimony according to which the witness confirms by their physical presence, their person, the link between the event and the testimony. In this case, the link would derive from fiction: at best, from a theatrical representation; at worst, from deception, if confusion were to arise between witness and mediator[56]. The substitute's words would thus be the simple manifestation of a text whose link to its universe of production should be explicitly vouched for and made visible by the institution, so as not to appear dubious; however, in any case, the relationship that the receiver has with the event through the witness, insofar as the latter is the presence of the event in the present, would be broken, the representation then having more importance than the link to the situation of origin. This is why the solution generally adopted is the playing of a recording of the documentarized testimony, the origin of which is vouched for by the institution: a heritagized testimony, in the sense that, as Rautenberg (2003a, p. 152) writes, "heritagization, in "reviving" the object in a new form – written, exhibited, ring-fenced by protection... – marks a rupture – which we call a heritage rupture – in representations of the past". The receiver will then be able to grasp the indices of patrimoniality, both through the content and through the marks of enunciation[57]. We have thus transitioned from testimony in the exhibition to the

56 The institution would no longer guarantee witness status. On fiction, see Schaeffer (1999). As part of her doctoral research, Émilie Flon studied the interplay between representation and fiction focusing on the recitation of a letter from the water curator at the *Musée du Pont du Gard* (2005, reprinted in 2012, pp. 181–200).

57 On the importance of these marks, let us refer to Idjéraoui-Ravez (2012, p. 126).

exhibition of testimony. It is this "cooling" through heritagization that Jean-Claude Duclos used in the reactualization of the museography of the Musée de la Résistance et de la Déportation de l'Isère, following the disappearance of witnesses and the evolution of historical knowledge, as demonstrated by Marie Lavorel in her doctoral research (Lavorel 2014; Cendoya et al. 2015, pp. 190–209).

The third use of memory is indirect. It is reconstructed from written or recorded memory, or historical knowledge, which is "revitalized" in an archaeological reconstruction in the form of a stage-set of the past. This is probably the use that is closest to the production of "simulacra" that Jeudy criticizes. This reconstruction activity takes two forms that are more or less complementary. The first is the use of the memory of practices – or rather the reference to this memory based on knowledge – by social museography in its most spectacular dimension, that is, one of the most "culturalized" forms of "viewpoint museology" (Davallon 1992): namely, either reconstruction, which Montpetit (1996) describes as "analogical museography"[58], or, more generally, recontextualization through the "immersive scenography" analyzed by Bealën (2002), forms that Chaumier (2005, 2013) has, for his part, dismantled and produced the criticism. The second form inverts the process, since it is based on the production of know-how (a form of memory) through experimentation in order to reconstruct practices that have disappeared. The best-known form of this kind of memory reconstruction is the cutting of flint or obsidian tools in experimental archaeology. However, there are also other experiments on habitat, boats and navigation, lifestyle (e.g. Vikings in Norway, or in Normandy, with Ornavik Park)[59]. In France, the best-known experiment is the construction of the Château de Guédelon using techniques and materials to understand the construction processes in medieval times (late 12th to early 14th centuries)[60].

3.3.3. *From collection of testimony to exhibition of the witness*

"Museography of time and space", this is how Duclos (2005) describes his practice of exhibition creation at the Musée Dauphinois using the memory of various collectives and based on the establishment of a relationship with the population that pursues, in an original way, the ecomuseum model of social heritagization. The resulting format of the exhibitions remains officially that of culturalized museology, but an analysis of these exhibitions reveals that their communicational organization

58 One of the emblematic examples of this type of museography is probably the *Plimoth Plantation* (renamed, in 2020, *Plimoth* Patuxet), which reconstructs the habitat and way of life of the first settlers from Plymouth in the 17th century.

59 For Ornavik Park, see: www.ornavik.fr/7-archeologie-experimentale.

60 Available at: https://www.guedelon.fr/fr.

mobilizes the particularities provided by the gathering and processing of testimony, and that they are therefore based on a heritagization of memory.

It is clearly not by chance that this kind of museography was developed at the Musée Dauphinois. First of all, even though ethnology museums no longer occupy the place they had won in the 1980s due to the changed position of localness as a source of identity (Mazé et al. 2013), the role that they attribute to memory in the interpretation of objects nevertheless remains a reference. In this context, the originality of the Musée Dauphinois, as a regional museum of ethnology, lies precisely in having established a new relationship with the population, as was the case since the exhibition *Le Roman des Grenoblois* (1982), in which Jean-Pierre Laurent sketched a portrait of the population of the city of Grenoble. However, a new threshold was crossed when Jean-Claude Duclos gave his series of exhibitions based on presenting an exhibition of the memory of social groups of Iserois inhabitants from Italy, Greece, Armenia and the Maghreb, using their "testimonies"[61].

It should first be clarified that Jean-Claude Duclos was very clear in formulating the close link between ecomuseums and the taking into consideration of the population by what we call "society museums". Already in the 1990s, in an article in which, after outlining the context of the creation of the name "ecomuseum" and the contributions of the first ecomuseums (natural parks and Creusot), it replaces the ecomuseum as an institution in the evolution of museological thought in order to highlight its major contributions: population participation, the museum's contribution to a collective reflection on development, and the use of interdisciplinarity. These are three principles, he explains, which have helped define the role of society museums: comparing, providing landmarks, reporting back, accomplishing "works of mourning" and giving recognition; this is the specification we can now give to the museum's action (Duclos 1995, p. 296). In an article published 10 years later (Duclos 2005, p. 146[62]), he refines these contributions and adds a fourth principle: "The development of a museography of time and space, focused on a given reality (a territory, a way of life, a population, a practice, a group, a corporation, etc.)". Therefore, recalling the importance of Rivière's reflections on the boundary between ecomuseums and ethnology museums (including the National

61 These are mainly exhibitions that associate various groups who immigrated to Grenoble: Italians in *Corato-Grenoble*, Greeks in *Des Grecs: Les Grecs de Grenoble*, Armenians in *D'Isère et Armenia: Histoire d une communauté*, Maghrebians in *Pour que la vie continue: D'Isère et du Maghreb, memories of immigrants*, repatriates in *Français d'Isère et d'Algeria*. It should be noted that the permanent exhibition of the *Musée de la Résistance et de la Déportation de l'Isère*, directed by the same Jean-Claude Duclos, was also initially based on the memory of the resistance fighters.

62 So, a few years after the exhibitions that will be discussed.

Museum of Arts and Popular Traditions), he consequently introduces the notion of "museography of time and space". These are already the principles of this museography that he implemented in the many exhibitions he designed and produced at the Musée Dauphinois and the Musée de la Résistance et de la Déportation de l 'Isère. These are the principles that are, in my view, pushed to the extreme, in a quasi-experimental way, in the series mentioned above, and especially in the last two, which I will spend some time on, since they are so revealing of the social, symbolic and media issues of the treatment of testimony by a heritage institution.

On the subject of these two exhibitions – *Pour que la vie continue: D 'Isère et du Maghreb, Mémoires d'immigrés* on the one hand (October 1999 to December 2000) *and Français d'Isère et d'Algérie* on the other hand (May 2003 to September 2004) – Idjéraoui-Ravez (2004, 2012), in her doctoral research on both the work of memory collection and the exhibition of testimonies and witnesses, showed the innovative and "unprecedented" character of an approach

> where it is no longer objects from collections that are at the center, but individuals whose testimonies profit from the favors of a museography putting in representation the lived experience, but also the point of view of a community touching upon sensitive historical facts. (p. 120)[63]

By their exemplary character, these two exhibitions constitute an enunciative event that formalizes the possibilities and constraints faced by those who collect memory in order to exhibit, in the form of testimonies, a memory treated as the heritage of a collective, especially when it concerns "sensitive" memories, due to being linked to a greater or lesser extent with situations that have profoundly marked their lives. I shall begin by summarizing the main characteristics of these exhibitions as emblematic media texts of social heritagization.

The first, essential characteristic is the introduction of a form of presentation of memory in the exhibition that is completely original: the *quotation*. There is no shortage of examples of exhibitions that present testimonial traces, but the novelty is in the modification of the format of the exhibition as a result of replacing the usual use of memory as a source of knowledge by a direct quotation from it; this is in the form of an exhibited document. This modality of putting memory on exhibition will have crucial consequences for the mediatic functioning of the exhibition.

The second characteristic arises from the first: the *complementarity between memory functioning and heritage functioning*. Against the temptation to seek prevalence either of memory, as a place of memory would do, or of heritage on the

63 In the following pages, I rely heavily on the analyses and results presented in the book entitled "*Le Témoignage exposé*", which picks up his doctoral research.

grounds that heritage objects are a tangible trace of what has been, the analysis of the two exhibitions reveals, rather, a hybridization between memory and heritage, and between documents and objects. Memory having been collected through interviews and workshops in the form of documentarized testimonies, the "heritage rupture" (Rautenberg 2003a, p. 152) that ensues takes this memory out of the control of the past and "cools" it; it can then serve as a tool for the interpretation of material objects, which are also cut off from their world of origin and become simply traces. However, to this use, which after all is common in society museums, are added quotations from this memory with the result that, by appearing as such, the memory introduces a specific link to the past, even underneath its interpretative function. Even when it is heritagized, the link between the witness and what they have seen or lived through remains sufficiently vivid to give presence to the human context of the objects presented, to "revitalize" them, or to "memorialize" them, so to speak. Conversely, the materiality of objects provides factual landmarks to what memory is saying: by situating it materially in past action, it reinforces its informative capacity, but above all the link between the document that presents it and the memory bearer, who is sometimes even present in effigy in a photograph. Therefore, the document in which memory is heritagized, linked to this representation of the witness-bearer-turned-witness, acquires the status of a textual monument, which will make the lived events all the more meaningful because of the novelty of its presence in an exhibition and thus confers a primacy of memory over objects.

The third characteristic also arises from the use of quotation. In the usual case of memory used as an informational resource so as to construct the theme of the exhibition, the museum is the sole enunciator of the media text that constitutes the exhibition. Therefore, the testimony is only indirectly present, since it is in a way "absorbed" in the staging of the exhibition (Idjéraoui-Ravez 2012, pp. 95–98). While the informant undeniably has witness status with regards to the institutions that collect, archive and use the memory, they still have no auctorial existence in the exhibition, and still less as a witness in the public space; the use of only the content makes the act of testifying invisible. On the other hand, "quoting" the memory, as is the case in the two exhibitions concerned, also exhibits the witness position. When the bearer is identified as the author of the citation, they appear as a subject enunciating their memory; their status visibly changes from that of simply witness bearer to that of witness, and even more so if they are present in effigy through photography[64]. *Two enunciators* are thus identifiable by visitors and not just one: the museum that produces the exhibition through the exhibition's theme (title, staging of the exhibition and comments), but also the witness through quotation and in a certain way by signing it.

A fourth characteristic, quotation is not exhibition: the fact that the museum "exhibits" both the testimony and the witness means that, as an institution, it

64 I refer to the discussion of this process in Idjéraoui-Ravez (2012, pp. 98–101).

accredits, on the one hand, the link between the enunciated memory and the witness (the bearer who enunciates it); and, on the other hand, the link, constitutive of the memory (and therefore of the testimony), between the witness and the situation they lived through. This double accreditation forms the framework from which *visitors* will be able to give meaning to what is shown to them and what is told to them about the situation to which the memory relates. The museum, as a *heritage institution* quoting the witness, thus occupies the position of "third-party mediator" between witness and visitors[65]. However, in doing so, it de facto publicly certifies the testimony; in accrediting it, it makes the link between itself and the witness visible to some extent. As Idjéraoui-Ravez correctly writes: "In the case of testimony in the museum, the witness, and the museum that authenticates their words and exhibits them as flagship objects, make a joint commitment on the trustworthiness of the narrative". And the author concludes: "In this case, testimony as a museum object goes beyond what is uttered, on the one hand, by firmly bonding what is related to the person relating it; on the other hand, by firmly bonding what is related to the museum" (Idjéraoui-Ravez 2012, p. 86). The main consequence of both this accreditation of the testimony by the museum and also the recognition it thus grants to the witness is that it invites visitors to exercise their empathy toward the latter and more broadly to in some way adopt a *position* that recognizes the witness community by trusting it as the museum itself does. This way of proceeding will work on the division inherent in social heritagization between collective and audience. It shifts this division by bringing face to face the witness bearers on the one side and the visitors on the other, who then form one whole, whether they belong to the audience or are members of the collective. Whatever the meaning that each of these visitors assigns to the content of the memory, they are invited to share a compassionate attention to the testimony and to the witness. This communication is therefore neither only participatory and restricted to members of the collective nor is it only transitive toward visitors forming an audience; it can be said to be *shared*, in the sense that it legitimizes and formalizes, by operationalizing it, the trust that the receiver of the testimony "naturally" gives to a witness. In other words, it mobilizes what Dulong calls the "natural institution" of testimony (Dulong 1998, p. 66) as a consequence of which "the social dimension constitutive of the testimony is its reception: the generally tacit mechanisms by which an audience accredits a narrative as a deposition, an autobiographical utterance as testimony, its enunciator as a witness".

The fact that witness and museum "make a joint commitment on the trustworthiness of the narrative" has another consequence, which constitutes a fifth characteristic of these exhibitions: their *political significance*, successor to

65 "Through the intercession of a third-party mediator, testimony is transformed into an object of communication turned towards an audience that will in turn interpret it" (Idjéraoui-Ravez 2012, p. 115).

ecomuseology, which cannot but bring to mind that of community museologies. The enunciative accreditation of the testimony and the status of the witness is necessarily also an accreditation of the value of the testimony – and consequently, of the witness. Indeed, if the narrative of witness is *worthy of being made public*, that means it is significant by virtue of the remarkable character of the situation to which it refers (Dulong 1998, p. 44). In the present case, the value of the situation that warranted the testimony was made visible by the hostile reactions toward the museum relating to the exhibition *Mémoires d'immigrés*, with some rejecting the recognition by a museum of not only the memory but also the witness and the collective – and rejecting perhaps even more the principle of a shared interpretative experience that puts immigrants and visitors on an equal footing[66]. It is the symbolic power of this accreditation that likewise had the effect of motivating the request, from the *Maison des rapatriés d'Algérie*, for a "right of reply", which was at the origin of the second *Français d'Isère et d'Algérie*[67]. This request indicates a sideways shift in terms of the object that the testimony is supposed to be about. Although in the museum project, the testimony is supposed to give an account of the immigration experience, those who protest against it are focusing on a different fact – on colonization or, worse, on the witnesses' presumed status as foreigners, indeed inferiors – superimposing an interpretation inherited from the war of decolonization onto the experience of immigrants in France in order to delegitimize this experience, indeed to impose the values of colonization; in short, a hierarchy of values is created between themselves and witnesses, and beyond, with the immigrant community. The second exhibition moreover bears traces of this position, since a part is devoted to colonization, life in Algeria, the war of independence and the return. From the point of view of the relationship between the museum institution and the two communities, we observe in *Mémoire d'immigrés* a "cooperative logic" of "summoned witnesses"; and in *Français d'Isère et d'Algérie*, a "logic of control" by "self-proclaimed witnesses"[68], counter-witnesses. From the point of view of reception, the cooperative logic "generates a close, strong, sometimes almost intimate relationship from the visitors to the testimonies", while the logic of control – in which the witnesses of the social group, controlling the memory collection and its archiving, will sometimes find themselves, as enunciators of memory and a constituted group, in competition with the institutional enunciator, that is, the

66 A commando, "armed, identified as belonging to a far-right group, invaded the museum in a threatening manner" to protest against the exhibition and attack the institution by putting an end to a lecture by the Algerian historian Mohamed Harbi (Idjéraoui-Ravez 2012, p. 19, 43).

67 A letter of protest from the *Maison des rapatriés de Grenoble* criticized the fact that an exhibition such as this could be presented in a so-called "French" museum and claimed, moreover, a right of reply. This right of reply was to be embodied, as per to their request, in a museographic *dispositif* at least equivalent to that of *Mémoires d 'immigrés*: the exhibition, at the Musée Dauphinois, of their own testimonies (Idjéraoui-Ravez 2012, p. 44).

68 On the conflict between these two types of witnesses (Idjéraoui-Ravez 2012, pp. 81–82).

museum, the latter being unable to accredit certain elements of memory – "seems to generate a more distant relationship" (Idjéraoui-Ravez 2012, p. 117).

To summarize, the power of the social heritagization mode specific to the museography of time and space could thus be said to lie in the fact that it makes it possible to simultaneously "cool" the vivid and direct character of the testimony, while at the same time preserving it. Once memory has been "disattached" from its social origin by archival processing, the testimony becomes a source that can be compared to other sources. However, having been intentionally produced, it nevertheless still possesses a special status that makes it symbolically more than simple knowledge, more than a simple "index" to be interpreted, and, moreover, quoted, it represents the witness insofar as the latter is a component of the past situation in the present. This museography uses testimony for what it preserves of the strength of memory: to bring knowledge about the original situation and to be the point of view (in all senses of the word) of a witness whose legitimacy and dignity the museum recognizes, but without attesting that what the witness says is true.

In order to operate a recognition that is at the same time a distancing, that is, this interplay between memory and heritage, the structure of the exhibition gives the historical framework that serves as a reference point in order to give an account of the person who was made spokesperson: a witness who is identified as such. Ultimately, heritage objects and memory form one entity, one representation made of *traces* of the past, material and memorial, that have acquired the symbolic status of things to be kept under the effect of the heritagization process.

3.4. In conclusion: social heritagization, a new reference

The examination of the different modalities of this new mode of heritagization that makes use of population (the collective) and memory – modalities which constitute so many variants and sub-variants – first of all confirmed the hypothesis that "transmission in action", as identified by Jeudy, could be considered as the most radical variant of this mode: this is the case in the ecomuseum and community museologies, in which the collective can be at the same time the actant who will replace the material trace to be heritagized (actant-object), one of the operators of heritagization (actant-subject) and the initiator of the assuming of responsibility for its destiny through the museum institution (actant-sender). Additionally, this can be in the case of the sociological intervention that approaches heritage as an experience where the actant-subject bases their action on their attachment to the heritage traces and on the experience they have of them. The other variant is characterized by textualization, operated and guaranteed by an institution, which produces a representation of past memory. This textualization can take the form of archived and heritagized testimony (in oral archives); the form of a mediation with exhibition visitors (for example, in museums or at industrial sites), or the form of testimony itself put on exhibition (in the museography of time and space). Since memory is an emanation of the population, its being taken into consideration has even brought about a revolution in the conception of the receivers of museums or exhibitions; since these can be addressed to members of the group or to audiences perceived as customers of a cultural enterprise. Hence, the question of the relationship between members of the population and members of audiences that are strangers to it (in culturalized museology).

The conclusion of this study is that the social heritagization mode undeniably introduces a new type of relationship to the past; on this point, we can only agree with Nora. This is moreover proved by the reactions to its arrival by proponents of the traditional, administrative political and legal conception, centered on monuments and works of art. Three crucial changes mark this "turnaround".

The first is indeed, as Nora criticized, the arrival of the collective ("groups") as the main actant that triggers, and in part carries out, heritagization, particularly through associations (Glévarec and Saez 2002; Glévarec 2003) or even more radically with ecomuseology or community museologies. The instance of this repository of traces of the past, that is, the one that considers itself to be responsible for keeping them, being their beneficiary and transmitting them, is no longer the state as the representative of the body politic, but the collective (referred to above as "population", in order to distinguish it from the "audience") who share an origin, a culture, an organization or simply a way of life.

The second change is the redefinition of traces of the past. These can be objects or places, but also practices, know-how, or memory. Their meaning and value are linked to the memory that the members of the group have of them; a memory that in some way replaces the national institutional memory. Hence, a change in the program of heritagization gestures. It is no longer the administrative legal declaration that is the pivot of the program of transformation of traces into heritage objects, but the interest of the collective, the knowledge at the center of which is memory (scholarly knowledge playing a secondary role); finally, publicization in which the staging of the exhibition is constructed on this memory. Hence, as a second consequence, the shift away from what used to be a scholarly interest toward an interest of the collective, whose origin is its attachment to these traces insofar as it see itself as depending on them and their world of origin, as shown by Tornatore (2010b, 2014). The construction of the relationship to the traces of the past will therefore happen through contact, what has been lived, experience, rather than through an interest based on the knowledge of scientists or experts. The criteria for recognition of what constitutes heritage will thus be disseminated in society, publicization resulting in them being passed from collective to audience, in the form of interpretive schemes.

The third change is a profound change in temporality: the introduction of the collective and its memory entails ensuring continuity between past and present. It is this continuity that Jeudy, one of the first to have studied the phenomenon through the case of industrial heritage, describes as "transmission in action". He rightly notes the existence of a process of representation of the past in the present, but he reduces it to the production of a simulacrum because he does not take into consideration the transformation of the symbolic status of what comes from the past, as operated by heritagization. He does nevertheless identify the rupture of this kind of continuity with traditional heritagization. However, it should be noted that the *dispositif* for the production of knowledge about this past, which is located in the present, objectively introduces a dividing line between humans of the past and humans of the present; even if the two belong to the same collective; or, to take it to extremes, even if they are the same individuals, having moved from a past situation to a present situation, who will then be in a reflective position. It is incidentally in order to abolish this dividing line that *places of memory* generally presuppose an intentionality of transmission from those who have lived the past to those of the present who have not lived it. This is the very principle of the place of memory (understood in the current sense of the term). In the opposite way from the latter, collecting and archiving testimonies introduces a *heritage rupture* that distances (cools) the relationship to what has been lived in the past, and revitalizing this relationship will then require a cognitive, affective and social reinvestment, which is the very nature of *commemoration*, but which, being ephemeral, presupposes periodical repetition.

We could say that the full symbolic force of social heritage is in the crossover of these three processes: continuity of the group in time (through testimony), reduction of the past to knowledge thereof (through the heritage rupture) and revitalization of the relationship to the past through knowledge and objects (through commemoration). All these changes are accompanied by the emergence of a new reference science: ethnology[69], which does not prevent the traditional heritage mode from continuing alongside the new.

69 With a certain degree of disconnect, geography will intervene, in relation to the territorial and managerial dimension of urban heritage and natural heritage. However, especially under the effect of the development of intangible heritage, ethnology still remains today the dominant reference science. We will have to observe whether bringing together the heritagization of nature (living things) with protection of the environment will change this state of affairs.

4

Heritagizing Social Processes

It was in the 1970s that the question first arose of exhibiting, not objects, but social processes and scientific knowledge. When the places presenting this kind of exhibition had no collections of technical objects, then they used to be regarded not as (science and technology) museums, but as science centers[1]. But the question was not asked in terms of "intangible heritage"; this may have been because the necessary presence of collections in museums implied that heritage had to be constituted of material objects. The terms of the debate began to change slightly with the advent of ethnological research on know-how, elements of culture, memory and also practices (Chevalier 1991). But it is, without a doubt, the 2003 UNESCO Convention (UNESCO 2018) which, by introducing the category of "intangible cultural heritage" into the public, political and scientific space, constituted a major discursive and institutional event, affirming what was then part of the oxymoron of the "intangible trace", a kind of contradiction in terms. Although in theory it could be seen as a second stage that complemented the 1972 World Heritage Convention[2], it put forward the existence of a heritage made of ephemeral traces. While social heritagization had experienced rapid growth throughout France in an environment that was, as we have seen, particularly conducive, the affirmation of the existence of intangible heritage objects burst into an environment that on the contrary was not expecting it. It is therefore not surprising that this event, which introduced a new disruption into the way of thinking about "traces" of the past – which, in a way, were not even "real" traces – after having aroused the skepticism of many heritage professionals, was discussed in an abundant anthroposocial literature, very predominantly in ethnology, which, as I mentioned in the previous chapter, had in the meantime become the reference science of social heritage. Just like history

1 Although, as far back as 1968, Cameron (1992 [1968], p. 262) introduced the concept of "kinetifact" to refer to the movement of Foucault's pendulum "and other similar phenomena".

2 Convention for the Protection of the World Cultural and Natural Heritage (UNESCO 2021).

before it, ethnology was henceforth made responsible for both producing the knowledge to be used for the inscription of heritage objects and studying the way in which this heritage functioned as a social being, in particular the creation of the category by UNESCO or the modalities, as well as the effects, of the heritagization of certain elements of culture for the social groups concerned.

I will give a swift review of the literature concerning this new type of heritage object. Faced with an abundance of ethno-anthropological publications, I will nonetheless confine myself to mentioning a few books which, in developing a wide range of questions and examples, illustrate this position as reference science.

The collection edited by Laurajane Smith and Natsuko Akagawa, *Intangible Heritage* (2009), while mobilizing a number of disciplines in order to provide a comprehensive discussion of the creation of the category, policies, practices and situations, conveys an anthropological approach (see in particular the text by Laurajane Smith).

Three collections that, in addition to presenting well-researched introductions discussing the chief questions posed by the category, bring together many of the anthropologists who work on this type of heritage or on situations characterizing it: one edited by Nicolas Adell Yves Pourcher, *Transmettre, quel(s) patrimoine(s) ? Autour du patrimoine culturel immatériel* (2011), another edited by Chiara Bortolotto, *Le Patrimoine culturel immatériel. Enjeux d'une nouvelle catégorie* (2011), and the last, presenting the proceedings of the 2012 Cerisy conference, edited by Julia Csergo, Christian Hottin Pierre Schmit, *Le patrimoine culturel immatériel au seuil des sciences sociales* (2020).

Two original texts: Valdimar Tr. Hafstein's thesis, defended in 2004 and recently published, *Making Intangible Heritage: El Condor Pasa and Other Stories from UNESCO* (2018), and Barbara Kirshenblatt-Gimblett's groundbreaking article, "Intangible Heritage as Metacultural Production" (2004).

To these ethno-anthropological publications, I will add a historical article and a geographical thesis that bring a point of view complementary to the previous ones: Dominique Poulot's article *"Le patrimoine immatériel en France entre renouveau muséographique et "territoire de projet"* (2009) and the thesis by Wladimir Mejía Ayala, *Les problématiques et enjeux de la mise en tourisme du patrimoine culturel immatériel en Amazonie. L'exemple du peuple indigène Sápara d'Équateur* (2014).

Box 4.1. *Some benchmark articles on intangible heritage*

Within a few years, this abundant literature constituted a rich collection of media texts confirming the existence of a specific mode of heritagization. But the manufacture and functioning of *the intangible heritage object*, as a trace and a being of language – I will say, its *poïetic*[3] – have, on the other hand, been little studied by the anthroposocial sciences, and even the infocommunicational sciences have only recently begun to examine them[4]. Therefore, my topic here will primarily be to understand how these *social processes* that affect practices, knowledge and representations are translated into beings that are both social and meaningful, having heritage status.

4.1. A new category of cultural heritage created by UNESCO

On reading the Convention (UNESCO), three observations must be made. The first is a very extensive usage of the notion of heritage in that it is, very nearly, coextensive with that of culture[5], and yet without this usage has not been explained or even asserted as such[6]. This first observation has two more as its corollaries, which are, for their part, explicitly present in the Convention. The first is that, officially and on principle, the choice of the element that constitutes heritage is a matter for the collective (community, group and, where appropriate, individuals),

3 In using the word "poïetic", I seek to stress the act of production of the characteristics of the semiotic and symbolic form of the heritage object.

4 An early approach to this poïetic can be found in Chapter 4 of Jeanneret's (2019) book, *La Fabrique de la Trace*, and also in the doctoral research of Pianezza (2017) or Brito (2018). On the use of digital tools in intangible heritage, see Severo and Cachat (2016). There is also an edition of the journal *Communication & Langages* (No. 211, 2022) on *Le patrimoine culturel immatériel en pratiques. Écritures, médiations et appropriations*, in which moreover an early version of this chapter appeared (Davallon 2022).

5 Here I mean "culture" in the classical sense usually used in anthropology. Yves Winkin follows Ward Goodenough in proposing a slightly more restrictive definition, usually serving as a basis for the anthropology of communication: "The culture of a society consists of everything we need to know or believe in order to behave in a way that is acceptable to the members of this society, in any role that they would accept for each of their own" (Goodenough, cited by Winkin (1996, p. 9)). A general discussion of the usage of the concept of "culture" can be found in Brumann (1999).

6 For the record, I will repeat in full the definition given in the Convention: "The 'intangible cultural heritage' means the practices, representations, expressions, knowledge and skills – as well as the instruments, objects, artefacts and cultural spaces associated therewith – that communities, groups and, in some cases, individuals recognize as part of their cultural heritage. This intangible cultural heritage, transmitted from generation to generation, is constantly recreated by communities and groups in response to their environment, their interaction with nature and their history, and provides them with a sense of identity and continuity, thus promoting respect for cultural diversity and human creativity" (UNESCO 2018, Article 2.1).

which "recognizes" it as being a part of its cultural heritage[7]. According to the second, the element of culture that constitutes heritage is said to "manifest itself" more particularly in certain areas of social life, mainly: oral traditions and expressions, the performing arts, rituals and festive events, knowledge and practices concerning nature and the universe or traditional craftsmanship[8]. It objectively defines, within a culture, the domain of intangible heritage.

Incidentally, it should be noted that the splitting off of this domain goes hand in hand with the fact that the Convention presents a redacted, not to say "euphemized" version of the elements of culture that have, in a way, the right to claim the title of "intangible heritage of humanity", since this status can only be claimed by elements ("practices, representations, expressions, knowledge and skills") that respect human rights, relationships between communities, groups or individuals, and also sustainable development[9]. There may therefore be a greater or lesser difference between the intangible cultural heritage of humanity, on the one hand, and, on the other hand, what the collective recognizes as defining it as a community.

I will develop the idea that these various observations are the index of a heritagization process that operates a *translation* from element of culture (a social process) into heritage object (the element inscribed) without saying so; an operation of translation whose sociosemiotic analysis will *reproduce* the functioning. In order to do this, I will begin with a rereading of the Convention, as the operative text of

7 For convenience, in the rest of the text, I will continue to use the designation of "collective" (or "community" when it is approached from the point of view of culture) for "communities, groups and, in some cases, individuals". Maguet (2011, p. 50) calls this expression a "triple syntagma" designating a somewhat imprecise entity. Although receptive to the author's arguments in favor of using only the notion of "community", which probably corresponds better to the sociological or anthropological reality (Bortolotto also uses it), I will for my part confine myself to a more generic notion that retains a trace of the linguistic imprecision of this "triple syntagma".

8 "The 'intangible cultural heritage', as defined in paragraph 1 above, is manifested *inter alia* in the following domains: (a) oral traditions and expressions, including language as a vehicle of the intangible cultural heritage; (b) performing arts; (c) social practices, rituals and festive events; (d) knowledge and practices concerning nature and the universe; (e) traditional craftsmanship" (UNESCO 2018, Article 2.2).

9 "For the purposes of this Convention, consideration will be given solely to such intangible cultural heritage as is compatible with existing international human rights instruments, as well as with the requirements of mutual respect among communities, groups and individuals, and of sustainable development". As for "safeguarding", it covers "measures aimed at ensuring the viability of the intangible cultural heritage, including the identification, documentation, research, preservation, protection, promotion, enhancement, transmission, particularly through formal and non-formal education, as well as the revitalization of the various aspects of such heritage" (UNESCO 2018, Articles 2.2 and 3).

heritagization, in order to formalize the two scenarios present therein, which I will then clarify from questions formulated by the most significant ethno-anthropological texts, especially through summarizing the critical discussions of the notion of intangible heritage set out by Bortolotto (2011) in the theoretical and problematic framework introducing the work published following the seminar she led with Sylvie Grenet at Le Lahic[10].

The text of the Convention is composed of two types of articles: *inaugural*, which structures the representation of what intangible cultural heritage is, announcing its interpretive scheme, and *procedural*, which is related to the implementation of the Convention (the production of heritage). The former include the definitions that outline the objectives and frameworks of operations, the role of states, or what specifies the intervention of UNESCO (the lists), and also the relationship between UNESCO and states[11]. The procedural articles for their part set out the practical implementation of operations and take on their full meaning through the clarifications and developments contained in the *Directives and Regulations* (UNESCO 2018). I will show that in reality, it is not just a technical division, but two ways of approaching intangible heritage: *ideal*, in response to principles, and *operational*, with each of these ways sketching their own particular scenario for inscription of the element[12].

4.1.1. *The ideal scenario for inscription of the heritage element*

The pivotal article is undoubtedly to be found within the inaugural articles of the Convention, article 2 ("Definitions"). Its strength lies in conforming with a particularly robust textual form (that of the definition that equates a term with the enumeration of its characters), but which, in addition, implies another communicationally very effective form (narrativity), since the definition is constructed from a series of actions that announce a program. It thus sketches

10 As well as the introduction, which does in fact present a literature review, I will refer, when necessary, to all the contributions included in the book, arranged in three parts that discuss the notion of intangible heritage, a series of cases and finally the connection with ethnology. See also the report by Heinich (2012).

11 Essentially, articles: 1: "Purposes of the Agreement", 2: "Definitions", 11: "Role of beneficiary States Parties", 16: "Representative list of the intangible cultural heritage of humanity", 17: "List of intangible cultural heritage in need of urgent safeguarding" and 32: "Ratification, acceptance or approval". A third list of "good practices" has recently been added, but it is more normative than properly constitutive of inscription.

12 What I here call "scenario" corresponds to what Ricœur (1983) calls "plot", especially in the configuring function of the action. But in the present context, I prefer "scenario" to "plot", which could take on negative connotations here.

the structure of a *narrative of inscription of the element*, which will then be operationally developed in the procedural articles.

This preliminary narrativization mobilizes three actors around the actant-object – the "element" – to compose the *ideal scenario* for recognition of the element as "intangible cultural heritage of humanity": the collective, the state and UNESCO[13]. Thus, starting from article 2, the subtext sketches out in the background the ideal history of successful recognition of an element as intangible heritage of humanity that brings it to the final status of being an element that is safeguarded and transmitted. Such a scenario, characterized by a focus on the history of the element, is fairly classic in the case of an object of material heritage, especially an archaeological object (Hammad 2015, p. 172): it is then processed as a quasi-subject that will undergo qualifying tests (in reality, the issue is a recognition of these different values) until there is acknowledgment of its new heritage status, which culminates not in the restoration of its initial state, but, on the contrary, in the transition to a new life, opening up a new narrative. As for the other inaugural articles, they mainly provide a bridge to the procedural articles.

The scenario as a whole is divided into two phases that mobilize the three actors around the actant-object, acting successively during the first phase (the tests), then jointly during the second (the new life). The tests in the first phase do not a priori[14] change the element itself, but consist of an examination intended to determine its value and assign a status to it. These are therefore essentially cognitive tests (i.e. concerning knowledge, not doing): (i) the first is fundamental, since it concerns recognition by the collective of the element's belonging to its cultural heritage, thus confirming its heritage status; (ii) the second test is a "recognition" of the element (a kind of official recognition of the recognition by the collective) on the part of the state of the country where the collective lives, which will culminate in the inscription of this element in a national inventory and a request to UNESCO for

13 We will see shortly that these three actors are in reality actants in their own right, since they are defined by the function they occupy in the scenario. I will reiterate that the concept of "actant" should be understood as a function contributing to the program (the "scenario") and *accomplished* in this case by what are commonly called social "actors". In reality, in the ideal scenario, actants and actors are designated by the same name. The actors in the real scenario will appear in the following articles of the Convention, and especially in the documents gathered together in the *Basic Texts* or in the application texts enacted by the states; or else in the descriptions and anthroposocial analyses.

14 A priori, because in practice (therefore, a posteriori), the change in status can have consequences on what constitutes the object itself, as we will see later. In any case, the intangible object preserves its social existence, unlike the archaeological material object, which goes through a phase of "dereliction", since it abandons its function of use and reaches the stage of being debris.

inscription on either the "Representative List of the Intangible Cultural Heritage of Humanity", or the "List of Intangible Cultural Heritage in Need of Urgent Safeguarding"; (iii) the third test consists of a third level of "recognition", by UNESCO, both of the element itself and the previous operations, which will be concretized in the inscription of the element on one of the two lists, with the obligations associated with its new status; the obligations will moreover be subject to periodic reassessment (with the risk of deregistration). With this process being essentially cognitive, this phase of heritagization consists, in essence, of a performative operation of language. The second phase involves the implementation of a certain number of social processes arising from the element's new status (see the first article): chiefly to preserve its viability by restoring the mode of transmission between generations prior to its examination and inscription (including its permanent recreation of which this mode is the bearer), with the aim of mitigating the risk of disappearance of the element. This is a pragmatic and no longer cognitive program that will be contingent upon the procedural scenario[15]; I will therefore examine it a little later. It should also be kept in mind that the first phase of the scenario is a cognitive detour whose purpose is to maintain – or, if necessary, restore – a continuity of practices; in other words, the continuity of a pragmatic order that appears to be under threat.

The Convention presents the initial recognition, which is operated by the collective, as something that can be taken for granted, defines the initial status of the element in this way and cannot be left unquestioned. Indeed, the fact that this recognition is reduced to the observation of the element in question belonging to the intangible cultural heritage of the group assumes that the collective knows (in other words: has already partially defined) what, among the various social processes present in the society in which it lives, is eligible as heritage of humanity, which, as we will see later, will inevitably give preponderant weight to the experts working with the collective. This is the condition for it to be able to not only declare that the chosen element is indeed part of it, which also means announcing its value, according to the place it holds in what the collective regards as its heritage, but also prove that the choice complies with UNESCO's criteria. Moreover, in its application, this collective must declare that "this" element has indeed been chosen from all the elements of its heritage, not only because of a simple sense of belonging, but for a number of characteristics that give this element its value. It is indeed an operation that Bortolotto (2011, p. 31) rightly calls "selection". And it is this selection that UNESCO will eventually "sanction" positively when it inscribes the element; "sanction" being understood here in the sense of the narrative semiotic of a recognition of success in the tests for which the reward is the new status of "intangible heritage of humanity". We thus see that in their relationship to the

15 I say "rather", because we will see that it is also contingent upon another scenario: that of transmission.

element (actant-object of the program), in reality the collective and UNESCO are defined, one by the operation of selecting the element from its heritage (as actant-subject) and the other (as actant-sender) by the sanction of the element's success in the tests. The state, for its part, seems at first sight to occupy a secondary function of mediator between the collective and UNESCO. To adopt the rather old-fashioned categories of narrative semiotics, we could say that it is an "adjuvant" of the collective, bringing the power of doing and know-how to it. In reality, its function is quite different, significantly more important and decisive than it seems at first glance, insofar as it is this that establishes the value of the object and the collective. What seems a simple mediation covers, in reality, a dual operation: on the one hand, it "sanctions", by validating it, the recognition of the element operated by the collective and thereby its legitimacy, when this is not the existence of the collective; on the other hand, it submits (or not) the application for inscription to UNESCO. It is therefore, ultimately, the actant-sender that is the master of the pragmatic program with which the collective must comply in order for its application to be submitted to UNESCO. This role is at the heart of the procedural articles.

4.1.2. *Multiple operational scenarios*

The procedural articles develop the normative rules that provide a framework for the operations falling within the competence of the state and UNESCO, governing the relationships that these two actors have, both with each other and with the collective. Supported and developed by the *Directives and Regulations*, these articles thus extend, by clarifying and giving guidance on them, the "interpretative hypotheses" (as Umberto Eco would say) prepared in the ideal scenario for these bodies (particularly states). The pragmatic analysis (in the socio-anthropological sense, this time), put forward by Bortolotto (2011), not only confirms the strategy of "openness" of the text, thus preparing ample room for interpretation by the reader-state, but above all shows how this connects to the very functioning of UNESCO, and also to the relationship between UNESCO and the member states. For this reason, this analysis constitutes a major enunciative event in the mode of heritagization of social processes.

Because of its "openness", the Convention offers a "*prise*" to states[16] that leaves them the possibility of interpreting the text by filling in the interstices and

16 Following Heinich (2009) and like Tornatore (2011, pp. 215–218), I adopt this concept of "*prise*" from the sociologists Chateauraynaud and Bessy to describe the way in which the French State will apply the category of "intangible cultural heritage", particularly in its relation to ethnology. However, it seems to me that this analysis can be generalized to all transpositions of the Convention by states parties.

disambiguating the wording. This is at least what emerges from the analyses of Bortolotto (2011): for example, in the point devoted to "the 'participation' of 'communities'", she explains how these two key concepts were the subject of bitter negotiations by states and how they are translated in reality into practices that not only can be very different from one state to another, but will be to a greater or lesser extent in line with UNESCO's ideas and objectives (Bortolotto 2011, pp. 31–36).

She develops and illustrates her argument in two, more recent contributions (Bortolotto 2013b, 2014). In particular, she shows how, upstream, the text of the Convention results from a compromise between the countries' representatives, the scientists and the objectives set out by UNESCO (which may also include nuances contingent on the components of the institution); and also how, downstream this time, states transpose the Convention according to their conception of heritage and the existing administration which is made responsible for the selection and management of the elements. However, in some countries (such as Venezuela), the proposal of an element may come from a community; in other countries (such as France, Italy or China), the selection is operated by scientists and/or by the administration, when not directly by policy (Bortolotto 2014, p. 112)[17]. In this case, the famous "recognition" by the collective (of the belonging of the element to its intangible cultural heritage) is limited to a "consent" enunciated by certain actors from this collective – and what is more, who are representative to a greater or lesser degree, or politically authorized and legitimate to a greater or lesser degree, and are sometimes quite close to the state instances that operate the selection (when not affiliated with them).

Seen from a sociosemiotic point of view, this "*prise*" happens according to specific *operational scenarios* which are closer to (or further from) the ideal scenario of inscription, and which, in any case, involve subactants (or, to be more precise, specific "actantial roles") that have very different ways of implementing all the "recognition" operations within the ambit of both the collective and the state.

4.1.3. *A critical analysis of the establishment of an ambiguous category*

I will spend a little time on the ethno-anthropological analyses that reveal how the schematic presentation of the definitions and procedures of the ideal scenario

---

17 As was the case in France for gastronomic heritage (Tornatore 2012), where we can moreover wonder of what collective it is actually the heritage. In a completely different context and relevant to any object, Hafstein (2011, pp. 75–81) gives the extreme example of an instrumentalization of the intangible heritage of the indigenous peoples of Bolivia by the military dictatorship, and Aykan (2014) of the same phenomenon in Turkey.

actually covers up the complex interplay between actors (organizational and political) and power strategies between collectives, states and UNESCO[18], made possible by the operational scenario.

In her “pragmatic analysis of the establishment of this new heritage category”, Bortolotto (2011, pp. 21–29) shows the link connecting this interplay and these strategies with the Convention’s assimilation of culture to heritage or with the place that the Convention assigns to communities. She sees in the fact that the notion of heritage “overlaps” with that of culture “the culmination of a progressive anthropologization of the notion of heritage within UNESCO” (Bortolotto 2011, pp. 22–24), and she shows how the notion of culture has been appropriated by social actors[19]. “Analysts”, as opposed to “heritage activists”, she explains:

> Denounce in this “showcasing” a form of taxonomic taming and cultural standardization (Nas 2002; Noyes 2006) which, by extending museological processes to the social living space of individuals, puts them in a new relationship, no longer cultural, but metacultural, to the cultural practices that were previously “taken for granted” (Kirshenblatt-Gimblett 2004). (Bortolotto 2011, p. 30)

She points to the “sugar-coating standardization” effect of this “heritage objectification”, mentioning in this context the “loss of symbolic strength” noted by Fournier (2011) in his contribution on the heritagization of the Tarasque to raise the issue of this transformation of culture into a tool:

> ICH is defined as the deliberate selection, by a group, of elements that can give visibility to the group’s culture. Heritage creation thus becomes possible when culture is taken out of everyday life and when its bearers, who then claim to be its holders, have developed a distant relationship with this “culture” which is henceforth used as an identification tool, a tool for the production of a collective identity. (Bortolotto 2011, p. 31)

She then alludes to the “new, more active role in the actions previously reserved for heritage specialists”, which would henceforth be vested in communities, groups

18 To the analyses already mentioned above by Hafstein (2011), Tornatore (2012) or Aykan (2014), we must add those collected by Bortolotto relating directly to the involvement of the collective: Laurent-Sébastien Fournier (2011) on the Tarasque, Dorothy Noyes (2011) on the *Patum* and Ignazio Macchiarella (2011) on the *Canto a Tenore*.

19 On this point, see also Bortolotto (2007). I will have the opportunity to come back to this question of the borrowing of the scholarly framework of thought in the operation of reflexivity.

and individuals (Bortolotto 2011, p. 32), noting from the start the tendency to extend the principle of community participation. This principle, she continues, present at the outset in terms of environmental protection and human rights and now extended to cultural policies as shown by the Council of Europe's Framework Convention on Heritage (the Faro Convention), makes "explicit the essentially political scope of heritage interventions" and raises the question of the legitimacy, the interests at play and the intervention of this or that group in the choice to "transmit certain cultural elements to the detriment of others". In reality, she concludes, these "heritage interventions, often regarded as highly technical or scientific, condition the identity representations of social groups and reveal their social and political dimension" (Bortolotto 2011, p. 33). A tendency, let us not forget, which is inherent in social heritagization.

However, she adds, the principle of selection by the community of the elements of culture constituting heritage presupposes that this community has itself been recognized as such. This kind of recognition of communities faces many problems. The first are political: specifically, the opposition of states to a recognition that seems to undermine their prerogatives or that seems to call into question national unity[20]; an opposition that translates, in action and in practice, into the preponderant role that states retain in the application of the Convention by selecting communities that are not in conflict with them[21]. Therefore, a distribution of functions between heritage "value" and heritage "status" is carried out:

> Although the attribution of heritage value to a practice is now the responsibility of its actual practitioners (the "communities"), heritage *status* is still assigned to it by government institutions, which alone remain in a position to establish a policy of heritage safeguarding and authorization, via, for example, the proposal for inscription on the two international lists of "intangible cultural heritage of humanity". (Bortolotto 2011, pp. 33–34)

---

20 I will return later to the example, which she gives in another contribution (Bortolotto 2014), of the adaptation of the Convention to the situation in France through the role given to the mission of ethnological heritage and the redefinition of community according to the principle of the national community. We should note in this regard that, according to a tradition of the state deeply hostile to regionalism (to the "hydra of federalism", to use the Jacobins' expression during the Revolution), and even more to the particularities of communities, of whatever kind they may be, Hottin (2012, 2020 [2012]) confirms the application strategy of the French conception of sovereignty to intangible heritage.

21 The example of Corsica's *Cantu in Paghjella*, which will be discussed later, inscribed on the emergency list with the support of the Corsican collectivity, was the affirmation of a community, in spite of the state, at a time when such an approach was still possible.

The second type of problem arises from the conceptual ambiguity of the very notion of "community", the definition of which in practice refers back, in a way, either to the definition given by states or to the performative effect of heritagization (the community is then defined in relation to the element that has been recognized)[22].

Moreover, contrary to the reference often made to "the community", communities are in reality complex and conflictual social systems[23], raising the question of authority – hence, of power and appropriation – such that in the end, "the association between a practice and a community, as it intervenes in this heritagization process, raises the question of the ownership (legal or symbolic) of this cultural expression" (Bortolotto 2011, p. 36).

These factual explanations and discussions, beyond the knowledge that they bring, above all invite us to unfold the ideal scenario, and, in terms of our topic, to reconsider the operational scenarios from the point of view of the heritagization operations (gestures), asking ourselves whether the definitional uncertainties, the paradoxes and the strategies might be at the root of a dispersion of the heritage operations that render them invisible, finding this dispersion at odds with the operational scenario.

4.1.4. *Heritagization gestures dispersed among several actors*

The easiest gesture to identify, because it is the most obvious, is that of the official declaration of inscription by UNESCO on one of the two lists of the "intangible cultural heritage of humanity". Someone will no doubt tell me that this is not a declaration of heritage status, only of the element's inscription on a list of heritage elements[24]. This seems a priori consistent with the fact that the element is said to have already been recognized by the collective as being a part of its cultural

---

22 On this last point, Bortolotto cites the "unofficial, but very revealing" definition posited by experts ((UNESCO-ACCU 2006), Tokyo meeting, May 2006): "Communities are networks of people whose sense of identity or of bonds is born of a shared historical relationship, rooted in the practice and transmission of, or an attachment to, their intangible cultural heritage" (Bortolotto 2011, p. 34, 2013a). For a more "sociological" and more critical analysis, see Noyes (2011).

23 Commenting on the analysis of the *Canto a Tenore* presented by Ignazio Macchiarella (2011), she explains: "As ethnologists observe them on the ground, communities are never homogeneous and consensual groups, but indeed complex and conflictual social systems crossed by specific interests and subject to a distribution of power that is not always democratic" (Bortolotto 2011, p. 35).

24 A discussion of these criteria can be found at Skounti (2012).

heritage. Nevertheless, in practice, this recognition by the collective can result either from a kind of inventory conducted by the collective itself of what, for it, constitutes cultural heritage, or from scientific research, or from expert advice, when not from political desire. However, without even trying, for the moment, to clarify or discuss what the truth of it is, let us note that, at least in some countries, such as France, nowadays it is the inscription on the *National Inventory of Intangible Heritage* that officially declares the element's heritage status (Hottin 2020 [2012])[25]. This declaration procedure is therefore very similar, when it is not a simple repetition of the institutional scenarios, to the administrative scientific declaration following the Inventory model.

As the UNESCO inscription is subject to the double condition of consent from the collective and inscription on the national inventory list, it can thus appear in some way as the simple confirmation of the element's intangible heritage status, and as an acknowledgment of the commitment of both the collective and the state to treat it as such. Conversely, it is UNESCO itself that defined the category of intangible heritage, and that, in *inscribing* the element, grants it the status of intangible heritage of humanity (which is the first category), and therefore enunciates the most integrative status; this has the effect, in a way, of legitimizing, after having made it obligatory, the *declaration* by the state and in consequence the *scientific recognition* by researchers and the "recognition" by the collective. If we follow through the process starting from the collective, this latter "proposes", the state "applies" and UNESCO "inscribes". The operations are in effect shared out among four actants – the three from the convention to which we must add the scientists – who each carry out a part of the work, but it is certainly the state and UNESCO that confer upon the element an administrative, political (sometimes legal) status of national and international heritage: the state decides on the status of both the collective and also the element, but, in doing so, it is only implementing the program defined by UNESCO.

The "gesture of rediscovery" presupposes the existence of a *social interest*, especially on the part of the collective, based on a recognition of the value of the thing, which involves keeping it safe (Davallon 2006, p. 120). It is, in essence, this interest, more than a declaration, that the Convention subsumes when it speaks of the collective's recognition of the element's belonging to its heritage. Although, in reality, there is more to it: ethno-anthropological analyses show that the process is far more complex than the theory suggests, because most of the time it results from an interaction between a great many actors: members of the collective, and scientific, technical or political instances. Certainly, if we look at the operations in

25 This inscription is indeed essential in order to make an application to UNESCO. Available at: www.culture.gouv.fr/Sites-thematiques/Patrimoine-culturel-immateriel/L-inventaire-national-du-PCI/Inventaire-national.

themselves, the collective is indeed the starting point, in that, at one moment or another, it will enunciate through its representatives in what way the element has value for it; the scientists tasked with the research will then confirm this; then the state will, for example, enunciate in what way something that has a value-for-the-collective also has a value-for-the-sociopolitical-whole that it represents (or, at least, that the former is legitimate in terms of the latter). Finally, UNESCO will declare that the element presents an interest in terms its own objectives of establishing a heritage fund of humanity[26].

However, the unfolding of the process does not follow this path in a linear way from the bottom to the top: not only do the tests create barriers, but the unfolding of the process, which can last for several years, is made of closures between levels, of stops and starts, and it mobilizes social, scientific, administrative, economic actors, experts and different policies in greater or lesser numbers before the inscription can be successful (when it is successful). The consequence of this is that the interest does not manifest itself through a sense of rediscovery, as can be the case for archaeological objects, because the recognition of the value of the element is dispersed among all these actors; nevertheless, it is indeed necessary for all of them to be enlisted in order to construct a sufficient interest, a condition for the success of the heritagization program. Indeed, in reality, no matter who the protagonists, assigning a value to the element begins a transformation of its representation of the social and symbolic benefits (sometimes monetizable, incidentally) that the object is likely to bring. In this sense, this awareness of the element's value actually is something of a "rediscovery" that will trigger the production of the *representation* of this element.

As – and perhaps even: more than – in the heritagization of the material object, the second gesture, the *production of knowledge*, plays a preponderant role, because the resulting scholarly representation will serve as a bond between actors and between levels, as it allows the element to be "circulated" (in representation, of course). The knowledge comes from the memory of the collective from scientific research and recordings, but, whatever they are, they have to be organized and put in the form of documents and an administrative format by an authorized instance: specialists or scientists[27], who will finally say what the element consists of[28].

---

26 The result, according to Noyes (2011, p. 144): the community, "itself a threatened heritage, is not in a state to renew its traditions, it is barely the 'bearer' of an inert ICH. The state that represents the community must safeguard its heritage in its place. On both sides – the community reduced to impotence and the State declared incompetent – the gesture constituted by the celebration is erased; only management remains".

27 UNESCO's rules (and therefore, to some extent also, the choices of states) are to a very large extent based on the recommendations of professional and scientific experts specialized in the domain of culture (in the anthropological sense) and heritage. This entails an expert writing format for the dossiers.

On balance, the operations characterizing the first three gestures of heritagization – the assignment of a value to the element, the production of knowledge and the declaration corresponding to the assignment of heritage status – are indeed present, but they are nonetheless dispersed between a number of actantial roles: each operation requires a number of actors, while one actor intervenes in a number of operations and gestures. This dispersion means that the element of culture, in order to become an element of heritage, will be transformed into a complex object that is a mixture of texts, practices, objects, discourses, conversations and documents. This kind of interweaving partly explains the existence of practical scenarios that differ according to states and situations. This just goes to show that "recognition" proves to be of a different complexity – not to say of a different nature – than is shown by the ideal scenario when it makes intangible heritage what "communities, groups and, sometimes, individuals *recognize as being part* of their cultural heritage". But, however dispersed the fragments of operations and whatever the plurality of actors, the fact that we can still find the gestures of heritagization means that the operational scenarios, as varied as they are, correspond de facto to the implementation of a heritagization program, that is, the production of a heritage object. The conditions of possibility for such a program having been established, we are now in a position to examine its operation.

4.2. Translating the cultural element into a heritage object

Intangible heritage is sometimes referred to as "living" heritage. This expression means that, unlike material heritage (and especially archaeological heritage), there is no visible rupture between what we might call the first life of the element prior to the application for inscription, in which the object's usage fulfilled a function, and its second life as a heritage object which happens after a loss of the initial use and following the acquisition of a status and a symbolic usage (a "semiophore" existence, as Pomian would say). It is difficult a priori to dispute such an obvious fact: even after its inscription, the element continues (and I will add, must continue) to exist socially in its old form. Nevertheless, as soon as the application for

28 In France, it is the *Département du Pilotage de la Recherche (DRPS)* (Department of Research Steering and Scientific Policy) advised by the *Comité du Patrimoine Ethnologique et Immatériel* (Ethnological and Intangible Heritage Committee) (composed of representatives of the administration, scientists and elected officials) and assisted by the *Centre Français du Patrimoine Culturel Immatériel (CFPCI)* (French Centre for Intangible Cultural Heritage) on safeguarding that has this power to certify (see Hottin 2012). This role of scientists seems just as preponderant and debated in Brazil (Sandroni 2011) or in the Balkans (Nikočević et al. 2012).

inscription is made, the element undergoes a significant change: a tipping in the administrative political world of the Convention, that begins with the construction of its *scholarly representation*. However, the production of this scholarly representation – as in any heritagization – presupposes defining the element, describing it, knowing its origin, which in this case can only be done indirectly based on the various *practices* and *memories* that manifest it.

Pierre Livet noted the paradoxical character inherent in this latter constraint. On the one hand, these manifestations "are regarded as reappearances of the same intangible entity, which preserves its identity through these various occurrences"; on the other hand, since the element cannot be reduced only to a reappearance or to a medium, it is therefore necessary "that social activities should reanimate these traces and bring these remnants and media back to life" (Livet 2018, p. 62). However, these activities, Livet continues, are social processes and, as Heraclitus had already mentioned, the ontology of a process "implies that no process may be a reappearance of a determined process. Processes seem to be reduced to their occurrences and appear unable to result in types, except when we satisfy ourselves with mental representations of past occurrences". He therefore sees a conflict here:

> Between a heritage element's claims to be able to constitute a type, identical across its various manifestations – and which fully accords with its classification as intangible since its mode of existence is ideal – and the need, in order to restore full existence to this element, to give it its mode of existence as a process. (Livet 2018, pp. 62–63)

This paradox, enunciated above from a philosophical point of view, can be reformulated sociosemiotically through the following question: how can we reconcile the fact that the element retains an identity over time with the fact that the manifestations present themselves as a series of independent social activities? While this reformulation considers that there is indeed an actual process, that process corresponds to the element of culture, and not to the observable activities (these, with their well-defined characteristics, are the materialization of the social process). For Livet, as we have just seen, such a question makes no sense, since the observable social activity cannot be a reappearance of what it previously used to be through the intermediary of common entity (a type); in the rest of his article, he thus endeavors to show how the maintenance of this heritage element can (at least partially) remove the paradox. On the other hand, the sociosemiotic approach to heritagization invites us to turn the subject around completely: to examine the proposition according to which the division between activities and entities is precisely an effect of heritagization. The initial element of culture can be said to be a *social process* that is part of the collective's habitual ways of doing and thinking: an identified practice, certainly, but triggered in certain well-defined situations, in conformity with rules and procedures, mobilizing certain objects; all conditions that determine its implementation in order to obtain the expected results and effects. In other words, this element conforms to the logic of practices described by Bourdieu (1980). Heritagization would thus have the effect of analyzing (in the primary sense of dissociating) this element of culture from, on the one hand, an entity (the process apprehended from the *characteristics* that define it); and, on the other, from the activities that *manifest* this process, in order to translate it, to produce it as a heritage element. Since it has to satisfy the principle of *safeguarding* the element, it will have to reconcile two logics: the *logic of writing*, with the production of the scholarly representation, and the *logic of practices*, with the continuation of the traditional practice. In this reconciliation lies the originality of heritagization of an element of culture which is a process: it will make the process both readable and visible. This is, in a way, the form that the hybridity of the heritage object takes on between scholarly representation and object – except that the roles are reversed.

4.2.1. *Writing the scholarly representation*

It was while studying the particular case of the *Cantu in Paghjella*'s inscription on the UNESCO emergency list that I was actually confronted with the crucial question of the difference between the heritagization of intangible objects and that of material objects[29]. The principle of a translation of material "things" into heritage objects was established at that time: the interpretation of these things as traces endowed with values, by making them traces bearing indices to their world of origin, ipso facto guarantees their authenticity and leads to their being attributed a status of things to be kept (Davallon 2000, 2006). The production of knowledge in this case brings knowledge of the history of the production of the object, which, if we apply the criterion that Goodman (1990 [1968]) devised a long time ago for works of art and that Genette (1994) took up and discussed[30], makes it possible to classify the mode of immanence of this object (and therefore its mode of social existence) as being of the "autographic" type. Given the fact that heritagization of the element of culture (in this case, *Cantu in Paghjella*), on the one hand, essentially involved a production of knowledge that was essential for its inscription, and, on the other hand, that the practice manifesting it could renew itself without thereby creating a new object, it seemed to me that intangible heritage corresponded to an "allographic" mode of immanence. This led me, initially, to distinguish two *regimes of heritagization* producing two types of *heritage objects*, tangible and intangible[31], referring, in order to do this, not directly to the distinction between autographic

29 Following a study day on this intangible heritage organized by Michele Guelfucci-Glinatsis and Dominique Salini.

30 I discussed in detail this borrowing from Genette, and thus the transfer of the analysis from works of art to heritage in Davallon (2019b).

31 See, for example, Davallon (2012a, 2015, 2018 [2014]).

works and allographic works operated by Nelson Goodman, but to the interpretation given by Genette (1994), which distinguishes physical object and ideal object[32]. I then discovered that I was not the first to have operated this homology between, on the one hand, the two types of heritage, and, on the other hand, the two regimes of immanence of works of art, autographic and allographic. In an article published in *Museum*, Kirshenblatt-Gimblett (2004, p. 60), referring explicitly to Goodman, was even then comparing, on the one hand, the sustainability and authenticity of material objects (the impossibility of them being counterfeited) with, on the other, the ephemeral character of intangible heritage (especially when this involved a performance)[33]. Comparing intangible heritage to the performing arts, she argued that it possessed two of its characteristics: "intangibility and evanescence". For Goodman, she explained, painting and sculpture were autographic, because "the material instantiation and the work are one and the same", while "representation (performance) (music, dance, theater)" was allographic, because the work is completely distinct from its instantiations that are made through representation (Kirshenblatt-Gimblett 2004, p. 60)[34].

A study of the Inventory led me to re-examine this comparison between the two regimes. Indeed, as I discussed in Chapter 2 (section 2.3), all material heritage objects are composed of a trace (the object in its physical reality) and a scholarly representation, in varying proportions according to the modes of production of these objects; they are therefore both autographic due to their materiality, and allographic due to their scholarly representation; except that it is the scholarly representation that confers heritage: by making this object exist in the world of language, on the one hand, it will serve as the basis for the declaration of heritage status, and, on the other hand, it will textualize the material trace, which will henceforth manifest it. So, what about the intangible heritage object? Is it only allographic?

One fact is certain: apprehending the scholarly representation as belonging to the allographic mode of immanence enables us not only to answer the question of "what constitutes" the heritage object (taken in its immateriality, as I mentioned just now),

32 Genette introduces the notion of ideality to free herself from Goodmanian nominalism. A discussion of this posture can be found in Martel (2005 [2003]) and Pouivet (2012).

33 She has taken it up, almost unchanged, in Kirshenblatt-Gimblett (2006) and had advanced its premises (including the affirmation of the "metaculture" dimension of heritage which will be discussed later) in Kirshenblatt-Gimblett (1998). I later realized that this homology had been mentioned – or discussed in more or less depth – by other authors, such as Heinich (2009, p. 24).

34 What she calls here "instantiation" corresponds, it seems to me, to what Genette calls "manifestation".

but above all, to understand the way in which this scholarly representation is constructed from observables. This explains the absence of an autographic material trace. It is, essentially, this construction of the scholarly representation that enabled me to understand the case of the *Cantu in Paghjella*: how the desire to maintain the existence of this element did, in fact, lead some of its practitioners, with the support of university researchers and of the Territorial Collectivity of Corsica, to observe, describe and analyze the practices (enactments of songs, their usages and the learning of them) in order to establish what this element of culture consisted of with the objective of applying for its inscription by UNESCO[35]. We should note that this heritagization of an element of culture of this type presupposes that it is still sufficiently alive for there to be in existence a transmissible practice and/or memory. It can be said that when all that remains in the group is knowledge about the element without the practices, the only trace that remains is its memory; when all that remains is the practices that manifested it without the memory – if a situation of this type can exist other than as a case study – then all that would remain would be an original "cultural" work (in the sense of an artistic creation and a spectacle); and when neither either observable practices or memory remain, the element has disappeared, because its presence as a social process no longer exists in society. The social existence of the element in the form of practice and memory is therefore the first condition – necessary, but of course not sufficient, as we will see – in order for an ideal object to be established, ensuring the immanence of the element. Establishing this ideal object means modeling the element of culture (the process), which, since it belongs to the *memory* of the collective, is transmitted between generations and manifests itself at the time of *enactments* (performances, practices, interpretations, usages, etc.) and/or in *(de)notations* (practices of transmission by learning, but also rules, principles, assessments, descriptions, etc.). This establishment of the ideal object corresponds precisely to what Genette calls "reduction": it consists of identifying all the constituent properties present in the manifestations to establish what the process (the ideal object) consists of. The function of the scholarly representation, as summarized, for example, in the presentation provided to UNESCO[36], is to conduct a review of the ideal object, accompanied by the description of characteristics relating to the circumstances.

In the paper presenting the first study of the application of Genette's model to intangible heritage (Davallon 2012a, p. 52), I had schematized the relationship

35 An illustration of the operation of reduction (in the sense of Genette) can be found in the broadcast FR3 Corse (1994) devoted to *Cantu in paghjella*. Available at: www.youtube.com/watch?v=GjdHqnaxPKg [Accessed 3 April 2020].

36 This presentation of the "*Cantu in paghjella* profane et liturgique de Corse de tradition orale" can be found on YouTube. Available at: www.youtube.com/watch?v=Oxd6YNBtT4M [Accessed 3 April 2020].

between ideal object and manifestation as follows, proposing this as an example of the heritagization regime specific to intangible heritage (see Table 4.1).

Purpose	Generic object	Individual object
Ideal	*Cantu in Paghjella*	*Versu*
Actual		Enactment and denotation (mass, concert, transmission by learning, etc.)

Table 4.1. *The heritagization regime of the Cantu in Paghjella*

The factsheet on the UNESCO website, summarizing the scholarly representation, presents the findings of this establishment[37]. Here, we see the two forms that the manifestation takes: performance and transmission. Performance is an enactment that takes place in the context of a social situation bringing together part of the social community (a mass, festival, concert, café, a more or less private enactment such as in madrigal form, etc.). As for transmission, this is a practice aimed at learning for the purpose of reproduction of the element by new performers; it aims at a memorial inscription essentially through imitation between practitioners. In both cases, the objective is to obtain, more or less, a replica of the combination of "ideal object + performance". More or less, because, in practice, the ideal object is not immutable, as it can be in a literary work, a musical work with a score, or in a classical ballet; nor can performances be considered as simple multiples (replays or even interpretations) that are absolutely identical to each other[38]. In fact, the variety of types of secular or religious songs, the existence of versions dependent on the village, dependent on the circumstances, but also dependent on the individuality of the enactors, mean that the manifestations vary from one village to another, from one group to another and from one practitioner to another. The performances are therefore not all perfectly identical: in addition to the variations in interpretation (specific to the enactment) depending on the singers, the circumstances, the mood, etc., there are variants within the ideal object (*versi*); without the performance seeming, however, to derogate from what the element *consists* of, that is, the generic ideal object of the *Cantu in Paghjella* and the individual ideal object of the sung variant. I will come back to these issues when I look at the extension of the usual functioning.

37 The factsheet for the UNESCO list is available at the following address: www.ich.unesco.org/fr/USL/le-in-paghjella-profane-et-liturgique-de-corse-de-tradition-orale-00315 [Accessed 3 April 2020].

38 It is in this that they are "replicas", not "duplicates" (Eco 1978, p. 143, 145).

Establishing the *scholarly representation* of the element means introducing a new mode of social existence of it. This constitutes a new form of manifestation of the ideal object – a documentarized form – manufactured from the collection and study of already existing forms (practices and memory); it inscribes these two forms in *the permanency of textualization*. This new written form, which combines an imitative, or rather a *reproductive* mode (recording enactments or comments by practitioners) and a *descriptive* mode (for example, description of the musical technique or the circumstances of enactment), therefore has the advantage of being both disseminatable outside the collective and also recognized by it as being both accurate and sufficiently complete. Its function is thus not to replace the already existing forms, but to take charge of them. One of the consequences, which will be of great importance, is that in bringing the element into a world governed by writing, this identifies it as a singular social process, differentiating it from the flux of social life. My hypothesis is that the scholarly representation, by "writing" and "documenting" the chief constituent properties of the social process, constructs a way of looking at it that makes it its own content, so that henceforth it will exist socially "in representation" alongside the flux of social life in the form of *media texts*. However, to understand this relationship (and thus, validate this hypothesis), we must look back in more detail at the necessary maintenance of the functioning of the social process.

4.2.2. *Maintaining the practice in its usual functioning*

The functioning of the element as a *social process* never appears in the Convention, because this element is presumed to exist and to continue; this, basically, is what is being shown by the fact that the collective has recognized it as forming part of its heritage. This recognition is the trace of its existence. The Convention, which only considers the element of culture in its guise as an element of heritage, rejects the social process as outside its field of competence, not only because it is the object with which it deals institutionally, but above all because it has already referred its usage, below certifying its existence, to the collective and the state.

4.2.2.1. *A functioning ruled by the logic of practices*

The analysis of a few contrasting examples, both elements and documents, especially documentaries, makes it possible nonetheless to grasp in broad outline the logic that governs the continuation of the usual functioning. Neither in the reactions of some of the members of the community involved that can be found in the documentaries[39], nor in the ethnological descriptions, does the process seem to be

39 The study of publicization helps us to see the point of using of documentaries as media texts.

split between entity and activation[40]. Most of the time, the persons who are questioned simply assert (or even sometimes assume) the existence of an element of culture that belongs to their way of living, thinking and imagining the world. The process is here a social reality in itself; named, memorized and transmitted very often orally by memory and practice, it very specifically makes us think of those "mental representations of past occurrences" of which Livet speaks, but which are here socially regulated and produced according to a logic of practices; this is in the same way as the gestures, techniques and operational scenarios that perpetuate it. In fact, this transmission does not function according to the logical models of scholarly objectification (as opposed to scholarly representation), but according to the logic of the "practical sense" as understood by Pierre Bourdieu when he writes, in the preface of his eponymous work:

> Coherence with no apparent intention and unity with no immediately visible unifying principle of all the cultural realities which are inhabited by a quasi-natural logic (is that not what constitutes the "charm of Greek art" of which Marx spoke?) are the product of the millenary application of the same patterns of perception and action which, since they are never constituted in explicit principles, can only produce an unintended necessity, therefore necessarily imperfect, but also a little miraculous, and very close in this respect to that of the work of art. (Bourdieu 1980, p. 28)

Or, according to the logic of "creative recall", characterizing oral societies, which Goody (1994 [1977]) distinguishes from "mechanical remembrance", proper to societies that use writing.

While this practical logic consists chiefly, as we have seen, of reproducing what we know we habitually do in a situation similar to the present situation, it also serves to adapt it to the constraints of the present. It thus sheds light on the paradox, inherent in the Convention, of a continuity by transmission that is at the same time accompanied by a recreation. The rules that allow recognition of the circumstances pertaining to the present situation, the reasons for making use of this practice and this knowledge, the procedures and techniques mobilized for the implementation of this practice, are all those that are transmitted (chiefly through memory and learning from generation to generation), which means that this practice is de facto part of the maintenance of the society[41]. There then emerges, with the selected texts, a third

40 In the following pages, "activation" is to be understood in the sense given to it by Cometti (2012) when he speaks of "activation" of a work of art.

41 In the sense that Livet and Nef (2009) give to "maintenance". On intangible heritage, see Livet (2018). An example of the ethnological discussion of this maintenance can be found in Noyes (2010).

scenario, alongside the ideal scenario and the operational scenario of heritagization, the *scenario of transmission* by continuation (often called "tradition"); this is the central operation of which is the usage of practice and memory, which makes possible the ordinary functioning of social life. This scenario is built around the *practitioner* actant-subject, possibly accompanied by one or more *participant* actant-subjects who are beneficiaries of the operation. This practitioner actant, implemented by a greater or lesser number of concrete actors, appears to be endowed with an *authority* linked to competences acquired through the transmission of knowledge and know-how, but which also results from a legitimacy recognized by the collective, due to their competences, but also to social factors such as their position in society, their initiation, their training, etc., or their personal qualities. This authority has a counterpart in the collective nature of memory that gives the beneficiaries the means of recognizing the qualities of the practice such as, for example, its "faithfulness" to what is done in these cases and its effectiveness. The transmission is therefore not based solely on the memory of the practitioners, but on a collective memory of the group, possibly including memory traces, such as objects or writings. The members of the social group assuming this actantial function of transmission thus possess the know-how (memory) and the power to act (position and social competences in particular) that allow them not only to enact, but also to transmit to future generations, which is necessary in order for the process to continue. They know (explicitly or intuitively) what is faithful to it and what is not, in terms of what has always been done, the spirit to be preserved, the objectives to be met, the practices to be enacted, the circumstances present or the resources mobilized in order to achieve it, and thus to be taken into consideration. They therefore mobilize memory traces while at the same time operating an *adjustment* of what has been transmitted to the situation being lived, adapting the practice to the constraints of the latter, in order to safeguard what is to be obtained or transmitted[42]. For example, in the above-mentioned two documents presenting the *Cantu in Paghjella*[43], we see how the adjustment of what is transmitted is operated in accordance with the constraints, possibilities, circumstances specific to the situation in order to preserve the spirit, principles, rules, technique and procedures that constitute the singing. Ultimately, the adjustment safeguards the logic of the element by reactualizing it if necessary: the element remains as it is in itself, while at the same time being attuned to the present time, generating the feeling of conformity

42 There is clearly a utilization of the memory of past situations. An excellent illustration of such adjustment and preservation strategies is provided by the leaders (*punake*) of the *Lakalaka* ceremony in the face of current developments, and especially the presence of tourists and the ethnologist in Tonga (Condevaux 2014, pp. 49–58, 2019).

43 The video presentation of the *Cantu in Paghjella* connected to its inscription on the list (available at: www.youtube.com/watch?v=Oxd6YNBtT4M&pbjreload=10) and the earlier broadcast, entitled *Paghielle*, in FR3 Corse's series *Territoires*, hosted by Pierre-Jean Luccioni in August 1994.

among the members of the group[44]. This process of adjustment is clearly illustrated by the analysis performed by Pianezza (2017, pp. 477–485) on the experience of the recording by young Guaranis of what the group's elders say about two rituals that these young people are tasked with documenting through video recordings[45]: these elders are able to say what is faithful and what is not, thus helping the younger ones to understand what the ritual consists of; in this way, they discover the practices and language transmitted within their culture[46].

4.2.2.2. *Maintaining the identity of the process by adjusting it to the present*

As a social process, the element consequently has a double particularity: as a social fact, it is synchronously inserted into the current functioning of society; as the form in which it appears as a result of being borne by social memory, it is the actualization of a practice that is diachronically iterative. Consequently, in the present, we perceive only the latest state of practices or knowledge that nevertheless seem to have been "always" repeated this way in the past; even if, seen from the outside (where possible[47]), these practices and knowledge can be seen to have been transformed over time, and sometimes even to have been created only a few generations ago, that is, in a past that is relatively recent from a historical point of view. This is because they are perceived from the present according to a logic of memory that functions on a simple distinction between present and past. Therefore, the maintenance of the process is stabilized in its identity and simultaneously in its adaptation, so that the memory to be mobilized or transmitted is always up-to-date in the present, while at the same time conforming to the old practice that it continues to make exist: since the element is being presented as it always has been, it gives the feeling of a past continued in the present. What has (always) been continues to be: *it is*. Hence, the observation of "authenticity" of the practice noted by Bortolotto (2012, 2013a, 2017, 2020) and Condevaux (2014).

This functioning, which mobilizes memory while at the same time adjusting it, is, however, very sensitive to any change in the social context in which it occurs, and therefore to any change in the way of life of the collective. It follows that, contrary to how it may seem at first glance and to what the Convention seems to assume, the continuity of existence of the element does not depend solely on its own

---

44 What is preserved corresponds to what the ideal object will be when the element is described. However, the need to adjust clearly does not rule out possible conflicts of interpretation.

45 And also acquiring a better perception of their own position in the group. The two ritual practices are "the *nhemgongarai* and the *nhemongeta*, but with a transversal attention to the Guarani language" (Pianezza 2017, p. 136).

46 While at the same time revealing the factors behind its possible disappearance.

47 This is precisely what is made possible by writing, especially through recording (Goody 1994 [1977], 2004).

maintenance, but above all that of its context; especially of the state of the collective, or rather here of the "community", in the sense given to it by Maguet (2011): of its capacity to maintain itself in its relationship to its environment (political, economic and ideological), and, within itself, of the relationship that its members, especially when they are bearers of the memory, maintain with the social process, with their culture and with the other members – or even with "others", outside the community, who can have an influence on the future of the process. Without this continuity of social life, the element risks losing its raison d'être, and unless it can find another, it has every chance of disappearing. However, the discussion of cases such as that of the *Tarasque* (Fournier 2011), the *Patum* (Noyes 2011) or the *Lakalaka ceremonia* (Condevaux 2014) confirms the extent to which inscription de facto produces a change in the social context by introducing new issues and new actors, the effects of which can run counter to the survival of the element[48]. Although the situation is not always so extreme, these examples draw attention to the ambiguity of the situation in which, following inscription, the mode of transmission finds itself, caught up as it henceforth is between the injunction to continue and the probable change in functioning of social life, particularly in relation to the fact that heritagization symbolically isolates it from its social context in order to universalize it[49].

What we know about an item emerges with what has been written about it; in this case, everything seems to start in the present, at the moment when the item is going to become a candidate for inscription. This of course does not mean that this element did not exist previously, but that it silently belonged to the flux of ordinary life. However, with heritagization, the past is in a way condensed into the element itself – and more especially into the practices and memory that manifest it, as they have been transmitted by social memory – thus modifying the order of time.

4.2.3. *The observer position between practice and scholarly representation*

When the members of a collective deem it necessary to safeguard a process, whether on their own initiative or on external recommendation, they inevitably adopt a "metacultural" position, because, as Kirchenblatt-Gimblett (1998, 2004) has shown very convincingly, this position is inherent in heritagization. For the members of the collective, observing, thinking or talking about the action, in order

48 Following a process similar to what happens for inscription on the World Heritage List, as is shown in Chapter 5.

49 This is what Kirshenblatt-Gimblett (2006, p. 170) notes when she writes that UNESCO converts "selected aspects of localized descent heritage into a translocal consent heritage – the heritage of humanity". Tauschek draws conclusions from this, to which I will return later.

to define the process and give it a value, amounts to exercising a *reflexivity* not only about the functioning of this process, but also about the relationship that the process has with its social environment, past and present. This reflexivity, which involves a cognitive decentering vis-à-vis the ordinary functioning of social life, is the principle of how the members of the collective are going to look at the element of culture, thus starting the production of the scholarly representation. From a communicational point of view, this reflexivity corresponds to the *observer position*, which, with the semiotician, we can consider as a "disengagement" from the communicational situations that traditionally used to characterize the transmission and implementation of the process and used to give meaning to it. It is thus the principle of the production of scholarly representation. Two types of consequences will ensue: the first concerns the relationship between practical logic and writing logic; the second concerns communicational functioning.

4.2.3.1. *The logic of practices managed by the logic of writing*

The consequence of the relationship between the practical logic and the writing logic consists of the former being managed by the latter. Before heritagization, the element of culture (i.e. the social process), to make it possible for it to be implemented, had to be transmitted according to the procedures described above; and while the practices of transmission and implementation did indeed have a meaning, it was clearly not for the purpose of formalizing this meaning in a media text. With heritagization, the objective changes. Since the scholarly representation of the element is affected through interpretation of the manifestations of the ideal object, as soon as they are recorded, these become *traces* of the social process. In other words, for the members of the collective, the scholarly representation is nothing but the writing of knowledge about the process produced by reflexivity. On the other hand, for members outside the collective, the implementation of the process, which, without this representation, seemed to be only an event made of gestures, acts and formulas, will become comprehensible through the enunciation of its characteristics and its organization. Therefore, they will have the impression that through the manifestation, it is possible to understand the very being of the social process, since this enunciation has made the practice into a "readable" text. From that point on, the manifestation puts the scholarly representation before their eyes: it renders it "visible", making it *presentify* the process.

In addition to that, by translating the social process, the scholarly representation will make it exist in the media and social space of the heritage world. All the true and legitimate knowledge that it constitutes, because it is attested by its authors (members of the collective, but also scientists or experts from administrations that have carried out surveys, collections or studies), will be transcribed onto proformas and presented in ancillary documents, whose format has been predefined by the heritage administration for which they are intended.

Consequently, written and inscribed in this way, this scholarly representation, in a way *concretized* in a form and on a documentary medium[50], becomes disseminatable, searchable, debatable, evaluable by other scientists, by other specialists or by experts in the field. It enters into an institutional communication *dispositif*, where it can be treated according to the scenarios of the application for inscription, in conformity with the constitutive and normative rules of intangible heritage: there it circulates and is interpreted, taken up, sometimes rewritten, serving as a basis for interactions between the various operators who participate in the scrutiny, the discussion, the decision on the nature and the heritage value of the element of culture (i.e. of the process).

The effects of this writing of the scholarly representation in heritage format take on another new dimension when the decision on inscription is positive. The first form of mediatization, internal to the heritage *dispositif* of the state and UNESCO, will then be followed by the "inscription" – in the strict and strong sense of the word – of the element on one of the two lists on the dedicated site, and in the various official documents. This inscription gives visibility to both the existence of the heritage object and its interpretation; it makes of it an *interpreted trace*. First of all, by publishing it, it *gives it visibility* as a trace: on the basis of its capacity as a media *dispositif* to make the stored information perennial and to facilitate access to it by an audience, it presents not only the name given to the process by the collective, but also a summary of the scholarly representation accompanied by reproductions[51]. Then, it *interprets* this trace according to the institutional format of UNESCO: that of a list made up of a great many elements classified by states and organized according to a scheme that defines the value and status of each of the elements. While the list is only a series of scholarly representations classified by country, a system of sorting from metadata – especially those that are titled "Concepts"[52] – nevertheless makes it possible to compare them. The scheme that structures the list corresponds to the actual way of thinking symbolically and mediatically about intangible heritage from UNESCO's point of view (i.e. of humanity, rather than collectives). It therefore establishes a universe of reference that defines intangible heritage and also its effectiveness. It draws an "imaginary matrix" (Jeanneret 2008, p. 193), forming the *interpretive scheme* of this type of heritage that the inscribed elements in some way illustrate. This scheme first characterizes the five types of

50 Its textual enunciation is, as I said in the Introduction, linguistic, editorial, documentary and mediatic. It is "concretized", in the sense that the theater spectacle is said to be the "concretization" of a text (Pavis 1983, pp. 51–88). For an example of UNESCO proformas, see: www.ich.unesco.org/fr/formulaires [Accessed 18 July 2022].

51 I refer to the lists on the UNESCO website. Available at: www.ich.unesco.org/fr/listes [Accessed 26 January 2023].

52 Available at: www.ich.unesco.org/en/lists?multinational=3&display1=inscriptionID#tabs [Accessed 21 December 2020].

objects (practices, representations, expressions, knowledge and know-how) present in the five domains (traditions and oral expressions, including language as a vector of intangible cultural heritage; performing arts; social practices, rituals and festive events; knowledge and practices concerning nature and the universe; know-how related to the traditional craftsmanship) enunciated in article 2 of the Convention; however, as we shall see, in practice (particularly since UNESCO places itself in the point of view of humanity as a whole, and not only that of the collective concerned), it applies much more broadly to any form of intangible heritage corresponding to the four constitutive criteria: the object is a social process, it comes from the past, it is recognized as such by the community and is the subject of a desire to safeguard and transmit. This scheme therefore plays a crucial role in the structuring and interpretation of texts and documents circulating in the media relating to intangible heritage and will thus contribute to forming the cultural being of the latter; but also relating to heritage in general. Locally, as a result, for a given element, scholarly representation and continuation of practice are reinforced: insofar as the "writing" of knowledge, in particular with the establishment of the ideal object, results from the observation of occurrences of practice, which in principle aim to be faithful to the culture, these manifestations appear both *faithful* (to the practice transmitted) and *in conformity* (with the description of the properties, i.e. the ideal object). Differences (there are inevitably always some) are generally the result of circumstantial adaptations specific to each situation, in particular as a result of the evolution of society. As for the two logics (writing and practice), there is no conflict between them, since the logic of writing, by translating that of practice, rises above it. The practice (i.e. the activation of the process) then appears as the implementation of a specific element of the culture (an entity that, in its narrowest form, corresponds to the ideal object) because the scholarly representation, by "describing" the process under these various facets, provides an *interpretation* of it that, making it intelligible to those who do not belong to the culture, enunciates its value. In addition, the scholarly representation confers upon it a new mode of perennial social existence by manifesting it in the form of *documents*. Essentially, what the Convention terms "heritage element" corresponds to the element of culture thus translated.

4.2.3.2. *From the participatory communication circuit to the generalized circuit*

I now come to the second consequence that has to do with communicational functioning. For the members of the collective, adopting the observer position opens up the possibility of establishing interactions and building alliances with scientists and experts who belong to heritage institutions. These experts, who are skilled in the written formalization and the interpretive scheme of intangible heritage, will be in a position to help the members write the scholarly representation in accordance with the rules, procedures and expectations of heritage institutions[53]. From this moment on, the group's internal communication linked to the process, often largely based on orality and experience, will be taken over by a wider communication network and inserted into the network of heritage institutions; and, in a second phase, into the potentially tremendously open network of media circulation. This network – a *generalized* communication circuit contrasting with the *participatory* communication circuit of the collective – now links its members (at least, the members engaged in heritage) with external actors around a sharing of the scholarly representation according to the heritage scheme legitimized by institutional standards. This generalized circuit aims to be a consensus (giving a "discourse, now authorized, on intangible heritage", we could add, paraphrasing Laurajane Smith)[54], whereas the participatory circuit essentially resulted from the organization of the community and had been constructed within it layer upon layer over time.

The arrival of this generalized circuit is very likely to complexify – if not modify – the ways of doing and even of thinking about the place of the process within the community, insofar as the members who have adopted observer positions are in the role of intermediaries between two social and symbolic universes that were hitherto separate. Anthropological research has pointed out the chief social consequences of this "metacultural production of heritage" (to use Barbara Kirchenblatt-Gimblett's expression): for example, the transformation of the point of intersection between the two universes into a "heritage arena"[55], or the effects of "showcasing" denounced by some anthropologists, as summarized by Bortolotto (2011, p. 30). To confine ourselves here only to the heritagization aspects, we can note that the reflexivity experienced by the members of society who are then in the observer position will mean that as a process, manifested in the present and interpreted by the scholarly representation as having come from the past, the heritage object (i.e. the element of heritage) will be *the trace in the present of what it has been in the past*: a social practice belonging to the culture of the collective, which is in this way certainly continued, but which is also accompanied by its translation into the present according to the interpretive scheme. It is customary to retain from this situation only the keeping up of the usual functioning of the process as a continuation of what "goes without saying", which is why the inscription procedure, or the publicization that follows, is mostly reduced to the simple observation of such a continuation.

In conclusion of this point, we may wonder whether, in some cases, it might not be the disconnect that makes visible the reflexivity between the process and the present society – hence, the fact that this process no longer seems at all self-evident – that lies behind the observer position and that then triggers in some members the

53 For two examples of such a conformation, see Tauschek (2012).

54 On the notion of "authorized heritage discourse" in intangible heritage, see Smith (2014).

55 For example, Givre (2012) presents a detailed analysis of a "heritage arena" case.

awareness of a value of the cultural element that lies behind the motivation to act to safeguard it. If this were the case, the showcasing of the element, discussed by Bortolotto, would perhaps not only be the a posteriori effect of heritage institutions, but rather that of a meeting between the heritage interpretive scheme (a way of perceiving things that are in danger of disappearing) and the division that the collective is de facto operating between, what it still wishes to keep from the past and what it is abandoning or has already abandoned[56]. The answer to this question lies in the field of ethno-anthropological research, not that of communicational research.

4.3. The intangible heritage object, a media text

Reviewing the process of translating the element of culture into an element of heritage – the process operated by heritagization – showed that this translation was done according to two connected but hierarchical processes: the continuation of the process in the social life of the group and the inscription of its scholarly representation, which makes it exist in the reference universe of the heritage institution and the universe of generalized media communication. The consequence of this is that safeguarding the process as a heritage element requires maintenance of this conjunction and hierarchy. The objective is thus to sustain the dual mode of social existence of the characteristic element of any heritage object: as a trace of the past and as a scholarly representation. It is clearly not a question of recommencing the inscription process, but rather of activating the process as in the past and at the same time mobilizing the scholarly representation in order to reaffirm the continuity of its heritage status (its patrimoniality) from the observer position. We should note that this kind of functioning is not so far from what happens for the material heritage object. However, with one great particularity: in the latter, it is the materiality of the object that makes it a trace, forming the link to the past, while the scholarly representation confers patrimoniality; in intangible heritage, since the trace of the process constituted by the enactment is ephemeral, and the perennity of the heritage object depends on transmission of the process within the group or on its transformation into a media text; the first is the cultural side, and the second is the heritage side for the reasons stated above. The fact, as we have just seen, that the second is in charge of the first is not a self-evidence because it modifies the temporality of the process. Indeed, it is no longer about a simple going back to the past from the present that is made possible by the material trace; there is therefore a different process at work, the principle of which must be clarified first before we look at its modalities.

---

56 This would amount, albeit somewhat iconoclastically, to extending to intangible heritage what Debary (2002) says about social heritagization: in the sorting and recycling of remains, preserving "what we cannot destroy" because we are attached to it.

4.3.1. *The production of the process as a heritage trace*

We know that the going back to the past from the present, made possible by the material trace, means that once the origin and authenticity of the material heritage object has been established through interpretation, its temporal perspective is that of an object from the past that is datable, with admittedly a greater or lesser degree of precision, but which is at least situatable on the timeline. There is nothing like this for intangible heritage; the perspective of an "identifiable" past becomes lost, so to speak, in the "dark abyss of time", to use that wonderful expression of Olivier (2008), which generally translates into the observation that it has "always" been there. Synchrony dominates diachrony. Therefore, we often remember the present in which the process appears, with an effect of focusing on the collective, to the point of doubting the heritage character of this element of culture. Speaking about a heritage object would then be, even more than in the case of social heritagization, only one of the means mobilized by the collective to present itself to others, a mark of identity, when not reduced to a simple question of ownership (Bortolotto 2011, pp. 36–38)[57]. This phenomenon of "presentism" is further accentuated by the presence of a future that potentially bears the risk of the element's disappearance (Hartog 2003)[58]. Hence, the question of knowing how the objective of sustaining the hybridity of the intangible heritage element, between trace of the past and scholarly representation, can be effected by the transmission of the process within the group and by the transformation of this process into a media text.

First of all, if we want to answer this question, we must abandon a chronological temporal perspective – past, present and future – and turn to a properly *aspectual* approach to time. We know that thinking about time from the aspect, and not from chronology, helps to understand the temporal perspective that characterizes any heritagization – both of material objects and social processes – since the features of patrimoniality are constructed in the present of the interpretation of that to which we give the status of trace. Indeed, in the case of the material heritage object, understanding the aspectual dimension makes it possible to grasp the temporality inherent in the semiotic relationship of recognition (Eco 1978) between, on the one

57 She notes the aberration posed by a reformulation in terms of ownership of the relationships belonging to ontologies other than naturalistic (see also Kirshenblatt-Gimblett (2006): *in fine*). On this question, we will look at the contributions of Cornu (2020 [2012]), Cominelli (2020 [2012]) and Benhamou (2020 [2012]). On the use of social heritagization, see Rautenberg (2003a, pp. 154–155).

58 Even if the author does not specifically develop the question of intangible heritage, unlike environmental heritage, he clearly poses the question of the "future perfect" (Hartog 2003, p. 127), but drawing consequences more from the point of view of the chronological dimension, the relationship between past, present and future (Hartog 2003, p. 210), than from the aspectual dimension whose decisive role I am intending to show here.

hand, what the object is today (here and now), and, on the other hand, what it was at the time of its creation – or, where appropriate, at other key moments in its existence (in days gone by). In other words, to understand how it has become (from our point of view as interpreters in the present) the *trace* of itself in time – what Riegl calls its "seniority value". The gap between what it has been and what it is, here and now, constitutes the foundation of the heritage value we recognize in it, of our relationship to it, even though the interpretation will then, obviously, make use of historical knowledge to construct a representation[59]. It is by comparison with this mode of recognition and interpretation of the heritage trace that the trace of the process seems ephemeral in intangible heritage, since it only lasts for the time of an enactment or only exists in memory form in the present of its transmission. The challenge will therefore be the production of perennial traces that will give the process a tangible and continuous mode of social existence. It is toward the decisive role of the interplay between the inscription of the scholarly representation in media texts and the continuation of the usual functioning that the aspectual approach directs our attention.

Jeanneret (2020, pp. 213–218) addresses this issue in his discussion of the case of the exhibition *"Une mémoire inachevée, le patrimoine juif en Galicie orientale"* at the Krakow Museum. It must first be clarified that this discussion took place as part of a large-scale project: an examination of "the trace of the mediatized social fact" that led him to define the relationship between "trace" and "aspect", especially in photography, basing himself on Barthes' formula of "it has been" (Jeanneret 2020, p. 49). The aspect, he then notes, "is a grammatical category that qualifies the predicate of sentences (to simplify what is happening) not according to chronology, but by the way the action or event is viewed and manifested" (Jeanneret 2019, p. 48). Before enunciating what he means by the *it has been*:

> This is what we call the perfect aspect. It has disappeared in itself, but something of it has remained present and thanks to this presence we seek to return to this disappeared past to restore, if not the existence – forever gone – at least the way in which it existed. The trace must therefore be somehow returned to the past to become a sign. To be clearer, reading a Peircien object as a trace is to make it a sign in the present related to the past. (Jeanneret 2020, p. 48)

Such categorization clearly applies to heritage in its various forms. An event that "took place in the past" is indeed always looked at from the present of the encounter

59 For more details, I refer to the *Don du patrimoine* (Davallon 2006), especially the third chapter, under *"Construire la continuité patrimoniale"* (pp. 114–126).

with its trace. However, there is even more to it when a *writing* of this trace occurs. Thus, in the heritagization of the intangible, Jeanneret writes, it is a matter of writing "in the future perfect, which consists of creating media text production devices in the present to provide traces in the future, rather than collecting information only *after the event*". (Jeanneret 2020, p. 222)[60]. While this writing is characteristic of "what is now called the *traceability* of society", and while it is also entirely emblematic of "heritage media writing", it plays a particularly crucial role in the heritagization of the intangible. Pointing to the main changes introduced by the development of intangible heritage – chiefly: a collective writing; the centrality of the documentation; specific modalities of collection, elucidation and recording – Jeanneret (2019) actually classes this heritagization as "a collective writing process of a form of invention of today's heritage for the future" (p. 227). He concludes:

> The shift from *it has been* to *it will have been* is therefore, fully accomplished when the present is transformed into the world of origin of the project to perpetuate a cultural form. It is also the ultimate achievement of the *emerging* trace of the text, since the trace, traditionally the initiator of the heritage process, is in this case its outcome. (Jeanneret 2020, p. 228)

However, we must not misunderstand; Yves Jeanneret is not inviting us to unquestioningly assimilate "writing in the future perfect" with a discharging of the relationship to the past, done for the sake only of a present that is anticipating a future relationship, because it is the effect in the present of the *it has been* that is thus written. Therefore, he goes on to explain, writing the schema of the trace in the future perfect (*it will have been*), means to "work in the present, through the production of media texts, so that it may yet be possible, in the future, to be confronted with the presence of traces of the past" (Jeanneret 2019, p. 213). He concludes with this definition of *writing in the future perfect* about intangible heritage: "A media production system intended to allow us, in the future, to still see that *it has been*, to avoid the fact that the past is, at best only fantasized, at worst totally erased" (Jeanneret 2020, p. 214).

While I fully subscribe to this analysis, I will nevertheless complete it in such a way as to reveal the link between the continuation of the process through the collective and its scholarly representation. To do so, I will use the approach of forming the "image of time" developed by the linguist Gustave Guillaume (1973 [1933], 1984) concerning the verb, applied to the documentary *The Inca Grass*

60 Jeanneret is referring here to Tardy (2010).

Bridge of Q'eswachaka[61], in which it is relatively easy to identify the interplay of the various temporalities. I will begin by setting aside the temporality of the watching of the documentary by the spectator, which is made up of isolated moments focused on the construction of an understanding of the content of the documentary. (i) The first temporality is that of the *documentary*. Throughout its own duration (i.e. 43 min), it is composed of sequences that take the form of a succession of operations that culminate in an outcome. Its form is thus that of a narrative, communicationally very effective and enabling an easy understanding of the *operative chain*. It is a distillation of the various sequences in the construction, the bridge accompanied by interviews with informers. Its duration is therefore very far from that necessary for the enactment of the bridge, which stretches over three days and ends with a day of festivities, to which must be added the weeks, even months, needed for the cutting of the grasses and their braiding into basic ropes by the villagers; ropes designed to be twisted to make the cables of the deck and the two handrails. (ii) This actual duration of the construction of the bridge is a second temporality that belongs to the world represented by the documentary: the temporality of the *practice*. It inserts its internal organization as an operational chain onto the horizontal axis of time (of diachrony) in the form of the four days of construction on an exact date specified by the project manager, during the second half of June. To the spectator of the documentary, the practice presents itself as a *present* that is situated between the past and the future on the horizontal axis of time. (iii) To these two temporalities is added a third: the periodicity of implementation of the practice, since rebuilding the bridge is an *element of the culture* of the community. This annual reconstruction is presented as a major event that traditionally marks out the life of the latter; it is also the occasion for religious rituals for harvests and festivities. (iv) This element of community culture that is one of its raisons d'être referred to by the title of the documentary. Another temporality then emerges that goes back from that of the practice to the designation of this practice as an element of culture. This fourth temporality corresponds to what, in terms of verbs, Gustave Guillaume calls "chronogenesis", which goes from the nominal form (the infinitive) to the modal forms (present, past and future in the various grammatical modes). The moment of insertion of the present of the construction, defined and diachronic, can be said to correspond to what he calls

61 Inscribed on the UNESCO representative list in 2013. I had used Gustave Guillaume's approach, for the first time, in the analysis of Poussin's *Orpheus and Eurydice* (Davallon 1990, p. 509). Then I repeated it in *Le Don du Patrimoine* (2006), where I alluded to it (too briefly) in the Introduction (p. 26) before applying it, without further explanation, to the way in which heritagization operations construct a relationship to the past from the singularity of the time of the heritagization process; specifically, a way that differs from that of history or memory (pp. 125–126).

"*temps expliqué*"[62]. Conversely, the moment of designation of the element of culture itself would correspond to "*temps impliqué*". Relatively global and imprecise, it would be roughly the same as the evocation of time borne by the name of the element of culture: in the event, a long time, most of which stems from an activity dispersed throughout the whole community, followed by an intense phase of three days of construction and one day of festivities. While the moment of *temps expliqué* corresponds to the *realization* of the practice, the moment of *temps impliqué* expresses a *virtual state*, which belongs to the very being of this practice. This approach makes it possible to identify a third moment: between the first two, virtual and realized, the moment of *actualization* is the moment when the "aim" of realization (we will express it in the optative: "may the construction be") is *concretized* in the choice of the precise date of the three days of construction by the project manager (Victoriano Arisapana).

These four temporalities thus correspond to four positions occupied by singular actants: the first is, as we have seen, the *documentary* as publicization; the second is the *practice* carried out by the community, and, in the documentary, particularly the project manager; the third is the temporality of the *cultural element* within the community; the fourth is the *scholarly representation*, or in other words, of the translation of the cultural element into an element of heritage such that *analysis* makes it possible to bring it to light.

Let us resume. While in the temporality of the practice (ii), the reconstruction actually has the status of an event; this is anything but the fruit of improvisation: the reconstruction of the year, as it is shown to us, must on the contrary remain faithful to those that preceded it, to what they *have been* previously, not only as an operational chain, but also as a social and religious fact (the periodicity of the element of culture). The transmission of the memory of the practice therefore plays a decisive role. This memory is chiefly borne by the project manager, who is moreover the central protagonist in the documentary: they are the custodian of an inherited memory (insider knowledge), which relates to know-how, but also to the successes and the problems, and consequently, the necessary adjustments when applying this memory to the present situation, as shown by the example of the decision they make on the number of ropes to be used in the cables. They are therefore the guarantor of the consistency of execution, a consistency that is indispensable to the success of the reconstruction. However, this memory is also partly borne by the whole community, be it in relation to the technical aspects (the cutting, preparation and braiding of the grasses), participation in the construction proper, the social and religious rituals, etc. It is therefore both the *memory* and at the same time its transmission that are the key to the iterativeness of the practice. In

62 I should explain that I am transferring to a social process the model that Gustave Guillaume constructed for the verb.

accordance with the logic of practices, it is this that allows each reconstruction to consist of a *replica* of those that preceded it.

The documentary, as a media publicization text, conforms to the principles of an investigation on the element of culture constituted by the annual practice of building the bridge, hence its heritage operativity. It is constructed from the observer's viewpoint on the memory transmitted, and on the practice that implements it, in such a way that understanding it amounts to operating, as an interpreter watching it, a concentration of the characteristics of this memory and of this practice that will establish the ideal object; that is, the mode of immanence of the element of culture. Since the media text is thus functioning as a form of scholarly representation, activating it by watching generates the operation that lies behind it: the translation of the element of culture into an element of heritage. An *iterative practice*, seen from the present of the reception of the documentary, is thus translated into *a definition of a continuous process* that runs from the past to the future. The writing of the scholarly representation has thus transitioned us from the position of the practice, with its iterative temporality, to the position of the observer, whose temporality is that of internal time, of a generic time, linked to the very being of the process.

The first consequence is that memory, practice and scholarly representation have now become inseparable. However, the logic is in a way reversed: the content of the memory is turned into a manifestation (in the form of a practice) of a process that is defined by the scholarly representation. The second consequence, which is of no less importance, is that the tensive temporality of the memory has been extended to the present. In terms of the present, the process (whose nature is "to be"), as it *has been* previously, falls within the category of what has been done since it belongs to a *time that is passing*, but it is also a know-how and a memorized knowledge that affects what will happen in the *time to come*. In Guillaume's words, the process is therefore in one way "in decline" (it *has been*), but through its effects in the present, it is in another way "incident" (it *will be able to be* again). That is not all, because in the documentary, it *will exist* in the future too. To understand the consequence of this, we must pause to consider the last sequence. During the festivities that bring to an end the (re)building of the bridge, the project manager announces (looking at the screen, therefore at us) that in a month's time, they will prepare themselves for the building of the new bridge, that they will busy themselves with the building of the next bridge which, as every year, will take place in the second half of June. We are very exactly at the tipping point between what is coming to its end and what is to come. From the viewpoint of the practice, they confirm that the year's building is part of the time that is passing, but they also open up the time that is to come. However, what is, from the viewpoint of the practice, the expression of an intended reiteration takes on, for the spectator, who is in the logic of the scholarly representation, the form of a closure of the present; that is, the end of one of the manifestations (enactments) of the continuous process of (re)building the bridge, and the announcement of the maintenance of this social process. In other words, the translation of the iteration into a permanence of the process means that one (re)building will succeed another. At this moment, the rebuilding that is coming to an end is becoming the *trace* of the process. As from the end of the festivities – or, for us, of the documentary – this trace, joining the previous ones, *will have been*. In the time to come, they *will have been* in the future perfect, as Jeanneret writes (2020, pp. 225–227), and I would add, it is no longer hypothetically but categorically.

The temporality of intangible heritage is therefore much more complex than that of tangible heritage: the element of social culture itself must continue to be replicated from memory in order to be potentially maintained, despite henceforth being in a textualized form that has been accounted for by the scholarly representation. It is in the light of these conditions that publicization should be approached.

4.3.2. *"In presence" activation*

The activation situation of the process is based on the presence of practitioners and possibly simply participants; the latter can be either members of the collective, or "others", non-members. This situation can, in theory, cover three different forms of communication, depending on the absence or presence of these non-members: be restricted to members of the collective, without the presence of non-members; be *mixed*, when it includes both members of the collective and non-members ("others"); address the members and the non-members *separately*. How does this work in practice?

For the first situation, it is a question of knowing how much it has to do with publicization. If the criterion is the presence or non-presence of non-members, there is no publicization. However, the situation is, in reality, more complex. Does the fact that there are only members of the collective mean that there is a simple reproduction of the functioning of the element of culture prior to inscription, or does the transition of some of the members of the community into the observer position, especially those engaged in the production of the scholarly representation in connection with the heritage institutions, not transform them into an audience? More radically still, we might wonder whether, prior to any heritagization, during the transmission or implementation of the element, some form of "publicization" is not present, due to the existence of a reflexivity that puts some members of the culture in a disengaged position. Let us start by examining this last question. It is certain that transmission and learning do indeed require the adoption of the observer position toward the element of culture in order to possibly sequentialize it to facilitate its memorization or to commentate on it to help its logic be understood; similarly, when implementing the culture element – what I call its activation – when it takes the form of a spectacle that requires specially trained practitioners, there are participants belonging to the community who attend and find themselves in position of being only "beneficiaries". So, is there then no publicization? If the receiver of the publicization (i.e. the audience) is an ad hoc collective that is formed and defined by the sole fact that it is the recipient of the *dispositif* through which it is being addressed, do the members of the group, whether they are receivers of the transmission or participants attending the enactment of an element of culture, not constitute an "audience", since the collective they form is only one part of a larger one, that is, the community; the transmission and the activation being only the implementation of the social functioning itself? The first question is thornier. As it is of a different nature, since we have seen that in effect, heritagization introduced a modification in the relationship to the element. From the constitution or the existence of the scholarly representation, undeniably, there is an adoption of the observer position, at least by those who are involved in the production of the scholarly representation. Does this transition to the observer position make these community members an audience? In support of a positive response, we can use as our basis the fact that, as soon as heritagization is initiated, the *activation* of the process (its implementation) acquires a new meaning and a new status in addition to that which it had in the past. It is thus characterized by a dual status: as an habitual *practice* of the process and, metaculturally – as a media text produced from this new point of view and because of this new status – as a trace of what this process was in the past: a heritage object to be preserved and transmitted. To summarize the argument from what we have seen about the *Q'eswachaka* bridge: the activation is not only a manifestation of the process (in the present), but also a trace of it (as it has been in the past). Cannot those who attend without taking part be regarded as an audience? The answer is to be found in looking at the diversity of situations: while indeed a sociosemiotic analysis shows that the key lies in the institutional definition of the position, only an anthropological analysis of these situations can say how it works in reality. Let us take two strongly contrasting situations: the bridges of *Q'eswachaka* and *Patum* (Noyes 2011) and the Tarasque (Fournier 2011). On this point alone, without prejudging the rest, the first situation can be more or less assimilated to the situation prior to heritagization, where, since all the members are supposed to participate, strictly speaking there is no "audience". In the latter two, power games have more or less "produced" fringes of members of the collective (the term "community" here being somewhat too strong) who find themselves relegated to the status of audience. This is one more step leading to a situation where the population finds itself reduced to the category of audience (as in some patronal festivals in the south of France or in the Nice carnival). We can form the hypothesis, still to be tested, that the members of the community of course do not become foreign to their own culture as such, but rather that it is the modalities defining the disengaged position of *observers* that vary; to take simple examples, between those who assimilate this position to that of participants in an "authentic" tradition, those for whom, this process belonging to the past, it can only be part of the "spectacle" and those who claim vis-à-vis the outside, in a logic of publicization, the position of *participants* (officiants or beneficiaries); and simultaneously that, disengaged, of *observers*, making this spectacle a trace with a particular meaning in their culture. In any case, the new heritage object tends to set itself apart, to become autonomous in relation to the rest of the culture in order to constitute what Pomian (1999 [1996], pp. 191–229) calls a "semiophore", without, usually, the members of the collective constituting an audience.

The second form, the mixed form bringing together members of the collective and non-members, often involves processes already covered by participatory communication, with the presence of participants who attend the activation of the process, as is the case for music or collective performances[63]. For members, welcoming non-members further reinforces the textualization effect that has a bearing on the activation – of its functioning as a media text – insofar as these "others" will necessarily have to resort to fragments of the scholarly representation in order to attribute meaning and value to what they discover. For them, the fact that the practice being implemented is more or less distant from their own culture will transform the enactment into an *illustration* of the scholarly representation, requiring them to adopt a position not of a reflexive observer, like that of the members of the collective, but "cultural". Moreover, in the absence of this cultural position, activation would remain purely a spectacle with no anthropological depth. Conversely, for the members of the collective, the observer position will thus be visibly "embodied"[64]. However, both will have to deal with the contradictions generated by this mixed communicational situation. Mejía Ayala (2018, pp. 111–121)[65] gives a very enlightening example of the negotiations that such a situation brings, with the gap between, on the one hand, the expectations of tourists who have come to discover the way of life of the Amazon Sapara, and, on the other hand, the imperatives of the culture of the latter who link the decision to hunt with the interpretation of dreams.

To avoid the difficulties inherent in a mixed communication situation and to preserve the relationship between traditional social practice and culture, the

63 Frédéric Maguet (2011, p. 60) points out that "85% of the representative elements present on the UNESCO website concern activities that imply the existence of an 'audience'" internal to the group.

64 An illustration of the complexity of such a relationship between two positions belonging to two cultures can be found in the analysis of her participation in the *Lakalaka* ceremony (Tonga) by Condevaux (2019).

65 Her thesis (Ayala 2014) also provides an overall discussion of the issue and the various tourismization situations that result from it (including the separation between process and spectacle). The possible consequences of this type of situation are discussed according to a different approach by Fournier (2011) or Noyes (2011).

community may offer to the "others" an adapted version, constructed especially for them. It then becomes a kind of staging that illustrates the process rather than an activation of the latter, strictly speaking. This strategy amounts to reconciling two imperatives regarded as equally decisive: the continuation of the practice for the community and, for economic reasons, for example, a concretization of the scholarly representation for the "others", possibly to the extent of this concretization taking the form of a dramatic text, as in the case analyzed by Condevaux (2014, pp. 49–58) of the decision of the Tongan officials (punake) of the Lakalaka ceremony (Tonga), faced with the current developments, and in particular the increase in the number of tourists, to offer them a kind of abbreviated representation of the ceremony.

4.3.3. *Mediatized publicization*

Mediatized publicization presents itself as a concretization of the observer position. By this we mean that, basing itself on the scholarly representation, it produces *media texts* that are a manifestation (of the allographic type) of the social process that accounts for the way in which it functions as an element of heritage. In a way, any media text aimed at an audience, as long as it has to do with the heritage element or its context, operates a publicized mediation. This is why the last two forms *in praesentia* (the mixed form and a fortiori the representation of the implementation of the process for a foreign audience), which require an organization of space and time, as well as of the practices and people involved (division between participants and public), are forms of publicization. However, we cross a new threshold when publicization takes the form of documents which, since they can be stored, circulated, interpreted, reproduced, revised, transformed, thus become part of what Yves Jeanneret calls "triviality"[66]. This is because they contribute to translating the relevant element of heritage into circulating cultural beings which will be able to be, in a certain way, "activated" in representation by the audience of these documents. This publicization can take several forms, depending on the communication circuit that it mobilizes; I will identify three of them here.

4.3.3.1. *An official form of publicization in the form of a dossier*

The first form is prototypical, if I may say so, in that it formalizes the scholarly representation in the form of documents that conform to the format of instances of the state, and especially of UNESCO, as part of an official recognition of the element of culture as an element of heritage. These are chiefly the production of

66 To reiterate the processes that guarantee triviality, "archival disciplines, alteration of objects, reconfiguration of communication situations in the discontinuity of mediatization" (Jeanneret 2008, p. 228), the issues of which are summarized in the general conclusion of the book (p. 229).

texts keeping a trace in the form of documents, through collection, description, recording, etc., of the enactments (i.e. activations of the process) and denotations (memory, documents and miscellaneous objects) relating to the process, as well as knowledge about its context. These documentarized texts are intended to be kept, for example, as part of an entry in an inventory of an administrative, scientific and archival type, and through them to keep a trace of the social existence of the process, by representing it (logic of definition, description, list, index and keywords). The media texts thus produced, especially in the form of documents included in the application for inscription of the element on the lists, constitute a *reference publicization* relating to the recognition of the heritage status of the process: they are aimed at an "audience" of experts and decision-makers belonging to the heritage world[67]. In addition to belonging to the set of scientific or administrative publications, these texts actually have a publicization function when they are made available to a wider audience through the intermediary of institutional websites, such as the UNESCO website, or when they are expanded upon by documentaries.

4.3.3.2. *More or less dedicated websites*

These websites, which in fact constitute the second form of media publicization of intangible heritage elements, are occupying a greater and greater place. Moving closer to the first category, they are of two types. On the one hand, there are a few websites devoted to a particular element, such as the innovative digital museum project Prometheus.museum, currently under development (Soulier and Roigé 2022) on *"Les fêtes du feu du solstice d'été dans les Pyrénées : Falles, haros et brandons"*, which are inscribed on the UNESCO Representative List of Intangible Cultural Heritage (2015)[68]. On the other hand, there are many websites devoted to inventories, the most notable of which is incontestably the UNESCO website, which relates directly to the very creation of the category of intangible heritage. However, I am not going to linger on this website, since it has already been mentioned above, but will simply note that its particularity is to allow the heritage elements inscribed to be compared in their aspect as productions of humanity. I will however dwell at greater length on the case of the French inventory insofar as it is an illustration of technical issues – above all scientific and above all political issues – raised through

67 The example is given by Tauschek (2012). It is particularly instructive on the way in which the recording of practices and commentaries makes possible a strategic writing of two applications for inscription (the Giants of Binche from Belgium and the Doina in Romania).

68 Presentation of the project available at: www.prometheus.museum/musee-virtuel/ [Accessed 7 July 2022]. The authors note the small number of *dispositifs* of this type, citing as examples: the Algemesí Festival Museum, the Carnival Museum of Cologne, the Virtualles Fastnachtsmuseum, the Kawkawaka Potlatch, the Gion Festival Digital Museum, the Patum, or the Memoriamedia (Soulier and Roigé 2022, p. 93, n. 48).

the use of mediatization on websites. Unlike the publicization of the General Inventory of Cultural Heritage, which seems rather to involve localized tourist promotion[69], the *dispositif* used to publicize the National Inventory of Intangible Cultural Heritage is deployed in cascade fashion on three different websites: first, the Ministry of Culture's website[70]; second, the PCI Lab platform, developed by Ethnopôle, Institut Occitan Aquitaine, under the project management of the Ministry of Culture; finally, in the form of a Wikipedia presence, including an entry on *"Patrimoine Immatériel"*[71], an entry on *"Inventaire du patrimoine culturel immatériel en France"*[72], and diverse entries relating to a number of heritage elements inventoried by the Ministry, together with a web portal: "*Portal: Patrimoine culturel immatériel*"[73]. The infocommunicational analysis of this mediatization *dispositif* by Severo (2022) brings to light two remarkable phenomena that raise the question of the issues and effects of this kind of publicization strategy.

The first of these phenomena is the transposition of UNESCO's conception of immaterial heritage. This transposition, which Bortolotto (2014) describes as a "transfer", of which she describes the characteristics and issues, and of which Noyes (2011, p. 143) notes "the oddity", can be seen on the Ministry's website through the fact that the "seven themes that ICH (*sic*[74]) covers" do not really pick up either the

69 The fact that this inventory depends on region (see section 2.1.5), while the destiny of the Intangible Heritage Inventory is linked to the mission of ethnological heritage, has probably had a great influence. Elements of a history of this inventory of intangible heritage can be found in Hottin (2012, 2020) and Csergo et al. (2020). On the advantages of using digital technology for the "saving", in the author's own words, of intangible heritage, see in particular Castéret (2017).

70 Available at: www.culture.gouv.fr/Thematiques/Patrimoine-culturel-immateriel/Le-Patrimoine-culturel-immateriel/L-inventaire-national-du-Patrimoine-culturel-immateriel. It should be noted first of all that, curiously enough, the Inventory of Intangible Heritage is autonomous from the General Inventory of Cultural Heritage. An explanation of this can be found in Hottin (2012).

71 Available at: www.en.wikipedia.org/wiki/Patrimoine_culturel_immatériel#cite_note-6 [Accessed 10 July 2022].

72 Available at: www.fr.wikipedia.org/wiki/Inventaire_du_patrimoine_culturel_immatériel_en_France [Accessed 10 July 2022].

73 Available at: www.fr.wikipedia.org/wiki/Portail:Patrimoine_culturel_immatériel [Accessed 10 July 2022].

74 Namely: "*Pratiques sociales et festives*"; *"Traditions et expressions orales"*; *"Pratiques physiques"*; *"Arts du spectacle"*; *"Jeux"*; *"Rituels"*; *"Savoirs et savoir-faire"* The formulation suggests that these themes did not come from intangible heritage, but that they pre-existed it.

five types of heritage of the Convention[75] or the five areas in which it claims that these types manifest themselves. The content and layout of the home page suggest that, in some way, the French Ministry has translated UNESCO's categories using its own nomenclature, apparently regarding the status of intangible heritage as a classification, recognized a posteriori, of certain social processes that were already known about due to having been studied by ethnologists, and not as the existence of heritage elements recognized as such by the collective, as the Convention affirms. If this is indeed the case, it would indeed be an actual translation of the Convention, all the more important since it serves as a basis for decisions on inscription – the Ministry speaks moreover of "inclusion" – of these processes in the Inventory[76]. For each of the elements inventoried under these themes, we are offered an inventory sheet[77] to download, while another link invites us – quite surprisingly at first reading, but I will come back to this – to go to "PCI Lab (Wikipedia) to contribute". This home page includes a link leading to the PCI Lab site and, finally, it invites us to download the list of elements included in the Inventory[78]. The second platform (PCI Lab) introduces itself as publicization (both mediation and mediatization) of the Ministry's institutional website. According to the Ministry itself, the particularity of PCI Lab is to offer "search and promotional features such as thematic queries, interactive mapping and also [to integrate] visualizations of associated videos". It should be noted that visitors' apprehension of the site is clearly very different according to whether they enter via the home page or land on an element's factsheet via that of the Ministry. In the second case, the element is described using the same headings as the Ministry's factsheet, but the information is supplemented by diverse

75 I call "types" the various items that the Convention groups together under the concept of intangible heritage: "The 'intangible cultural heritage' means the practices, representations, expressions, knowledge, skills – as well as the instruments, objects, artefacts and cultural spaces associated therewith [...]" (UNESCO 2018, art. 2, Paragraph 1).

76 Among others, an active intervention on the part of the former mission of ethnological heritage and "ethnology", then within the *Département du Pilotage de la Recherche, Direction des Patrimoines* (ethnological conception of heritage) and both political and administrative criteria (in particular, the French national conception of heritage, as shown by Bortolotto in his 2014 article). The pre-eminence of the ethnological conception over heritage recognition of elements by the collective is clear when comparing the Wikipedia entries "Inventory of intangible cultural heritage in France" and "Intangible cultural heritage". However, it should be remembered that the Convention only requires signatory states to draw up an inventory, letting them choose to operate it in their own way. Hottin (2012) presents a history of the construction of the policy transposing the Convention in which all these points are discussed from the point of view of ethnology and administration.

77 The so-called "standard form for the inventory of the intangible cultural heritage of France".

78 It should be noted that this list makes no mention of the possible inscription of elements on one of the UNESCO lists.

clarifications, and we are given an option to access a map, videos or images, or to view associated factsheets[79]. Entering through the home page, however, a study of the website reveals that the elements, which were classified in seven themes on the Ministry's site, are redistributed according to ten themes that essentially identify practices[80]. With the development of tools to present these practices, the logic of the inventory seems to tip towards an encyclopedia logic. As Marta Severo explains, this website acts as an interface between the Ministry's website, governed by a logic of scientific and administrative experts, and the Wikipedia website, governed by an "enthusiast" logic. Moreover, the Ministry announces that the PCI Lab website uses data from Wikipedia to enrich its database, so that by "joining the Wikipedia community, it is thus possible to freely and collaboratively enrich the encyclopedia in the field of ICH". This tells us that the partnership between the Ministry's communications arm and an online encyclopedia goes beyond the simple desire to set up a *dispositif* to address both specialist and non-specialist audiences to aim for the target of making it possible for the audience to participate in the enrichment of the Inventory's data.

As Marta Severo explains, the second noteworthy phenomenon is, precisely, an interplay between institutional writing and amateur writing. As she writes at the conclusion of her article: "What is exceptional in this example is undoubtedly the flexibility of the institution that is prepared to redefine the 'spirit' of the Inventory as an official document in order to reach out to the amateur enthusiast" (Severo 2022, p. 69). This is due to an inventory being basically a writing operation *that conforms to procedures and to a format*. Hence, the importance of negotiation in order to operate an adjustment to two modes of writing that correspond to two models of production, definition and legitimation of knowledge, and of which we find textual traces not only in the generic entries *"Patrimoine culturel immatériel"* (Intangible Cultural Heritage) and *"Inventaire du patrimoine culturel immatériel en France"* (Inventory of Intangible Cultural Heritage in France), but even more in the

79 The sections are broken down into *"Description", "Lieu d'exercice", "Apprentissage et transmission", "Historique particulier", "Mode de valorisation", "Action de valorisation", "En savoir plus", "Données techniques", "Contribuer à cette fiche sur Wikipédia"* and *"Fiches liées"* (Description, Place of Exercise, Learning and Transmission, Individual History, Mode of Promotion, Action of Promotion, Find Out More, Associated Factsheets).

80 *"Pratiques sportives"; "Pratiques alimentaires"; "Arts du spectacle"; "Pratiques rituelles"; "Pratiques chorégraphiques"; "Musiques et danses"; "Savoir-faire"; "Pratiques agricoles"; "Jeux"; "Pratiques festives"* (Sporting Practices, Food Practices, Performing Arts, Ritual Practices, Choreographical Practices, Music and Dance, Know-how, Farming Practices, Games, Festive Practices). The fact that "practices" are listed, further reinforces the hypothesis that "inclusion" on the list is only the categorization of a social process, and also that mediation aimed at a non-specialist audience puts more emphasis on the manifestation of the social process than on this process itself.

"Portail : Patrimoine culturel immatériel" (Portal: Intangible Cultural Heritage) (especially under the "Discussion" tab). On the Ministry's side, the issue is to produce a scholarly representation in the form of inventory factsheets designed to be put on a database: (i) based on a reference science, in this case ethnology; (ii) conforming to the documentary tradition of inventories; (iii) being located in a political institutional framework that is both strategic and symbolic. On Wikipedia's side, we are looking at a typical writing procedure of "triviality" (in the sense that Jeanneret understands this term and which I will echo in the following pages). A writing that "is bearer of a (descriptive) conception of what the circulation of texts is and a (normative) conception of what it should be", characterized by "the way in which a de facto critical authority asserts itself in the movement of writings", which is all the more powerful because it is not explained: "a vaguely defined theory, dealing with what constitutes the triviality of cultural beings, in this case consensus-building on acquired and cumulative knowledge" (Jeanneret 2014, pp. 42–43)[81].

The issue at play between these two phenomena is the question of *authority*: who possesses the knowledge, the power-to-be and also the power-to-do, as well as legitimacy to have the capacity to declare what constitutes heritage, so occupying the *authorial position*? This is an issue of the same order as that which, as we have seen, already resides in the Convention in its very principle with the paradox residing within it between what it asserts (the ideal narrative of a heritage defined by a community that recognizes it as its own) and what it does (hand in hand with the states, because it depends on them). The Ministry's translation of the Convention amounts to positioning itself as actant-sender, establishing the legitimate subjects of legitimate knowledge: whether that be the researchers who describe the process in its various components – including the members of the collective who implement it and who find themselves obliged to submit an application to see whether this process can be considered as a (national) heritage being – or whether it be the committee responsible for proposing a positive or negative sanction to the application. Only the Ministry ultimately possesses the *power to declare what can be included in the National Inventory* (the implication being, in the national heritage). From this angle, PCI Lab appears to be above all a mediatization tool. However, not entirely, because the truth is that while the confrontation with Wikipedia, and the negotiation that followed it, in no way changed the Ministry's position as institutional author with regard to declaration (hence authorship of heritage status), it did, on the other hand, boost the circulating representation of this declaration and it also initiated a breach in its institutional enclosure. At the end of

81 To illustrate the authority of the anchoring in triviality of this writing, which is only detectable through a detailed semiotic analysis, he comments (not without a certain irony, I think): "It is not without significance that in a documentary construction that defines itself as an encyclopedia, the dominion of quantity, the signposting system of red lights and hit parades embody the critical authority" (Jeanneret 2014, p. 32).

the day, this is the issue at play in the negotiation of the two writings: the recognition of the social legitimacy of the authorial position belonging only to the Ministry and its adjuvants to "write" what constitutes heritage, in the face of an authority that favors the circulation of representations as the principle of evaluation, and as the factor of legitimacy. When the authority founded on triviality recognizes the findings of the expert (i.e. scholarly and institutional) authority by publicizing them, it endows them with high visibility in the world of circulation of cultural beings, and this is basically essentially what the Ministry expects. From this point of view, for the Ministry, one of the secondary benefits of the existence of knowledge contributions by members of the contributory "community" of Wikipedians, seems to be a recognition of the elements of the Inventory – or improvements to the factsheets – coming from members that belong to a population that is not isomorphic with the necessarily restricted circle of the collective (and, perhaps, to partially free itself from potential claims from the community whose heritage it is), without affecting either the constitutive rules that it has itself defined or, seemingly, the principle of the Convention. If this logic were to be pursued to its full extent (but we would probably then be going outside the framework of the Convention), this contributory "community" could even function as a kind of placeholder of the "community of practice" (i.e. the collective). On the other hand, if the trivial authority constituted by Wikipedia claims to make the expert authority accept its own authority in terms of production or criticism, it seems to de facto call into question not only the functional position of author, as an actant-subject that consists of "declaring"; but above all, this declaration having a performative character, the very position of sender – in other words, the position of the one who holds the power to define the constitutive rules of the game. It will be interesting to follow what happens to the relationship between the two positions in the coming years. Wikipedia does not position itself as simply an author, in the sense of simply a collective actant-subject gathering information, but as a sender enunciating constitutive rules that govern the writing that has meaning in the world of circulation of "triviality"[82]. This gives members of this "contributory community" their own agenda. As Severo's investigation shows, this "community" brings together expert-authors, guardian-of-the-rules-of-triviality-authors, authors-belonging-to-the-community-of-practice, authors-from-the-audience-interested-in-the-heritage, etc.[83]; hence her final reflection on the place of these last two categories: "It remains to be seen in the long term how the ICH enthusiast, who makes their shy appearance on Wikipedia in the midst of a community of extremely active and extremely experienced Wikipedians, will be able to make a place for themselves in this already very structured context and how the institution will be able to support them in a

82 As can be seen when Wikipedia enunciates its own editorial rules or when it writes an article on "the inventory of intangible cultural heritage in France", shifting the Ministry from sender to subject of discourse.

83 More than positions, they are therefore actantial roles that can vary or combine.

triangulation with Wikipedia administrators" (Severo 2022, p. 69). Nevertheless, no matter what happens to the communities of practice, the fact remains that objectively arriving on Wikipedia gives further reinforcement to the importance of circulating interpretive schemes linked to the openness produced by the translation of the Convention. Appearing alongside other media texts and also other practices, the *interpretive schemes* thus generated serve to expand what can be included in the category of intangible heritage. This phenomenon takes on more importance because of its interaction with publicization through documentaries.

4.3.3.3. *Publicization through documentaries*

Especially during the last decade, another form of heritage publicization has grown tremendously, in which the intangible heritage scheme occupies a special place: *documentaries* aimed at a non-specialized "audience". These are broadcast particularly by cultural television channels, such as Arte, or by some France TV channels, either within the program schedule or on these channels' websites[84]. However, sometimes they are also produced and distributed by specialized institutions or associations. It is possible to distinguish two types: whether they deal explicitly with intangible cultural heritage or whether they limit themselves to mobilizing (or making reference to) its interpretive scheme.

The following are two examples of documentaries, in addition to the one studied above on *The Inca Grass Bridge of Q'eswachaka*, which deal explicitly with intangible heritage. First, the documentary *Saisir un patrimoine immatériel* is a particular example that perfectly illustrates the first type. It is "particular" in that it is presented as an educational tool designed to help people discover intangible heritage; it is therefore situated in the register of scientific publicization (or popularization, if you prefer). It targets an audience that is interested, but not specialist; it deals with the way in which we study this type of heritage, in view of its specificity, taking as an example a particular type of intangible cultural heritage, that of gastronomy (actually, gastronomic catering), which it consequently also publicizes. Second, the documentary *Bulgarie, danser sur les braises ardentes* is an example of a documentary presenting a ritual where initiates belonging to a brotherhood walk on embers with icons of saints on June 3, the day of the Orthodox feast of Saint Constantine and Saint Helena. This ritual attracts a great many visitors to the village of Balgari[85].

---

84 The case of the Arte website is an example of the construction of an original offering, complementary to the program broadcasts.

85 One of the villages where this ritual is promoted, inscribed on the UNESCO Intangible Heritage List in 2009 under the title *Le Nestinarstvo, messages du passé : Le panagyr des saints Constantin et Hélène dans le village de Balgari.*

Olivier Givre's analysis (2012) of the historical and social context, together with the political, cultural and territorial issues of this ritual, reveals how the central theme running through the documentary is the initiates' commitment to the ritual, especially that of one woman who is not a native of the village, notably in her relationship with the other members of the brotherhood and with their practices. The editorial structure of the documentary, concretizing the observer logic, is simultaneously narrative, descriptive and interpretive. In fact, it offers a discovery of the heritage element that meets the same rules of veridiction as the exhibition: it presents recordings (activation of the process in practices, extracts from social memory, types of places and objects, discourses on social subjects, etc.) and offers an interpretation based on knowledge borrowed from the scholarly representation, without, however, going so far as a review of the social and political context brought to light by Olivier Givre. Consequently, the documentary highlights what characterizes intangible heritage (in particular, the importance of the community in accordance with the Convention's ideal scenario), as well as its attractiveness for spectators who are members of the collective or non members. The number of this type of documentary entirely devoted to one element of intangible heritage remains relatively small[86].

While relatively few documentaries are devoted to one element of intangible heritage, the interpretive scheme of intangible heritage is, however, present in a great many of those that deal with subjects of an anthropological, cultural or heritage nature. Even though they may not mention the "intangible cultural heritage" category in naming, classifying and describing the relevant social processes, a sociosemiotic analysis of these documentaries shows that in them, social processes are dealt with, mentioned or simply evoked in such a way that, from the viewpoint of the external spectator, the audience, they seem (rightly or wrongly) to be derived from the domains of the Convention referred to[87]; and, in any case, to meet the four constitutive characteristics of intangible heritage: the process involved is a social

86 According to a first analysis of the offering, which would need to be carried out systematically and in depth, in addition to the examples above, I will mention *Les trésors du patrimoine équestre, Kihnu en Estonie : Une île d'un autre temps, Patrimoine immatériel : La peau rouge*, and sequences in *Kirghizistan, Au cœur des monts célestes*, which is devoted to learning and reciting part of the "Epic of Manas", the founding text of the Kyrgyz oral tradition; sequences that explicitly mention inscription on the intangible cultural heritage of humanity.

87 Five types of objects (practices, representations, expressions, knowledge and know-how) present in the five domains (traditions and oral expressions, including language as a vector of intangible cultural heritage; performing arts; social practices, rituals and festive events; knowledge and practices concerning nature and the universe; know-how related to the traditional craftsmanship) (UNESCO 2018, art. 2).

process; it comes from the past; it is recognized as heritage by the community: it is the subject of a desire to safeguard and transmit. Depending on the place they give to the presentation of these processes, we can put them into three categories. First come those that are entirely devoted to one social process while meeting the above criteria. Here are some evocative examples (taken from Arte's "Voyages et découvertes" section): *Le Carnaval de Schignano, étrange et archaïque*; *Frioul, une vallée tout en musique*; *Les Secrets de l'indigo*, and the 14-episode series *Rituels du monde* with anthropologist Anne-Sylvie Malbrancke. Second, we have the documentaries that, dealing with material heritage such as monuments, show how their study or restoration brings to light knowledge relating to the community that produced them. Even if, in a way, the reconstitution of knowledge relating to the social group is characteristic of the archaeological approach, it takes another direction here in that it is aimed at ensuring the continuity (or the resumption) of a practice. The most radical form of this aim is clearly experimental archaeology (such as *Guédelon II: Une aventure médiévale*, to stay in the domain of monuments), but it can also take the more limited form of a reconstruction from a group's memory of know-how. This is the case for the documentary that shows a reversal in the approach to the restoration of the Parthenon (*Les secrets du Parthénon*), the knowledge and technical practices of the builders (especially full-scale drawings or polishing for the assembly of marble blocks using butt joints) having been taken up based on an interpretation of indices of these practices and their transmission, or the documentary about the discovery of construction techniques during the ongoing restoration of Notre-Dame de Paris (*Notre-Dame de Paris, le chantier du siècle?*). The third category is composed of the multitude of documentaries in which elements of the interpretive scheme of intangible heritage are present in a partial or dispersed way, not because there is an involvement of this scheme, which would be secondary in relation to the dominant scenario, as in the previous case, but because they deal with, for example, complex heritage objects of the social heritage type. Heritage objects that are identified and defined less according to the criteria of specialists or scientists – for example, by types of heritage – than as cultural beings with relatively hazy outlines, more or less in symbiosis with their social and material environment, characterized by the need to be safeguarded. This is the case of entities such as landscapes, villages, specific social practices or ways of life[88] that combine material realities (some of which are already heritagized), memory, practices, scholarly representations and knowledge. It is precisely because the humans, who live in these places bringing together these components, regard these places as an

88 This scheme is thus present, more or less markedly (sometimes even with an explicit reference), in the majority of the documentaries under subsections *"Évasion", "À table !"* and *"Vies d'ailleurs"* of the *"Voyages et découvertes"* section, in the subsection *"Civilisations"* of the *"Histoire"* section of the Arte website; or in the broadcasts in the *Des racines et des ailes* series.

environment to be protected that they are heritage objects. A more detailed examination of this mode of heritagization will be the subject of Chapter 5.

4.4. To recap: producing traces to construct a continuity

Confining ourselves, initially, to the institutional functioning of this mode of heritage, we can first note that the intangible heritage object is hybrid in that it consists of a social process – an element of culture – and a scholarly representation. While as an element of culture, the present occurrence is nothing but the continuation of a social process ("we do as we always have done in these circumstances"), seen from the observer position, this occurrence is a priori nothing but a phenomenon, generally a practice, which only has existence in the present of its perception. A scholarly representation, on the other hand, is constructed from observation over time of the practices manifesting this process and from observation of the knowledge of the collective that serves to transmit it (its memory). And it is this that transforms what presents itself as a phenomenon into a *trace* of the process. This transformation makes the element of culture – in other words, the social process – an element of heritage, hence a profound change in the way the trace functions. This change had moreover been initiated by the social heritagization of know-how, in that the trace, which up to then was the foundation of cultural heritage insofar as its materiality made it perennial, is here a consequence of the production of the scholarly representation. To translate this change into more technical terms: the instantiation, which used to be based on the autographic mode of existence of the trace, is now dependent on the allographic mode of existence of the process; this is in the form of an ideal object established from implementations of the process. To put it yet another way, it is by being a *replica* of what it has been in the past that the present implementation may in the future be the trace of what it will have been. However, it is through the production of the scholarly representation that it can be, in the present, a replica of the past process (of which the collective memory has also kept the trace).

However, there is another particularity of this operation, certainly less confusing, and which is even so self-evident that we tend to forget it: the continuity in time of the collective, or society, that produced this process. This continuity makes of the process that has become heritage, what Quebec's public policies call a "living heritage"[89]. This particularity, also inherited from the social heritagization, helps to draw a clear distinction between this heritage and traditional material heritage,

89 Available at: www.patrimoinevivant.qc.ca [Accessed 18 July 2022].

especially archaeological (or even ancient) heritage whose creator, whether individual or collective, has in any case disappeared.

The consequence of this second particularity constitutes a third. While the relationship of those who discover a material object, be they the experts who heritagize it or the audience who then visits it, is established only with an imaginary being, the relationship of experts and audience of intangible heritage is formed, whether in person or through mediatization; this is initially with the members of the collective who still implement this heritage in the present and who recognize it as "their" heritage, because they have seen it carried out by the previous generation. However, the imaginary beings of the past do not disappear as such, since they are members of the collective who created the process in the "distant" past (i.e. several generations). Consequently (as a sort of derivative consequence), while within the collective, there is thus a continuity between present and past. Conversely, on the audience's side, an identification is established with the present members of the collective, insofar as the latter are themselves engaged in a reflective relationship with the process. However, for this audience, the process only exists *in representation*, whether in person (even though present, the audience is not part of the community) or through mediatization as in the emblematic case of documentaries. In inventories, it is documents that make it possible to perceive practices – for example, recordings – which make it possible to establish a relationship that goes beyond simply obtaining information.

More generally, the most surprising phenomenon is still the disconnect between the principle position, at the basis of the ideal scenario, of recognition by an actant-subject – whose vagueness, it is true, is in itself already indicative of antagonisms – of an element of their culture as constituting "their" heritage, and the practice of selection (when not of manufacture)[90] of an entity by states, but also in part by UNESCO, which excludes, which is understandable given its principles and objectives, certain forms of practice from possible inscription on the lists of the

90 France has two wonderful examples of such "manufacture", in a kind of retrospective historiography of national memory: *"Le repas gastronomique des Français"* (2010) and, more recently, the even more incredible inscription of *"savoir-faire artisanaux et la culture de la baguette de pain"* (2022). On the first, I refer to the article by Tornatore (2012). With regard to the second, not only can Jean-Louis Tornatore's analysis of the first easily be applied in many points (especially with regard to the French state's takeover of a heritage supposed to be in rupture with the traditional Eurocentric conception), but that UNESCO has allowed the use of the generic term "baguette", which (in everyday language – the only thing in this case that the French share) actually covers, for by far the most part, a highly industrialized product (even if artisanally baked). This dates, in its current form, to the post-war era; and which, what is more, contains far more than the four ingredients mentioned in the UNESCO factsheet, which leaves everyone quite lost for words.

Convention. However, in view of the dissemination of the notion of "intangible heritage", accounted for by the proliferation of interpretive schemes, we may wonder whether this disconnect does not have the advantage of freeing these heritage interpretive schemes from the straitjacket of institutional procedures in favor of a relatively open social usage, traveling via media texts such as documentaries, websites and posts on social networks. It seems, in fact, to foster a progressive empowerment of a heritagization mode based on an attachment of collectives that is notified, demonstrated and then shared, of which these media texts are the vectors, shaping it and amplifying it.

5

Heritagizing Complex Entities

The development of complex heritage entities, which has been observed in heritage objects since the late 1960s and early 1970s, is usually apprehended as a simple epiphenomenon of new modes of heritagization. So far, therefore, I have not paid particular attention to it. But when we analyze it, we find that these entities contain issues that deserve a more assiduous and more detailed examination.

We have already encountered this type of entity in previous chapters. In Chapter 1, with the archaeologists' proposal to protect Aboriginal engravings and publicize them by transforming the space where many of them are located into a World Heritage Park, we encountered a practice, which is moreover common in archaeology, of constituting one single heritage object from various elements. The same is true of the Inventory of heritage, which studies singular "works" in order to group them into families of specific heritage objects (families) in a given area. But since the arrival of social heritagization, the phenomenon takes on a completely different scale. It is no longer simply the production of knowledge of objects that makes it possible to form a single heritage entity, but as objects become more numerous and more diverse, the boundaries of one single complex entity become variable. For example, industrial sites consist of the traces of historic activity, both material (adapted workspaces or living spaces, various buildings, machines, archives, etc.) and immaterial (practices, know-how, ways of life, memories, etc.). A complex entity can thus be constituted of a number of sites and cover a wide territory, as is the case with mining sites in the Nord-Pas-de-Calais basin[1]. Ecomuseums are also defined by a wide territory and its population, and the objects therein may, in some cases, already be a part of heritage. Finally, the heritagization of social processes adds an extra degree of complexity when these processes are combined with material objects, the territory or the population (Skounti and Tebbaa 2011; Brumann 2013). Out of all of these examples of the production of complex

1 Available at: www.whc.unesco.org/fr/list/1360 [Accessed 8 July 2022].

heritage objects, a number of constants emerge. The most visible are as follows: the gathering of more or less heterogeneous components and the spatial base of the entity thus constituted. Other constants are less apparent, such as the relationship to a present, past or even imaginary population. Ultimately, one final constant seems to be decisive: the role of knowledge and its production as the starting point for a publicization that plays the leading role, with, for instance, the staging of exhibitions and documentaries.

But this still leaves the most difficult thing: understanding what constitutes the complexity, and in particular the relationship that can exist between the elements and the entity. However, in order to know whether there is a heritagization mode specific to these complex objects, it is essential to shed light on this point. To accomplish this, I will proceed in two stages. I will firstly explore how a collection of objects can become a single complex object through being put on exhibition, on the one hand, by using as a basis the museography of Georges Henri Rivière, and, on the other hand, by examining what characterizes the heritagization of this type of object with an example borrowed from urban heritage. Then, in the second stage, working from the basis of a sample of documentaries, I will seek to show how *a media heritagization mode* of complex objects, namely landscapes, is emerging today.

5.1. Understanding the production of complex heritage entities

To understand how a single heritage entity comes about from a number of elements, I feel it would be helpful to make a digression via a restricted situation where we can easily observe the production process of such an entity, so as to then apply the knowledge to the production of a complex heritage entity – namely the heritagization of a historical urban complex – the latter entity having the advantage of remaining, both literally and figuratively, quite contained in itself.

The first situation is borrowed from the museum, insofar as there is a tradition of gathering together material objects in the form of a collection; objects that can then be arranged so as to form a complex entity in the exhibition. Indeed, fortuitously, this situation exists (or rather, existed) with the former National Museum of Popular Arts and Traditions in Paris, now gone[2]. It constitutes a kind of "laboratory" situation that presents several examples of the structuring of heterogeneous heritage objects into a complex entity.

2 I'd like to extend my warmest thanks to Florence Pizzorni who allowed the Culture and Communication laboratory to set up a photographic survey, carried out by Cécile Tardy and myself, of the cultural gallery. Although the museum no longer exists, I use the present tense wherever possible.

5.1.1. *Producing heritage entities through the textualization of traces*

The extraordinary opportunity constituted by this situation can be ascribed to the indissoluble link that Georges Henri Rivière posited between the production of knowledge, the collection of traces of a way of life – or to be more precise, of a social fact (in the sense of Mauss) or a technical system (Desvallées 2009, paragraph 18) – and the modalities of staging an exhibition. Following the "museum-laboratory" principle that he shared with Rivet, Rivière produced exhibitions that were based, on the one hand, on a consideration of the relationship between artifacts and world of origin, making the former into witnesses of the latter, and, on the other hand, on a particularly innovative combination of these artifacts to produce an entity representing what characterized this world, in other words, based on a *writing* of these artifacts in the form of an exhibition.

In terms of the relationship of artifacts to their world of origin, field studies were there to produce knowledge about the way of life of the "inhabitants", in particular through the gathering of collective memory from individual memories, with this collective memory[3], which provides knowledge about the objects collected, then allowing the exhibition to present a display of "ways of being and of inhabiting the world", in the words of André Desvallées (1997, cited by Rolland-Villemot (2016, paragraph 14)), and making the objects into the "witnesses" of this relationship to the world. The *scholarly representation*, constituted from knowledge that was collected at the same time as the objects, and part of which brings back memory traces of the functioning of the group, would serve to "contextualize" these objects either within the exhibition by means of texts, maps, diagrams, photographs, etc., or more broadly through the actual construction of the theme governing the exhibition. On this aspect, Rivière progressively developed and experimented with *museological skills* on the method of creating exhibitions from the collecting of these material and memorial traces (Segalen 2005, pp. 147–183, 2013; Desvallées 2009; Calafat and Chevallier 2019). All of these points, which are nowadays well documented, led to the expography of the National ATP Museum providing examples of these experiments. I have selected three of these: three showcases that go from a simple combination of components to an entity that presents a piece of the world of origin as a heritage object, although they are rarely approached in these terms.

It is common practice to refer to Rivière himself in considering that this museum's exhibitions were composed of two levels: the objects, collected or chosen from the collection, and the sequence, which corresponded to the articulation of the message in accordance with what Duclos (2019, paragraph 2) calls the "discourse", which Rivière defined as "the ideological organization of the message to be

3 See, for example, the Aubrac survey (Segalen 2010).

transmitted"[4]. There is absolutely no doubt, Rivière poses the objects, collected and exhibited, as the basic unit of the museum, while the sequence is the execution, on a specific subject, of the gallery's organizational theme (the "program"), in different possible "variants" (Gorgus 2003 [1999], p. 186), the most famous of which is the description of a manufacturing process, of a "*chaîne opératoire*"[5] (Calafat and Chevallier 2019; Rolland-Villemot 2019, paragraph 16–17). It is "contextualization" that creates the link between object and sequence: in order for the object to be able to be a witness to its world of origin, the theme (borne by the sequence) must provide information on it. Contextualization therefore occupies a central place in the expology of Rivière, as pointed out by Jean-Claude Duclos (2019, paragraph 5) and also André Desvallées (2009, paragraph 11–19)[6], who explains that it was to play an increasingly important role from the 1960s. Rivière's approach is therefore quite radically communicational.

So then, why choose the showcase[7]? Because this is the media text created by conceptualization. In fact, while I am adopting the point of view of the visitor who must interpret the expographic *dispositif* (and therefore who, in reception, must carry out the proposed contextualization semiotically), the sequence is easily identifiable due to the information panel that gives it a title and lists its contents. Under the aegis of this information panel, the sequence is composed, perhaps of paintings or some other art object, perhaps of documents (images or maps) put on the walls, but above all of showcases of different sizes, some of which may occupy one, or sometimes several, sections of the wall. The showcase is thus a unit that initially appears to the visitor as a *spatial entity*, because it is circumscribed and enclosed. This entity is composed not only of objects (almost all of these are in fact in showcases), but also of linguistic texts such as a list of the objects and often an introductory text or at least a title, and sometimes script-to-visual documents

4 Duclos (2019) explains that "this stage [of the exhibition] is the 'program' stage which he [Rivière] often describes as 'scientific', the stage of reviewing the knowledge required to deal with the theme of the exhibition and the writing of the discourse that one wishes to propose to the future visitor".

5 At least for one part of the cultural gallery, the other being organized ("customs and beliefs") according to a collection principle that Martine Segalen calls "thematic" (Segalen 2008, p. 640). I will come back to that shortly.

6 Desvallées (1997, 2009) shows the importance of the contextualization operated by objects to putting on exhibition mankind's relationship to our environment (milieu, fellow human beings, daily life). An idea taken up and developed by Rolland-Villemot (2016) in terms of ecological units. On the characteristics of staging exhibitions in Rivière, we can refer to: *La muséologie selon George Henri Rivière* (1989), Gorgus (2003 [1999], pp. 167–192) and Segalen (2005, pp. 147–183, 2013).

7 I should explain that the word "showcase" here refers to the content: what is presented in the showcase-container.

(although the supporting materials and other presentational elements are purely functional components). The content of the showcase, as a part of the sequence, therefore constitutes a *text* in its own right offering a contextualization process. It is, in view of this, a particularly relevant unit when considering whether, and if so how, the expographic experiments that Rivière carries out in composing of some of them culminate in the construction of complex heritage entities.

I have selected three showcases that operate three different methods of constructing the entity: the mining and stone cutting showcase, the so-called herding showcase and finally the showcase of the shepherd's hut from Aubrac (the *buron d'Aubrac)*. Their heuristic interest becomes visible when we compare them to the more "traditional" showcases, from which they stand in strong contrast. This is true, obviously, for those that make up the popular arts section, in which the objects are "presented as works" (Gorgus 2003 [1999], p. 186) in thematic series according to the form, motif or style of the objects; but these three showcases stand out just as strongly from the showcases that present either a single object or mini-series to illustrate categories such as place (for headdresses or puppets), functions or uses (for fairground objects, games or musical instruments), etc. All of those "traditional" showcases are in no way intended to form a new unit of meaning, different from the sum of their components or from their simple grouping by type. Conversely, in the showcases chosen here, the objects are connected with each other and with the texts through a *writing*, in such a way as to give an account of a practice (and more broadly a social fact or a technical system) belonging to the world of origin of the objects.

5.1.1.1. *Contextualizing through exhibiting*

The first showcase, entitled "Mining and stone cutting in Villebois (Ain)", is the flagship showcase from the sequence "from quarry to edifice". It is constructed on the principle of the "technical *chaîne opératoire* as designed by André Leroi-Gourhan", which "goes from raw material to finished product, explaining the manufacturing sequences" (Rolland-Villemot 2019, paragraph 18). If we consider the objects solely from the point of view of the collection, this can be summed up as (or rather is reduced to) a presentation of the tools used for the mining and the cutting (both by hand) of the stone, because for each tool there is a number that refers to its identification and its provenance written on a frieze at the bottom and front of the showcase. This identification signage is however broken down into the various types of operations in which these tools are used. What could after all be a simple functional categorization here takes on a different meaning in its relationship with the showcase's other three types of components. (i) With, very specifically, a number of groups of blocks of stone illustrating, on the left, the mining operations, and, on the right, the cutting operations, together with, in the center, a presentation of the instruments used to lift and transport the blocks. The mining and cutting tools are "set in a scenario" (Duclos 2019, paragraph 5) of use or above the blocks in a scenario of waiting to be used. The identification labeling thus indicates which tool goes with which block and which operation. (ii) Labeling relating to, next, a second frieze, located this time at the top and back of the showcase, which describes these operations. (iii) To finish with (or possibly start with) an introductory text, placed at the beginning (in the reading direction) of the bottom frieze, which gives the list of tools its social context, tracing the memory of who they belonged to by the Compagnons who produced the stone blocks for the occasion, thus handing over the key to interpreting the showcase as a whole[8]. In addition, other tools (jig, plotting instruments, measuring instruments, storage instruments, handling instruments, etc.) are arranged between the top frieze and the tool block assemblies. Hence, when "reading" the showcase, through the visual and interpretative back-and-forth between the various components, the transformation of quarry stone into building material is shown in its diverse operations, in its actant subjects and objects, in its social context and its values. The objects are thus integrated into the *chaîne opératoire* represented by the showcase, which is in turn integrated into the sequence. This integration, specific to the writing of the staging of the exhibition, in return helps to contextualize them by making visible and readable the relationship they have with their world of origin, in this case: for the human aspect, that of the *Compagnons tailleurs de pierre* (stonemasons, as mentioned in the introductory text), and for the social activity aspect, that of the construction of buildings (as shown in the sequence as a whole).

Consequently, while the *contextualization* of the object in its world of origin is indeed a specificity of Rivière's expography, it appears that another specificity, complementary to it, must be taken into account: the *integration* of the objects by a higher textual unit, namely, the showcase, which is itself integrated by the sequence. Contextualization and integration are both based on knowledge produced from the field through a study that looks at the objects' world of origin. Contextualization equates to the scholarly representation of artifacts as they belong to their world of origin (i.e. the place they occupy in their mode of origin). Integration, arising from

8 I will give it in its entirety: "Joseph Dulaud, called La Gaîté de Villebois, Honest *Compagnon passant, tailleur de pierre du Devoir*, practiced the trade in the quarries of the deposit of Villebois (Ain). Grouped by operation, the tools we see here were used by him and other stonemasons, his friends. Serge Laisney, called La Fraternité de Geffosses, *Compagnon passant, tailleur de pierre du Devoir* worked under the supervision of Eugène Lamborot, called La Sagesse de Saint-Georges-de-Reneins, of the same order, from blocks of stone from a quarry in Porcieu (Isère), to illustrate the said operations. The objects in Collection 62.130 are from the donation of J. Dulaud. Mission A. Desvallées. Formed of hard limestone, used notably for many buildings in the cities of Lyon and Paris, the Villebois deposit runs along both banks of the Rhône, at the edge of the departments of Ain and Isère. [...] No. 1 Villebois deposit/map".

the ability of textual writing to generate units of meaning from the structuring, within it, of lower ranked units (Harris 1993), corresponds to the program of staging the exhibition of the sequence, which is also created from knowledge about the world of origin. In this way, the semiotic functioning of the showcase, combining contextualization and integration, makes it *an autonomous media text*.

On balance, through the process of integrating the showcase as text, what was the description of each of the objects and its context is extended to what this showcase is about. The scholarly representation of the showcase becomes the description of the practice (here, a technical *chaîne opératoire*), the context of the showcase being the transformation of stone as a "thing" of nature into a material of which edifices are built; a transformation approached here from the angle of *functional operations* that are carried out by a certain type of social actant: the *Compagnons*. For the visitor, the objects henceforth derive their meaning from their relationship to this context (and not only from their provenance or characteristics, nor even simply from their use). The analysis of the showcase shows how writing structures, by integration, the components (objects and text panels) into a media text whose context (the referent) is in actual fact an aspect of culture. The objects obviously retain a relationship to their context of origin: they are *traces* of it and have an autographic mode of existence. They thus confer part of their symbolic power as "real things" upon the showcase, but the showcase can only represent the knowledge of a process that belongs to a way of "being and inhabiting the world": the transformation of stone by the Compagnons.

In summary, the showcase demonstrates that the heritage objects it presents have the particularity in their world of origin of belonging to a cultural process, and at this point, two interpretations are possible (but not exclusive). The first takes into account only the communicational approach (in the usual sense of the word) that aims to render a process understandable. The second considers the entity produced by the showcase that has a tendency to turn the cultural aspect presented (the social process) into a heritage object. But this second interpretation is valid for us who have interpretive schemes to think about the heritagization of processes (intangible heritage). However, at that time, only material objects constituted heritage, the conservation of the process being taken care of through knowledge of it such as is produced by ethnographic studies; this process can therefore only exist in "representation"[9] in the showcase.

The next two showcases illustrate two attempts, taking two opposite directions, to play on the relationship between integration and contextualization. In one case,

9 The circularity between integration and contextualization makes the showcase a closed and self-referential *dispositif* of meaning that brings to mind what Louis Marin (1977, p. 29, 1980) calls a "representative system".

the showcase represents a practice that it seeks to constitute as a single object (what we term the herd showcase). In the other, it is made materially coextensive with what it represents (the showcase of the Aubrac shepherd's hut)[10]. What is the effect on the production of the showcase as a complex entity, or a heritage entity?

5.1.1.2. *The writing of a "complex heritage quasi-object"*

The famous "herd showcase" – or to give it its exact title: "The marching order of a large transhumant herd, between winter grazing site and summer grazing site"[11] – is presented visually to the visitor as the freezing of an action in the process of happening (i.e. the herd on the move): as a snapshot revealing the *ordering* of a herd in the process of making its seasonal movement, apprehended in its organization (the *functional character* of its composition) and in its meaning (its *ceremonial character*), as explained in the introductory text[12]. It is a picture, a hypotyposis of the herd that is all the more striking and powerful because it is a condensed version (a synecdoche) of it: it is in fact explained, at the end of the frieze, that it is "a shortened reconstruction of a whole, evocative of an immense herd". This power of evocation arises from the choice of objects and a staging that presents them "in operation", by using the nylon thread technique; this technique makes it possible to imply what is not there (the living) by putting before our eyes what "constitutes" a transhumant herd: not individuals that change, but sign-objects bearing witnesses to a generic social practice. The only texts in the showcase are as follows: the title, accompanied by an interpretation of what is represented; the identification labels grouped together on the frieze at the base of the showcase, which further reinforce the generic bias of the scene, because, with the same gesture, they identify the transhumants (their function) and the objects that represent them (cape, coat, collar, pannier, etc.), specifying the provenance of these, along with their noteworthy characteristics. The ensemble consisting of exhibited objects and texts supplies the contextualization: it takes up almost the entire showcase to represent the herd; for example, there is no secondary register of objects as in the mining and stone-cutting showcase.

---

10 Two directions that cannot but bring to mind the two methods of exhibition advanced by Rivière in 1949 are (cited by Desvallées (2009, paragraph 19)): the "systematic" method, in which "objects of all kinds are grouped by ideological sequences (nowadays we would say: "conceptual")"; and the "realistic" method where objects "having been culturally part of the same complex [...] are grouped and arranged, as far as possible, in the state in which they had been at a time of their "cultural existence"".

11 On the directional signage, the showcase is numbered "223.300 Sheep" and the sequence "223.00 Sheep farming".

12 "The marching order is functional: at the head and tail, the humans and animals in charge; in between, the mass of sheep and pack-bearing animals. The march is ceremonial: the pomp of a procession; the music of the bells; elegant collars".

Like the earlier showcase, this one operates through integration against a backdrop of contextualization. But the integration here is carried so far that the objects become simply components of the entity represented (the herd). Or to be even more precise: the context of the objects belonging to the world of origin is no longer introduced only by knowledge, it is also presentified in the very layout of the showcase (the herd in the way it is organized). The showcase, as a media text, no longer describes only a process, as the stone-cutting showcase did, but sets out to give presence to an entirely real situation captured at a given moment (the herd as organized for transhumance). It therefore moves toward producing a complex heritage object, as identified by the title, but without being able to completely achieve it. In fact, even if the autographic mode of existence of the objects endows the representation with its power of "evocation", its overall apprehension by the visitor remains first and foremost that of a text constructed a posteriori from traces of an aspect of material culture. We can of course discuss what status should be granted to the object thus constructed and wonder whether it really does constitute a heritage object on account of the heritage components being arranged to reproduce a complex cultural entity; but, whatever the answer (which, in my opinion, is quite indeterminate), what is however certain is that the showcase, taken as a whole, is the *exemplification* of a practice based on traces that manifest it. I would therefore say that it puts before the visitor's eyes a "quasi complex heritage object", which makes it, for us, as a representation of a process, potentially a metaphor for a manifestation of the intangible heritage of transhumance.

5.1.1.3. *Exhibiting the assets of a complex social entity*

The "Aubrac shepherd's hut" showcase[13] is one of the "ecological units" which consist of "sets of objects integrated into a medium, transferred from this medium to a museum environment, animated by audiovisual means" (Rivière (1971), quoted by Calafat and Chevallier (2019, p. 98)).

In one way, these ecological units mark a clear rupture with the two previous experiments, because (at least for the showcase) the integration seems to be entirely dependent on contextualization: to constitute text, they do not need to be staged as an exhibition, since not only the components, but also the syntagmatic organization itself come from the context of origin[14]. Visitors actually have the impression of finding themselves face to face with the original reality (a room with its objects). In this sense, the content of the showcase thus has a perfectly autographic mode of

13 Part of the sequence "421.00 A human establishment, Aubrac".

14 Rolland-Villemot (2016, paragraph 11) explains that the ecological unit is "a set of objects coming from the same context and arranged in a showcase as they usually were at the time they were collected in the field". But it also shows that the ecological unit actually belongs to the analog form of viewpoint museology, while the two previous showcases belong to its descriptive form.

social existence[15]. And in the present case, this impression of "reality" is further heightened by the absence of any interpretive device apart from an audiovisual device combining a recorded commentary and illuminated signage, and also by the presence of a window going down to the floor to separate the place where the visitor is from the room. However, these expographic devices, by their very existence, completely contradict the idea that every exhibition process has gone. Because what the visitor sees is the result of mounting an exhibition through collecting. It was necessary to disassemble and then reassemble the elements of which the original room was composed (walls, floor, ceiling, furniture and all of the objects) in order to produce a copy. Even if it were the most perfect copy possible, the reassembled exhibition unit has in any case still lost its original context, both material and social. In effect, the unit's autonomy isolates it from everything else and the components serve as context for each other. The unit's "realism" (the integration of the components on the original model) has the corollary of cutting it off from the outside world.

Hence an unexpected effect: the contents of the showcase can be defined as the *facsimile* of a complex entity, obtained from original components[16]. A kind of "false truth", dare I say, it is of course made from authentic components following the original layout, but its integration and contextualization depend, somewhat paradoxically, entirely on *know-how* about the relationship of the entity, and its components, to the world of origin, as well as *know-how* about the world of origin itself. It is this know-how, which must be precise and must be collected at the time of taking the unit, that makes it possible, in production, to reassemble the unit, and also, in reception, to understand it. In the absence of this know-how, composed of knowledge produced and of memory, the reconstruction would remain a perfect enigma, and its heritage character, as well as that of its components, would disappear (with the exception of their age value relating to their marks of wear and tear). To this know-how related to the original entity is added, on the visitor's side, the need for another kind of know-how, about the procedures that guarantee that what we have is a piece taken from the world of origin and not a copy, or worse, a piece of fiction. So here again, the relationship between contextualization and

15 These "integrated real things", as Rivière calls them, "bear witness to the environment with more intensity than do isolated real things" (Rolland-Villemot 2016, paragraph 18). This is precisely the characteristic of viewpoint museology that constructs the *dispositif* from the visitor position.

16 As Rolland-Villemot (2016, paragraph 62) notes: "The ecological unit corresponds to a thematic museography: the object is presented according to a given integrating theme". As a consequence: unlike the facsimiles of the ornate caves (Lascaux, Chauvet, Cosquer) whose components are themselves copies, they do not require the use of very specific technology and know-how to ensure, in the absence of authenticity, the greatest possible veracity and fidelity when reproducing the original entity.

integration is shown. In order for visitors to know that the "numerical identity" of the original has been preserved in what they have before their eyes, they need to know that it is an object collected by a faithful transfer of the original entity. In the case of the shepherd's hut, in spite of the disappearance of the initial mediation devices at the time of our survey, these conditions are nonetheless met, because the showcase belongs to the sequence devoted to Aubrac, which, with its texts and its other showcases about the inhabitants' way of life and practices, contributed to the knowledge that made it possible to understand, albeit summarily, the use of the utensils, the arrangement of the objects and the layout of the room. Another contribution to this was the existence of a calico on "Aubrac/how a shepherd's hut becomes a museum object, etc.", which had been added later, of the same type as those in other sequences, initiating a reflexive museology.

Comparing this showcase with the previous two makes it possible to clarify what characterizes a *complex heritage object* in the particular case of the museum. Unlike those two, which represent cultural processes by being written from their traces and from knowledge about them produced by the study, the "Aubrac shepherd's hut" showcase is first and foremost plainly a *heritage object*, since it was taken for this purpose, in its entirety, from its world of origin. But it is also a *complex entity*, because its heterogeneous components already possessed this complexity in their world of origin. They were not integrated in the staging of the exhibition in order to represent or exemplify through the use of knowledge produced by the study. They are an integral part of a piece of material culture that belongs to a certain way of "being and inhabiting the world". The flipside is that once taken, this piece of material culture is cut off from its original social context: it exists ex situ, being musealized. As Rolland-Villemot (2016, paragraph 52) points out, the ecological unit, as implemented in the museum, is a "situational exhibition [that] makes reference to a reality familiar to visitors". Consequently, once taken, "it can still only represent a very small part of the reality from which the musealized witnesses were taken. The ecological unit is a metaphor, like the isolated object in its showcase" (Rolland-Villemot 2016, paragraphs 62–63). Hence the necessity, when the reality is unfamiliar, or no longer familiar, of using the knowledge collected in order to interpret this complex heritage object as a trace of its world of origin.

This examination of Rivière's experiments to implement contextualization delivers three major lessons. Firstly, it shows that there are two forms of complexity: that produced by the writing of the exhibition text, and that of the material and social world. However, when I talk about a complex heritage object, we have just seen that it is not about the exhibition text, but about a piece of this world with its objects, its spatial base, its actors and its practices. Then, this examination confirms the essential role of knowledge and therefore of its production. This is what makes it possible to define what the heritage object will be, to draw its outline in space and in the flux of life, to describe and understand it. Because it is with this that we can grasp the origin of the components, but also what keeps them together. Finally, the limits imposed by musealization, which emerge explicitly with the taking of the ecological unit, lead the study toward the way in which a collection of objects becomes one single complex object with the examination of the production and functioning of the complex heritage object in situ.

5.1.2. *Heritagizing a complex social entity: urban heritage*

The notion of *urban heritage* was introduced by Gustavo Giovannoni, as Françoise Choay explains. It seems to be one of the first complex heritage entities to have been explicitly defined as such:

> The expression "urban heritage" that he invented describes the district's overall fabric as an entity *sui generis*, and no longer the sum of independent monuments which was, at the time (as often still today), synonymous with the historic city. (Choay 1998, p. 13)

In terms of the museum object or the historical monument, the change is therefore important, since we move from the individual object or the single building to a complex entity, due to being approached and studied as the sedimentation of historical layers that has made it inseparably a monument and a living fabric (Choay 1992, pp. 154–156). This change of scope in relation to the museum affects institutions, decisions, practices and representations. A change that is also partly in rupture with the traditional French mode of administrative legal heritagization focused on monuments and which was to result in the emergence of new heritagization practices.

5.1.2.1. *The emergence of the notion of urban heritage*

While the actual notion of "urban heritage" emerged with Gustavo Giovannoni after the First World War, it is the result of a gestation spanning nearly a century. Choay (1992, pp. 135–157) traces its various phases in her book *L'Allégorie du patrimoine*, especially in the chapter specifically devoted to "*L'invention du patrimoine urbain*". I will not go into the details of the analysis she develops, referring the reader to her book but will merely note a few points relating to the constitution of urban heritage as a complex entity in what she calls (Choay 1992, p. 139) the "figures" that govern this constitution.

The first figure, which she calls "memorial", features the fight by John Ruskin, and then William Morris, "for the life and survival of the pre-industrial Western city" (Choay 1992, p. 141). This figure was to bequeath two noteworthy imperatives that would serve as a foundation for the conception of urban heritage: on the one hand, taking domestic architecture into account as constituting the urban tissue, as opposed to taking an interest solely in monuments considered in their singularity; on the other hand, the imperative need to preserve this fabric (this texture) of the pre-industrial city as an entity.

The most important advocate of the second figure, called "historical", is Sitte (1996 [1889]), with *L'Art de bâtir les villes*. Based on a *morphological analysis* of a whole series of old city squares, it shows how the city, the center or the district impose themselves, "independently of their constituent parts" (Choay 1992, p. 149). While positing the need to introduce urban beauty into contemporary metropolises (urbanism as art), Sitte nevertheless supports the obligation to save *historic cities* from destruction insofar as they are, precisely, entities in their own right. Choay notes that this "museum" approach to the ancient city as something to be kept can be found amongst urban planners who have had an "ethnological experience" of cities in the Maghreb in the context of colonization (especially in Morocco). The distanced look that they then took at the cities of Europe is the opposite of the "*tabula rasa* ideology" of the CIAMs[17] and their emulators. In a way, it foreshadows the shift that took place (in France) with the creation of protected districts (1962), which recognized the heritage value of the old part of a city.

But it is with the "historial" figure that urban heritage was truly recognized. By operating the synthesis between the technical imperatives of travel, which characterize the modern city, and the preservation of the urban and aesthetic coherence of the old city, Gustavo Giovannoni simultaneously posits preservation (in theory and in practice) of the coherence of the latter and its interlacing with the contemporary parts that are complementary to it. Thus, "in the form of isolates, fragments, nuclei, they [i.e. the old centers] are able to find a topicality that Viollet-le-Duc and Sitte denied to them", since their scope renders them suitable to constitute units of daily life (Choay 1992, p. 153).

In terms of the complexity of urban heritage and the heterogeneity of its component parts, for Giovannoni, the monument, as Françoise Choay explains, is not a building independent of the built context of which it is part. It is the *ambiente*, that is, "the (historical) soul of places, embodied in spatial configurations" (Choay 1992, p. 155), that gives cities their nature; however, it arises from the dialectic between the major architecture and the peripheral fabric made up of minor architecture. The procedures for the preservation and restoration of this heritage must therefore be derived from that: in particular, respecting the scale and morphology of the old city, preserving the original relationships that linked plots

17 International congresses of modern architecture that deny any value (a fortiori heritage) to the old city.

and paths[18]. This involves the production of specific knowledge about this type of object. Giovannoni (1998 [1931], p. 40) defines this knowledge as "urbanism", the basis of the approach to urban heritage as a complex entity.

The conceptual definition of this complex entity illuminates its political and practical recognition.

5.1.2.2. *Recognition of urban heritage in France*

In view of the French heritage tradition that has long approached built heritage from the aspect of the monument[19] and urbanism from the aspect of renovation, the recognition of urban heritage required several stages that François Tomas (2004) has described as "temporalities of heritage and urban planning".

The first step was the promulgation of the law of August 4, 1962 (known as the Malraux law[20]), which created "safeguarded sectors" and their management through the establishment of a "safeguarding and development plan". It is generally accepted that this law introduced a rupture in a French heritage tradition, the only form of urban stewardship at that time being monitoring by the departmental architect of Bâtiments de France, of the works carried out within a 500 meter radius[21]. It is also accepted that it put an end to the domination of the CIAMs and the functionalist logic inherent in the 1958 decrees on urban renewal[22], which regarded old cities as having no value, apart from a few flagship monuments, and consequently as obstacles to progress that should be flattened and replaced by new constructions. But make no mistake. This creation of safeguarded sectors is very far from having led to old town centers being immediately, or even quickly, taken into consideration as urban heritage, first because of the administrative rigidity and regulatory constraints weighing on the implementation of these sectors – which greatly reduced their

18 "Two truths appear to us today: (1) a great monument gets its value from its visual and spatial context, amid the masses and colors where it originated or to which it has adapted, a value almost always preserved thanks to the persistence of the plan, as well as to the permanence of the urban characters of the space and of local artistic feelings; (2) the typical aspect of cities or villages and their essential artistic and historical value often lie primarily in the collective expression provided by the topographical plan, by the layouts of buildings and the architectural life that is expressed in the minor works" (Giovannoni 1998 [1931], p. 211).

19 See the Law of December 31, 1913.

20 Note that this law was introduced two years before the creation of the Inventory.

21 Law of February 25, 1943.

22 In the wake of the framework law of August 7, 1957 establishing urban renewal as a national obligation, rulings and a decree (itself followed by a number of decrees and circulations for their implementation) were published on December 31, 1958.

numbers (Tomas 2004, paragraph 11)[23] – and probably also because of the persistence of the traditional conception of heritage as a monument, and certainly also because of the lack of tools to think about the complexity of the old city, not to mention its relationship to the contemporary city. On the other hand, this creation did mark the beginning of a slow and gradual redefinition of the rules, practices and representations pertaining to ancient cities and the relationship between redevelopment and urban heritage.

This redefinition, some 20 years after the creation of protected sectors, was first translated, at the political level, into the possibility of creating, in parallel with these, "architectural and urban heritage protection zones" (*zones de protection du patrimoine architectural et urbain* (ZPPAU)). With the addition of landscapes, these ZPPAUs would become, 10 years later, "architectural, urban and landscape heritage protection zones" (*zones de protection du patrimoine architectural, urbain et paysager* (ZPPAUP)), which would be relaxed further, some 17 years later, with municipalities being given the option to create "architectural and heritage enhancement areas" (*Aires de Mise en Valeur de l'Architecture et du Patrimoine* (AVAP))[24].

It is thanks to this evolution that the complex nature of urban heritage would be gradually taken into account. But to understand this evolution of public policies relating to old cities, it should be placed in the context of the (partial) decentralization of the French state, which in this case took the form of a gradual reduction of the powers of decision-making, protection, management and promotion of built heritage by the state in favor of local territorial authorities

23 According to François Tomas (2004, paragraph 13), it is thus necessary to relativize the interpretation that makes this law seem "a kind of catalyst for transition from the culture of the isolated historical monument to that of urban heritage". "But there", he writes, "there seems to me a mistake of the same order as that which consists in claiming that the national natural parks, created around the same time, paved the way for the culture of the environment and sustainable development. In both cases, in fact, we are forgetting that the protection of the urban landscape or nature fits perfectly into the logics of functionalism and zoning, which have constituted the two basic notions of modernity in its most caricatured form, whereas they are incompatible with the cultures of heritage and the environment". The whole article is an illustration of the slow and progressive redefinition of policies and practices, and also an illustration of their limits.

24 The possibility of creating ZPPAUs dates from the Law of January 7, 1983, of creating ZPPAUPs from the Law of January 8, 1993, and creating AVAPs from the Law of July 12, 2010. For an analysis of this evolution, I refer to Tomas (2004).

(basically departments and municipalities), facilitating the reviewing and managing of the complexity of this type of heritage[25].

But what is of interest to us here is that with this evolution, historical urban complexes, which benefited from one or the other of these policies, ultimately acquired the *status of heritage objects*, even if this is rarely referred to in these words. In effect, these complexes, after having being identified as traces of the city of the past, have been interpreted and studied as such and a declaration made that they should be safeguarded and enhanced. Even though they are often presented as simple technical urban planning exercises, all of these operations are forms of heritagization gestures. However, in order be carried out technically, because of the complexity of the old city, they require the existence of knowledge about the latter that cannot be reduced to technical aspects alone. This knowledge, as Giovannoni had shown, results from the production – or at the very least, the mobilization – alongside technical knowledge, of historical and architectural knowledge, relating not only to the monuments, but also to the very fabric of the city. In practice, this type of heritage intervention thus calls for a trilogy of competences (technical, architectural and historical). Also, because this trilogy is often supplied by various professions or stakeholders – urban planning departments, architects, heritage experts (whether heritage specialists, learned societies, associations or sometimes academics) – there is a greater risk that it will spontaneously go against knowledge that integrates these various branches of expertise, even though it is this integrated expertise that will act as the *scholarly representation* of these old urban districts as complex heritage objects. It is this scholarly representation that will serve as a reference for the safeguarding, the possible restorations or the actualization, and then (often as a secondary stage) for publicizing them to tourists and residents.

To illustrate how a historic urban district can become a complex heritage object, I will take an example from the "Towns of Art and History" (*Villes et Pays d'Art et d'Histoire*), Annecy's old city center: as well as safeguarding and promoting and urban heritage, it also publicizes it explicitly to tourists and residents.

---

25 The urbanism of cities (in the administrative sense of the term) is now the responsibility of municipalities and although many of the monuments (with the exception of great national monuments) have been transferred to local authorities, the state has nevertheless retained some of its prerogatives for the protection this type of heritage, through an advisory role (with architects of Bâtiments de France or counsellors of the regional directorates of Cultural Affairs), also through a process of contractualization (especially with regard to funding) and incentives (through labeling); but sometimes, by means of direct control (as for national monuments or protected sectors).

5.1.2.3. *An example of a complex heritage object: Annecy's old city center*

Navarro's doctoral research (2015), dealing, in his own words, "with the circulation and mediation of urban heritage in the Towns and Lands of Art and History" (*Villes et Pays d'art et d'histoire*), provides three lessons on the construction of a complex heritage object in situ: firstly, the heritagization of this object, a historic town center (an "ancient city", in Giovannoni's words), unfolds over a long time and in stages; secondly, it involves the writing of the heritage object in a media text; finally, it must link this media text with the living urban fabric in order to make of it an integrated object[26].

In the first place, as we have seen in the previous chapters, as the object to be heritagized grows in complexity, the heritagization process takes place over an increasingly long period of time, in an ever-increasing number of stages, mobilizing more and more different actors for each operation. The example of Annecy confirms this observation. Indeed, for quite a long time learned societies and associations have been showing an interest in what they regard as traces that have heritage value for them. Moreover, since their creation[27], these learned societies have produced knowledge that makes it possible to interpret these traces of the past as potential heritage objects. However, for the heritage process to come to fruition, it is also necessary for this interest and this knowledge to be translated into a declaration, as well as into a desire for preservation and transmission. Such conditions, which also require a social context that recognizes the value of urban heritage, mobilizes a large number of social operators who may have divergent interests (or may be openly antagonistic on certain points), but who will have to coordinate their action to perform the operations that make it possible to heritagize this complex object, the historic urban complex. In Annecy's case, the study reveals the way in which the municipality gradually recognized, then declared, but also treated the historic town center as a heritage object (Navarro 2015, p. 173 and 229). The deputy in charge of heritage and urbanism, who was at the same time president of the Society of Friends of Old Annecy (*Société des Amis du Vieil Annecy* (Sava)), was its promoter. By coming to an agreement with the architects of Bâtiments de France and the chief architects of Monuments Historiques (the historic center actually included a listed historical monument, the Palais de l'Île, which falls within the competence of the latter), it was possible to develop a method for dealing with and promoting this city

26 All of the analyses performed in this point are based on Nicolas Navarro's thesis.

27 That of the Florimontane Academy dates back to the 17th century, for example, and that of the Society of Friends of Old Annecy (Sava), which was heavily involved in the project to preservation project the ancient city, to the inter-war years.

center on a similar mode to that of protected sectors[28], and, we could add, akin to the principles of urban heritage as a monument and as a living fabric, as set out by Giovannoni.

Lesson number two: the writing of a complex heritage object such as this in media text (staging its exhibition) is necessarily more complex than the writing of a showcase at the National Museum of Popular Arts and Traditions, and above all, it unfolds in situ, requiring intervention on existing material realities and the involvement of the population. As with any promotional activity, Annecy's historic center more or less consisted of putting it in the format of its scholarly representation, the historical and urban contextualization having served as a basis for the integration of its elements. This was done through three operations: (i) restoration, (ii) signage and (iii) a meta-communication *dispositif*.

(i) Without going so far as to believe, with Viollet-le-Duc, that the restoration of a monument consists of "restoring it to a complete state that may never have existed at a given moment" from its morphological analysis, restoration always operates a (relative) alignment with what we know – or what we interpret – of the state of the monument prior to its deterioration; in other words, with what analysis reveals as features that characterize it both in its type (i.e. its period and its function) and in its individual specificity (its scholarly representation). The same applies to the rehabilitation of an old city center[29]. In Annecy's case, beyond the general characteristics in terms of plots, the groupings of buildings, the layout of its streets and squares, or the urban landscape, which are common to all old cities, a certain number of specific features were retained, such as topographical and historical constraints, or uses and cultural traditions. These features have served as a basis for its rehabilitation: development of the canals and quays that characterize the urban landscape; little streets that have been repaved, in order to be pedestrianized; arcades in front of covered passages on the ground floor of buildings that have been partially restored and in some cases reopened; old facades either restored with

28 "The close links woven between Sava and municipal actions (see Navarro 2015, pp. 171–173) explain the major role of this personality: for several decades President of a learned society producing knowledge on heritage, and also municipal officer for culture and heritage for several terms, he takes action on institutional protection of heritage and its communication to the public. His recent withdrawal from political and associative life necessitated the establishment of heritage urban planning rules that are capable of replacing "his brain"" (Navarro 2015, p. 229). This regulation is that of the Architecture and Heritage Development Area (*Aire de Mise en Valeur de l'Architecture et du Patrimoine* (AVAP)).

29 I use the term "rehabilitation", usually used in architecture when referring to buildings, to refer here to an intervention and a town planning process (restoration, actualization or illumination) based on respecting the characteristics of the old city center, as revealed by its historical and morphological analysis (as opposed to renovation that aims to replace it).

colored renders (“Sardinian” colors, as they are called), or cleaned when they were made of cut (and possibly carved) stone for the well-to-do residences (palace, castle); old storefronts maintained; etc. To these restorations, we must add the rehabilitation of numerous buildings and houses, as well as various highway improvements that also act as signage, such as the removal of sidewalks, the demarcation of pedestrianized zones and zones for motorized vehicles.

(ii) Signage in its various forms is indeed an essential component in putting a historic city center on exhibition: whether it be directional signage effectively showing the special quality of the old town center (particularly the signposting places of note, and traffic segregation); nameplates of public spaces (streets, squares, etc.) or shops (restaurants, various shops, souvenir shops, etc.); interpretation signs (display stands commenting on viewpoints, wall plaques on buildings, “Heritage route” markers; etc.); not to mention what I may term “volatile” signage (documents with one or more sides, guides, smartphone commentaries, etc.). By offering a program of heritage use of the city center, these various signages effect an operational translation of the scholarly representation.

(iii) The meta-communicational *dispositif*, in this case the Centre for the Interpretation of Architecture and Heritage (*Centre d’Interprétation de l’Architecture et du Patrimoine* (Ciap)), was set up within the confines of the old city center itself, unlike other *dispositifs*, such as promotional documents, websites or guides; it offers a historical, architectural and urbanistic representation of the heritage object through the exhibition of documents (models, maps, views, sketches, texts, etc.). This media metatext is a form of mediation that can complement (or possibly replace) that offered in the guided tour.

Lesson number three: the exhibition of the old city center to its functioning as a living fabric presupposes its appropriation not only by tourists, but also by its inhabitants, and by the inhabitants of the rest of the city. However, such an appropriation entails the old city center not being reduced to a collection of material elements apprehended only for their value as traces of the past, nor to an entity resulting solely from combining these elements into a media text that will tend to become part of the decor. The challenge, then, is to know what constitutes heritage, with a consequent tendency to redefine the complex heritage object. Indeed, in line with the convention of the Towns of Art and History, there is no question of “museumifying” the historic center, nor of isolating it from the rest of the city. The department for the animation of architecture and heritage (*Service de l’animation de l’architecture et du patrimoine*) will therefore develop actions that are in accordance with an open conception of heritage created not only from an approach to objects, but also from the uses and practices of the city, landscapes and industrial and human memory (Navarro 2015, p. 76). It offers an apprehension of the city in its entirety according to various types of heritage, from the most traditional and established heritage, such as the Palais de l'Île (a listed monument), to what can become the subject of a heritage experience, such as the districts built in the late 20th century, via intangible heritage (which can extend to including local memory, city practices, even civic rights). One of the results of Navarro's thesis (2015, pp. 63–80) is thus to show how urban heritage combines what he calls "operators": the traditional form of the monument (the *unicum*), the historic city center as a "family" of urban components (the *typicum*) and the heritagized city (the *totum*). With the particularity that the third of these operators "thus stands apart from a purely material reading of heritage to add an immaterial dimension" (ibid., p. 77), such that heritage is then, as summed up by the curator of the museum-chateau, "above all a mentality; above all a state of mind". There is no clearer way of saying that it is henceforth essentially a "cultural being".

In conclusion, the old city center of Annecy corresponds, to a fairly large extent, to the definition that Giovannoni gives to urban heritage, it is composed both of heritage objects declared as such and of an urban fabric that together constitute as many traces of the memory and practice of the places, know-how and models that governed its production; and again, traces of ways of "being and inhabiting the world". It therefore results, as a complex heritage object, from the condensation of a number of heritagization processes scattered over time, ranging from "institutional" heritagization (in the sense of Rautenberg), including that of Annecy as a Town and Land of Art and Culture, to social heritagization, to which there is also added an intangible dimension.

In terms of heritagization of a complex object, the analysis confirms the decisive role of the production of knowledge, but also sheds light on the role of the instances that operated the heritagization. The existence of this object in fact depends on the capacity of the heritagizer actant – operationalized by a great many social actors (in reality, actantial roles), at the center of which in this particular case we have the municipality and especially one of its elected officers – to integrate components having varied statutes into a single entity. It is the scholarly representation that has thus constructed the unity of the new complex heritage object. It actually resulted from the knowledge produced through the various interactions between these multiple social actors, then formalized and conveyed initially by the various media texts mainly of the "operator" type (dossiers, reports, summaries, maps, decisions, technical documents, etc.). It is by retracing history by enunciating its characteristics that it has defined the old city center, enunciating its unity and identity and socially conferring upon it a heritage existence.

This scholarly representation has been operationalized in two ways. First, the description of the old city center carried out from the observer position set the direction for the work of restoration and promotion that integrated the various component into a single object. Second, this scholarly representation served as a basis for preservation and publicization. It is thus at the heart of the *relationship* that the residents and the audience (the "others") have with this old city center. In fact, it helps to shape the *heritage interpretive scheme* that will set the direction not only for the way the place is treated (urban planning) and the modalities for safeguarding it, but also the practices (mode of inhabiting and visiting), and more generally still the circulating representations (how this city constitutes heritage). With a role in the way of apprehending the city, inhabiting it, understanding it, living it – in short, experiencing it as a heritage object – this scheme is shared, albeit according to different modes, as much by the residents as by the "others", in particular the tourists.

Thus, everything suggests that with complex objects, a new *mode of heritagization* is emerging. Does the case of the landscape confirm this emergence?

5.2. Toward a new mode of heritagization: the example of the landscape

The landscape is a far more complex object than the ancient city[30]. While in everyday terms, speaking about the landscape pose no difficulties in comprehension, as soon as we try to define it a little more precisely "the word landscape is indeed a problem word", as Lenclud points out[31]. Therefore, while it is quite easy to recognize the way in which the city constitutes heritage, the complexity of the landscape makes such recognition far from self-evident. But, looking at it from a different side, this is what makes it an object rich in lessons.

5.2.1. *How do we study the heritage character of a landscape?*

The complexity of the landscape means that we should first take stock of what defines a landscape as heritage.

5.2.1.1. *The starting point: the landscape is a representation*

In the diverse texts dealing with landscape, it is almost unanimously assumed, if not explicitly posited, that a landscape results from the transition from an entity

---

30 In addition to the conceptions of the landscape itself that are summarized in Paquot (2016), there is a wide variety of approaches to its heritage complexity. On this question, see, by way of example, some previous studies (Dubost 1994; Tomas 1994; Voisenat 1995; Davallon et al. 1997; Briffaud 1998; Vivien 2005; Sgard 2008, 2010; Amboise 2012; Larrère 2018; Davallon 2019c).

31 "When we stop to consider its meaning instead of using it spontaneously in speech or writing, the word *landscape* is indeed a problem word. It signifies a category of thought that is determined culturally, historically" (Lenclud 1995, paragraph 7).

made up of material and living realities, studied through geology, geography, ecology, history, sociology, agronomy, law, etc., to a representational entity which, in addition to the previous disciplines, art history, ethnology, communication sciences, political science, etc., seek to grasp. This transition from what, according to the authors, is described as "environment", "space", "milieu", "territory", "country" and which I will subsequently refer to by the generic term "place"[32], to "landscape", has as its inaugural act, as Gérard Lenclud explains, the perceptual act in which we deem a place to be a landscape. The consequence of this act is that an "object perceived is an object redesigned, given already-known attributes. At first we see only what we recognize; seeing a landscape is recognizing it as such" (Lenclud 1995, paragraph 10). A landscape can therefore be defined as:

> A datum reconstructed by a perception that informs conceptual schemes including not only the representation of the landscape, but also the representation of what a landscape should be in the most diverse circumstances. The landscape can be said to be an existant, established as a function of an ought-to-be. (Lenclud 1995, paragraph 12)

This deontic modality, the semiotician would say, means that "any landscape is fixed as such by comparison with prototypical properties; these properties determine the reference" (Lenclud 1995). Recognizing a place as a landscape is therefore not reduced only to the perception that I may have had of it subjectively, but through the mobilization of preexisting *conceptual schemes*, presupposing that these have been formed earlier by other subjects and that they have been more or less transcribed into pictorial or literary works, scientific texts and mediated texts that have circulated them in society. This is why the schemes make it possible, from the moment of perception, to evaluate the *landscape value* of a place, that is, to see if it has characteristics that are recognized as landscape markers due to their presence in other landscapes. It should be noted that, socio-semiotically speaking, such cognitive activity equates to having adopted an observer point of view, not only if one is a researcher studying this landscape, but also (and which, in parentheses, poses far more difficulties to the said researcher) for anyone who perceives, observes and "landscapes" a place. This is precisely what Lenclud writes:

> In short, it takes an Other to confer upon my space the quality of landscape. In this hypothesis, the gaze that makes landscape, proceeds from an attitude of critical distance towards the sites observed; this critical distance calls for the adoption of *a viewpoint* (gaining

32 Which I mean in a much more comprehensive sense than that given to it by Michel Conan (1994), since he covers all practices other than those that, requiring cognitive experience, define "landscape".

elevation), a viewpoint without which, precisely, there is no landscape. Perception of landscape is part of a logic of social and cultural differentiation that depends on this movement of critical distance that causes one, or does not cause one, to climb a promontory and look down from above. (Lenclud 1995, paragraph 21)

The observer actant, who thus produces a representation of the place by exercising a perceptual act, can limit themself to this single program of action, but at that point, the landscape will be a purely mental representation that does not go beyond the framework of subjective experience. As Lenclud rightly points out, this experience will then be difficult for the researcher to apprehend. But fortunately, there are cases where the observer actant will themself give a social existence to this representation, by instantiating it in the form of a "document". According to the actor implementing the observer act, this document can be pictorial, literary, photographic, audiovisual, scientific, administrative, etc. But no matter who the actor, transitioning from the purely cognitive (introspective) act to an enunciative (communicational) act, by writing the representation, they will produce a *media text* that will deliver the *traces* of their experience and that can be accompanied by commentaries, and we will then be looking at a text that falls within the "documentary" genre. The experience of the landscape can now be perceived and interpreted by a different subject, and above all, it can circulate and be shared socially.

5.2.1.2. *Landscape conceptual schemes and heritage interpretive schemes*

The traces present in a documentary are of two types. First of all, traces of the landscape *conceptual schemes* mobilized by the maker of the documentary in the way the landscape is shown. Then, the traces of *heritage interpretive schemes* borne by different actors to potentially indicate in what way components, processes or the landscape as a whole constitute heritage. But we must add to these media traces that, just like the view of the landscape itself or listening to spoken commentaries on it, the act of watching a documentary (on television or on a website) can call heritage interpretive schemes to the viewer's mind, when they get the feeling that an element, an aspect, or the landscape in its entirety must be kept and transmitted because they have come from the past and are therefore endowed with values. We have already encountered such schemes, particularly with regard to documentaries on intangible heritage, but they take on a particular importance here.

To realize this, we only have to think about the somewhat paradoxical feeling that we may experience when, looking at an in situ or documentarized landscape, we ask ourselves the question of its possible heritage character. On the one hand, indeed, any landscape, since it was necessarily produced in the past (even if this past was the recent "time that is passing"), gives the feeling that it is heritage up to a point. But, on the other hand, it is difficult to enunciate precisely what makes it heritage. This paradox can be seen as being simply due to the disconnect between the recognition of landscape through the presence of prototypical properties, and the tradition according to which heritage objects are defined and delineated. The paradox would then come from the blurring of the boundaries of the landscape. However, this blurring is not a symptom of a defect in the landscape, but simply an indication that its recognition as heritage is more of the order of a *heritage interpretive scheme* than official categorization as heritage element. Hence, it is no longer the boundaries of the object that are decisive, but certain properties that will be traces of its heritagization: this can of course be an official declaration granting heritage status to the entity as a whole or to some of its components[33]; but also the decision or action to safeguard them; an interest stemming from the fact that they come from the past; not forgetting the production of knowledge. With the additional difficulty that quite often, the link between properties and entity is not immediately visible or even interpretable with the naked eye, and only emerges thanks to a scientific study, hence, the essential role played by knowledge, its production and then its publicization.

5.2.1.3. *Analyzing heritage traces in documentaries*

A documentary on the landscape can be defined as a *media text* writing the representation of a place in landscape. It bears *traces of the landscape experience* not of one single subject, but of several actants (and a *fortiori* a great many actors) with various, complementary, and in some cases antagonistic, competences in the semiotic sense of the term. It also helps to circulate, complete, modify and structure the landscape conceptual schemes and heritage interpretive schemes corresponding to these traces. All of these things make *television documentaries* good material for examining and discussing how landscapes become complex heritage objects. Since my objective is to illustrate the emergence of the heritagization mode of complex objects and not to undertake a study either of a landscape category or of a program genre, constituted a very restricted corpus of a few documentaries, in order to illustrate this emergence, a corpus that, in order to do this, satisfies two conditions: announcing explicitly that it deals with landscapes and presenting or calling to mind heritage interpretive schemes.

Among the impressive number of documentaries, presenting landscapes and meeting the two criteria above, that can be found on Arte – not to mention other channels, such as France TV (France 3, especially with *Des racines et des ailes*) – I selected the series entitled *Paysages d'ici et d'ailleurs* which explicitly focuses on

---

33 See on this topic the *Proceedings of the multidisciplinary seminar in Villefavard, Limousin* (Inventaire général du patrimoine culturel 2012).

landscapes[34]. From the 80 episodes in the series, I extracted a final corpus of seven documentaries to illustrate my topic. By cross-referencing the requirements – landscapes that are officially recognized as heritage, and others that are not; landscapes possessing a strong cultural dimension and others possessing a predominantly natural dimension; fully heritagized landscapes and others where simple use predominates – I selected the following documentaries: *Val d'Orcia, Cinque Terre, Agrigento, Alpujarras, Baie de Somme, Côte d'Opale* and *Bocage Normand.*

One of the advantages of this choice is that all of these documentaries are constructed in the same editorial format. This combines several levels of enunciation: that of the director (structure corresponding to a topic serving as a common thread enunciated at the beginning by the presenter-narrator); the commentary of the presenter-narrator throughout the documentary, giving it coherence; four or five conversation sequences with scientists, scholars, heritage specialists (curators or mediators) or residents[35]; views of the landscape from viewpoint relating to the topic, but also the discourse of the various enunciators. These documentaries thus have the advantage of asserting a declared point of view and at the same time being polyphonic, which increases the chance that there will emerge the mobilization not only of conceptual schemes emerge (even if only through the choice of views), but also of heritage interpretive schemes.

These documentaries thus deliver a form, albeit reduced and simplified, of the key elements of the scholarly representation of the landscape concerned. In this way, they facilitate the viewer's cooperation, in particular a bringing to mind by the viewer of conceptual and interpretive schemes from other media texts, especially other documentaries. Consequently, for this research, while they cannot constitute a primary source for a scientific study of the place, they do however contain indices on the construction of the cultural being "landscape as a complex heritage object", being bearers of information on the three major operations of the heritagization of complex objects: (i) on the role of knowledge in the definition of a landscape as a cultural being; (ii) on the modalities of safeguarding it; and (iii) on the process and role of publicization. But before beginning an analysis of the traces of these operations, I will quickly review the traces of official declarations of heritage status of these landscapes as they emerge in the documentaries.

34 Series broadcast between 2013 and 2015. According to the Wikipedia page on this series, 80 26-minute episodes were broadcast over three seasons. There is also a complete list. Available at the following address: www.fr.wikipedia.org/wiki/Paysages_d%27here_et_d%27ailleurs [Accessed 3 September 2022].

35 This editorial format is thus different from that of a documentary with a more explanatory or discovery-orientated intent, in which there is only one enunciator, such as the *Paysages d'Europe* series.

5.2.1.4. *Traces of official heritage status of the selected landscapes*

Two landscapes, Val d'Orcia and Cinque Terre, have been declared as heritage sites in their entirety since they are inscribed on the UNESCO World Heritage List[36]. The Val d'Orcia is the most strongly heritagized landscape in the entire corpus. The documentary also mentions a very large number of objects with heritage status – museums (including one in Montalcino, where one of the interviewees is a curator), monuments, churches, villas, gardens, the Via Francigena, etc. – to which must be added the urban center of Pienza, which was inscribed on the World Heritage List 8 years (1996[37]) before the Val d'Orcia landscape itself. All of these heritage objects are part of the landscape and contribute to its heritage status. For the Cinque Terre landscape, the review mentions only a few traditional heritage objects (churches and urban planning in the village) that do not in any case have declared heritage status, apart from the designation of origin of its white wine (especially Sciacchetrà). The only noteworthy declaration mentioned briefly in the documentary is the marine protected area.

For the Agrigento landscape, the archaeological site of the temples of the Greek city is listed as a World Heritage Site[38], but the documentary deals with the relationship between this site and the landscape around it.

The Alpujarras landscape is recognized as natural heritage, since it is part of the Sierra Nevada Biosphere Reserve declared by UNESCO and classed as a National Park of Spain (for its botanical and landscape riches)[39]. The documentary, however, as for the Cinque Terre, puts the main focus on the cultural dimension the place's redevelopment by man.

The Bay of the Somme and the Opal Coast are recognized primarily for their natural riches. In both cases, few heritage recognitions are mentioned. In view of the dates, the documentaries may have been filmed shortly before the official granting of the label Grand Site de France to the Bay of the Somme and the two capes, Blanc-Nez and Gris-Nez. But we also know that there were already multiple and fragmentary recognitions, of which the documentaries bear traces. For the Bay of the

36 Val de l'Orcia (2004). Available at: www.whc.unesco.org/fr/list/1026. Cinque Terre, (1997). Available at: www.whc.unesco.org/fr/list/826 [Accessed 12 March 2023]. On the World Heritage Convention, see UNESCO (2021).

37 Description on the UNESCO website: Pienza. Available at: www.whc.unesco.org/fr/list/789 [Accessed 25 September 2022].

38 Description on the UNESCO website: Archaeological area of Agrigento, 1997. Available at: www.whc.unesco.org/fr/list/831 [Accessed 25 September 2022].

39 Respectively in 1986 for the reserve and in 1999 for the park. UNESCO List of Biosphere Reserves. Available at: wwww.unesco.org/new/en/natural-sciences/environment/ecological-sciences/biosphere-reserves/europe-north-america/ [Accessed 25 September 2022].

Somme: an interview with the curator of the National Nature Reserve of the Somme and with the coordinator of the Somme Bay Railway between Noyelles and Saint-Valéry. For the Opal Coast, an interview with the "Nature" guides on the dunes and on the Bay of Authie, leading us to believe that one structure has a preservation strategy, as well as an entire sequence devoted to the South Opal Museum (a Musée de France) in Berck which has a collection of paintings and pochades[40] from the late 19th century.

Finally, I selected the documentary *Bocage Normand* as a counterexample, since this landscape has received no official heritage recognition[41].

In contrast to these official recognitions from heritage institutions, what heritage recognition of all or part of the landscape comes through the application of heritage interpretive schemes? In order to answer this, it is necessary to carefully examine the presence of these schemes or bring them to mind by the viewer through the traces of the heritagization operations of these landscapes: their production, their maintenance and their publicization as objects, as complex heritage entities. Identifying these traces requires a very detailed examination of the seven landscapes for each of the three operations.

5.2.2. *Landscapes produced by knowledge*

The documentaries underline certain elements that characterize the landscapes, such as the views, their selection and how they are presented, the comments of the various actors, which, being presented as having been created in the past, will become traces of heritagization of these landscapes, in other words, *heritage traces* that contribute to defining them. I will start with the documentary on the Val d'Orcia which presents the latter as the very prototype of the cultural landscape. Four documentaries, looking at the Alpujarras, the Cinque Terre, the Normandy bocage and Agrigento, illustrate a heritagization of landscape based on the practice of the place. Finally, the documentaries on the Bay of the Somme and the Opal Coast illustrate the specificity of the heritage future of natural landscapes and its challenges.

40 Color sketches executed with a few brush strokes, with the specific aim of capturing the light.

41 But we know, apart from this, that Domfront has a number of monuments classified as historic monuments or inscribed on the supplementary list of historical monuments, although the documentary does not mention them. Available at: www.fr.wikipedia.org/wiki/Domfront_(Orne) [Accessed 10 March 2023].

5.2.2.1. *A prototype of artialized landscape-heritage*

The documentary on Val d'Orcia presents the outcome today of the place's transformation into an "ideal landscape" operated in the Renaissance from a pictorial representation. This ideal landscape was therefore born from an in situ artialization, as Roger (1978, 1997) would say, which made the place – the "pays", to stay with Alain Roger's categories – what I will call a landscape-place[42]. As explained by the guide at the Val d'Orcia Park of Art, Nature and Culture (Valentina Perghidi), and the art historian curator at Montalcino museum (Alessandra Dani), the landscape's organization is the result of the application of a pictorial representation: from Ambrogio Lorenzetti's fresco *The Effects of Good Government in the Country* which is in the municipal palace of Siena.

The documentary gives the two processes that constructed the history of this landscape and that enabled it to come down to us today: a conservation of the landscape-place and a transmission of the ideal representation. The landscape historian (Ugo Sani) lists the reasons for its conservation: the continuation of sharecropping until the mid-20th century, the fact that it was bypassed by the motorway built in the 1960s, then the creation of the Val d'Orcia Park of Art, Nature and Culture which has in recent years enabled this landscape to be safeguarded and its evolution to be supervised. Even though it has changed over time[43], this historical continuity has made it possible to sufficiently preserve the main features of the original creation, which had resulted from an alliance between the producers (political authority, landowners and sharecroppers) and the observer actants (painters, landowners and walkers), or at least, the image we have today of this continuity. Hence, the importance of the transmission of the representation that translated the features into *markers* constituting the conceptual scheme of this landscape as a "Renaissance" landscape: valleys; type of habitat; divisions and crop

42 The documentary's presentation: "When we walk in the Val d'Orcia, we have an impression, a perception of singular harmony as if each tree, each path, each house was in exactly the right place. A feeling that we can fully understand when we know that the region as a whole was thought out, designed; a landscape policy that dates back to the 16th century, when the Republic of Siena was using its campaigns as a showcase to demonstrate its good governance to the whole world. The story I am going to tell you is the story of this ideal landscape".

43 As the historian Ugo Sani rightly says: "In reality, landscape is dynamic, it cannot be imagined that it has stayed the same. There have been periods when these lands were abandoned because they were difficult to work; it was very tiring, because these lands are clayey".

types (wheat, vines, meadowland); the planting of olive trees and cypresses[44]. These markers have made this landscape an entity by giving to it an overall coherence, so that it has become what Patrick Dal Cin calls a "de facto heritage"[45]. Nowadays, this landscape corresponds to conceptual schemes that pick up the characteristics of the place itself, but have also become heritage interpretive schemas, insofar as they have been historically transmitted through paintings, writers' texts, scholars, art lovers or art historians. And, we could add that they are now found in promotional discourses, or in tourists' comments on social media. Thus, the landscape that the documentary shows us results from the concordance between the "landscape place" (lived and perceived) and the scholarly representation transmitted: it is a "heritage landscape", a landscape that, even if has changed since its creation, seems authentic to us. A concordance that inscription on the UNESCO list has officially consecrated by declaring the heritage status of the landscape, and that the documentary, by publicizing the landscape's history, is continuing in its turn.

Two additional points, however, draw our attention. The first is the position claimed by the inhabitants interviewed in two sequences: one sequence in which a farmer's wife cooks and serves to tourists pasta made by hand according to the traditional method, with her husband explaining in parallel the contribution of agritourism to the continuation of agriculture. The second sequence sees a sheep farmer, whose herd preserves the openness of the landscapes, presenting his sheep's milk cheese (*pecorino*)[46].

These inhabitants thus pursue the historical hybridization of the landscape place, by participating in a practical way in its preservation and in its in situ heritage publicization. These two sequences amount to making the landscape-place a social heritage, but in accordance with the observation that "where a man lives out his work and his days, where he is attached, he has no thoughts of creating a landscape representation of the space" (Lenclud 1995, paragraph 20), neither the crop farmers nor the livestock farmer approach the landscape in terms of a representation, even

44 According to the historian Ugo Sani, the cypress, linked to mythology, has become a symbol of the Tuscan landscape that has been disseminated by cinema, advertising, television, and; for this reason, it is planted in abundance (often in rows, whereas previously it used to be more often planted singly since the Etruscans).

45 "We should recognize that the landscape is a de facto heritage and that it is defined as such when its analysis, with its multiple criteria (forms, colors, rhythm, functionality...), makes it clear that it is an entity of landscape coherence, but also of cultural coherence offering specific and noteworthy characteristics (modes of exploitation, division of the plots, habitat...)" (Dal Cin 2001).

46 The documentary shows vineyards, but does not dwell on the importance of the appellation contrôlée of the wine (*brunello*), which in fact complements the range of quality products.

though they adopt an observer position toward their practices[47]. The second point concerns the mention, by the park guide, of non-landscaped hills, and also the mention, by a historian from the city of Pienza, of Mount Amiata as a landscape counterpoint to the city[48]. These two traces of the place more than of the landscape – to which must be added the constraint of preserving the openness of the landscape (i.e. of preserving its character as a man-made landscape) with the sheep – sketch a division between the major part of the landscape that bears the heritage features and the fringes that can be said to be landscaped but without being (completely) heritagized[49]. These two points invite us to pay attention, in the other documentaries, to the fact that the landscape is not necessarily heritagized in its entirety, and to the need to verify the degrees of heritagization of the components according to viewpoint.

5.2.2.2. *Practical knowledge and scientific knowledge*

Unlike the Val d'Orcia, the landscapes of the Cinque Terre, the Alpujarras, Agrigente and the Normandy bocage were not produced from a landscape model, but acquired landscape coherence from the practice of the place. This has determined the way in which they constitute heritage.

This is obvious for the Cinque Terre, which, despite having been sung about by poets since the early 19th century, acquired the dual status of heritage and landscape among its inhabitants at the time of application for inscription on the UNESCO World Heritage List, which happened in 1997[50]. The (re)discovery of the place by its inhabitants thanks to writing of this application constitutes a "singular point" in the relationship to the place. The origin of this application was in fact the need to stop

47 It would probably be different with other publicization operators, such as professionals of trade, tourism, art publishing and websites, etc. – or even tourists themselves – who are not included in the documentary.

48 Located on the other side of the valley, it can be seen as the vanishing point of the urban setting: "The sketch of the urban setting has its vanishing point here, in this volcano opposite, called Mount Amiata. And the whole city converges towards the top of Mount Amiata, as if the city were a continuation of the valley. The nature that springs from Mount Amiata, like the water that descends, reaches the city. Thus, culture unites with nature".

49 Not entirely heritagized, since it is not impossible (in spite the distance introduced by the representation) for them to be partly so, being connected to the hills shown in the background of the illustrated scenes in the original picture.

50 A summary of the scholarly representation can be found on the UNESCO website. Available at: www.whc.unesco.org/fr/list/826 [Accessed 10 March 2023]. Even though the inscribed site includes not only the Cinque Terre, but also Portovenere and the three islands Palmaria, Tino and Tinetto, added when the application was submitted by the region, with the argument making reference to the criteria of Outstanding Universal Value reveals the specificity of the Cinque Terre.

its deterioration, which came about as a consequence of viticulture being abandoned in the 1960s. Moreover, this deterioration occurred after an initial period of exodus linked to phylloxera in the 1920s, according to the specialist in the economic history of the Giuliana Biagioli landscapes, and can be explained basically by the rural exodus as a result of an "economic boom in Italy, and hence additional work opportunities opening up either in the transport sector or in the industrial sector". Because of the rather atypical situation of a mobilization of the inhabitants led by the mayors, these inhabitants became aware of the characteristics of the place as well as its value: what was until then a source of problems turned into traces whose historical interpretation showed its exceptional value and revealed opportunities. This interpretation is presented in the first part of the documentary in which the historians (Attillio Casavecchia and Enrica Salvatori) explain how this landscape-place is the result of an interaction between, the inhabitants and nature, which has always taken the form of a struggle with it; growing vines has, in fact, made it necessary to build many dry stone retaining walls to achieve cultivable terraces. Nevertheless, if World Heritage inscription has granted the place the status of landscape-heritage, far from solving everything, it has, on the contrary, generated new constraints relating to the safeguarding of the place as a living heritage. When I come to deal with preservation, I will return in detail to the two conditions of this safeguarding: the ongoing restoration of the dry stone walls retaining the terraces, which presupposes the continuation of viticulture; and the management of the massive influx of tourists that came as a consequence of the site's inscription on the World Heritage List. But for the moment, I will mention that the analysis shows these dry stone walls are in actual fact a *semi-heritage object* (the thing to be kept) that conditions the patrimoniality of the entire landscape[51].

The Alpujarras landscape, which straddles the Andalusian provinces of Granada and Almeria, in the heart of Sierra Nevada (Spain), presents the same scenario: possessing a semi-heritage object declared explicitly in the documentary. This landscape was also produced as a place before being recognized as a natural heritage, initially as part of the UNESCO Biosphere Reserve, then as part of the Sierra Nevada National Park. But the documentary essentially deals with how the landscape-place came about as a result of Arab colonization in the Middle Ages. How it was shaped, on the one hand, by the creation, development and upkeep of a ring of canals (*acequias*) that made pasturing possible on the alpine pastures, the growing of fruit trees (olive trees, almond trees, etc.), food-producing crops (gardens) and human habitat (villages), and, on the other hand, by the transmission

51 I call this component a "semi-heritage object" because it does not have declared heritage object status. Even if it is explicitly recognized as imperatively to be safeguarded, its patrimoniality is derived from a heritage interpretive scheme.

of terrace farming techniques that make good use of this irrigation system[52]. The first part focuses on how man produced the place by making the canals, as explained by the National Park guide (Antonio Ramos) who takes the presenter on a tour of the mountain at the beginning of the documentary, and on the Arab inheritance in terms of water management with a professor of medieval history (José Martín Sivantos)[53]. The second part shows how the inhabitants still continue this way of life today, regarding the canals and water they carry as essentially utilitarian, they claim that they use them as their ancestors did. Even though they do not describe these canals as heritage, their comments cannot fail to bring to mind a heritage interpretive scheme. And the presenter's conversation with one of the farmers who runs a farmer-irrigator community (Modesto Alonso Alonso) serves to confirm that the canals do in fact constitute heritage – a semi-heritage object – that governs the whole environment. This is why, the farmer explains, irrigating with a system of pipes, which would supply water only to the fields, would be tantamount to destroying the eco-system.

The example of Agrigento is somewhat different, because we do not a priori have a component of the place that governs the heritage character of the entire entity as in the two previous cases. While only the archaeological area of the ancient Greek city, the valley of temples, has been listed as a World Heritage Site, the documentary, by dealing with the relationship between the city and the surrounding countryside, will mobilize a heritage interpretive scheme based on this relationship[54]. In the first part, the agronomy professor (Guiseppe Barbera) and the archaeologist (Maria Concetta Parello) explain that this relationship has existed since the consecration of the city to the Greek goddess of fertility, Demeter

52 Raphaël Hitier (village street and square, then aerial view towards the mountains): "Whether it is the layout of the villages, the architecture of the houses, or the white walls that we see here, the Arabs left a deep imprint on the landscapes of southern Spain. But their major contribution to the landscape is not necessarily there. If the Alpujarras are a fertile, cultivable, habitable land even today, this is thanks to the vast irrigation network they built in the Middle Ages at the very top of the mountains".

53 As the latter explained when being interviewed in the Alhambra gardens, in Granada: "In the Nasrids' reign in Granada, we are talking about an Arab cultural presence of more than 800 years. It lasts from 711, the year of the Arabian conquest, to the Christian reconquest in 1492. So it was a very important presence from a cultural and material point of view. It can be seen in the mountains, but also in Andalusian culture. We find this influence in the strategies for the use of natural resources; and concretely, in agriculture with the use of water".

54 As Raphaël Hitier announces in the documentary's opening sequence: "In the south of Sicily, men have for centuries combined their devotion to the heavenly powers with their love of the fruits of the earth. Blessed by the gods, the region of Agrigento has long been the granary of Italy. […] The image of paradise is omnipresent. Moreover, Sicilians have been convinced for centuries that the fertility of this land is a gift from the gods; and as you will see, it is this link between sky and earth that has made this landscape".

(here accompanied by her daughter Persephone), who gave agriculture to men. Throughout the documentary, Raphaël Hitier's commentary traces how, throughout the diverse events that marked the history of the city (the destruction of the city by the Carthaginians, annexation by Rome, Christianization, period of Arab occupation), the religious link between countryside and city continued, in various forms, up to the 1960s. The unity of the landscape between vestiges and countryside thus posited, the second part presents two uses of the countryside, one claiming modernity, the other tradition. But for the heritage interpretative scheme, this opposition is an example of the heritage character of the city-country unit. The changes in agriculture and in the city over the last 50 years (as mentioned by the professor of agronomy) are illustrated by a farmer who grows cereals as his ancestors did, but who would have liked to move toward intensive agriculture (but is unable to, because of the relief of the terrain), while at the same time recognizing that the soil is productive and the countryside is beautiful. In terms of the tradition of a relationship between the land and the supernatural, a citrus farmer (Guiseppe Lopilato from the Kolymbetra garden) aims to follow a logic of heritage preservation of the countryside in a domain situated in the valley that in the time of the Greeks was used as a vast water reservoir for the city; thus, in an area between the temples and the countryside where irrigation is made possible by the ancient water source (which seems a semi-heritage object for him). He concludes the tour of his domain by saying he has set himself the "mission to preserve the beauty of the island" and affirming his pride in preserving traditions with the hope that "the future of our land, of Sicily, will be a future of beauty". The relationship to the supernatural thus seems here to take the form of an aesthetic relationship.

As explained above, the documentary on the Normandy bocage does not deal explicitly with either heritage status or even the heritage character of the landscape place: it presents it as an organization of the place that produces a type of landscape whose folds "resemble those of the soul"[55]. The topic of the documentary thus puts the spotlight on a link between the bocage and a state of mind that could be said to characterize the inhabitants and in particular to make "the little shaded paths bearers of secrets and mystery". But from the very first conversation with an inhabitant engaged in the upkeep one of these paths, these inhabitants are to take on a landscape and heritage role that is indeed more or less expected by anyone who has had geography lessons on the bocage at a French school. This inhabitant shifts the subject toward the entirely practical *raison d'être* of these paths: their being dug down to the rock, the raised sides of the embankment thus formed and it being planted make it possible to not have to move around in mud and to be sheltered from

55 Opening sequence of the documentary by Raphaël Hitier: "The Normandy bocage, a fragmented territory of hedgerows and sunken paths. It stretches out to the west of Lower Normandy. With land consolidation and urbanization, the bocage is receding, taking with it a landscape that used to protect men and their secrets".

the wind and rain. Yet he does not completely deny the type of relationship between people that these paths may produce (being "connected to each other without being seen", he says, with the fantasies that this relationship can nurture). Then, a geographer (Armand Frémont) defines the bocage through the history of an appropriation of land by poor people, reflected in a close relationship between man and land he farms. This relationship to the land is illustrated by conversations with a photographer (Christian Malon, author of a book on a local farming couple). In reality, the tension between the bocage as a practical organization of the place so as to be able to live it (and live from it) and the bocage as a representation of a relationship to the place (with its social consequences) thus runs throughout the documentary. After a part devoted to the Roche d'Oëtre, the adjacent forest and the city of Domfront, the final sequence shows an encounter with two Calvados producers who are busy picking pears. One of the two summarizes and resolves this tension in his own way: he talks about the pride of seeing pear trees around a farm, especially when you have planted them yourself (he mentions that they live for three centuries, one growing, one producing and one dying); a pride that leads you to think to yourself: "We did something, didn't we, there is a heritage here!"; a "de facto heritage", to go back again to Patrick Dal Cin (2001), that is both transmitted as both an entity of landscape coherence (conceptual scheme) and cultural coherence (heritage interpretative scheme). In a certain fashion, the paths and the way they divide up the bocage, as well as the type of social relationship they produce (at the end the presenter comes back to the opportunities they provide for transporting alcohol without being seen by customs officers), while they are not quite, like the dry stone walls of the Cinque Terre or the canals of the Alpujarras, semi-heritage objects, a physical condition for the safeguarding of the place, they nevertheless appear as the symbolic condition; or at least, as the memorial condition.

To conclude, it can be noted that, as Lenclud (1995, paragraph 20) says, there is a difference in the definition of landscape between inhabitants and specialists. The former talk about what they do; they formulate in reality the *raison d'être* of their action. The latter are in an observer position creating a representation of the place. But talking about what we do is in itself the beginning of an observer position, perhaps less about the landscape than about what has been transmitted. What the inhabitants say about their actions emerges as focusing on markers (components, processes or memory) that prove to be the key heritage elements both of the activity to be preserved and of the coherence of the complex entity: the stone walls for the Cinque Terre; the irrigation canals (*acequias*) for the Alpujarras; the traditional citrus cultivation with its mode of irrigation for the Agrigento countryside; the paths, hedgerows and pear trees for the Normandy bocage. This set of documentaries thus shows the importance of the two functions (pragmatic and observer) which historically, as we saw with the Val d'Orcia, jointly shape landscape. It is this conjunction that invites us to interpret the markers mentioned by the inhabitants as heritage interpretive schemes.

5.2.2.3. *Knowledge of the functioning, at the foundation of the complex heritage entity*

In contrast to the previous documentaries, those on the Bay of the Somme and the Opal Coast are about natural sites: they illustrate the specific role that scientific knowledge plays in recognizing the heritage dimension of this type of landscape.

The landscape of the Bay of the Somme, a "landscape between land and sea", as the documentary's opening sequence announces[56], is at risk of disappearance[57]. In order to recover salt meadows for the livestock, explains the head of the *Promenade en Baie de Somme* association (Jean-Michel Dolivier), ever since the Middle Ages man has used embankments to reinforce the Bay's sand dunes, which are produced naturally by the sand left behind by the outgoing tides. In addition, the replacement, in 1912, of the wooden jetty on which the train traveled between Noyelles and Saint-Valery via an embankment, blocked the rising of the tide at the back of the bay. Because of this, silting increased significantly during the 20th century and is in danger of eventually filling the bay. This situation has a doubly negative impact: economic and ecosystemic. The former is described by the geographer in charge of depolderization (Vincent Bawedin): on the one hand, the difficulty of navigating between sandbanks hinders the Hourdel fishing boats on their way out and, on the other hand, the bay risks losing its tourist appeal. However, fishing and tourist activity are two major economic resources. The conservator of the Somme Bay National Reserve (Patrick Triplet) explains for his part the impact on the biotope: by transforming the beaches that are covered and uncovered by the tide (the foreshore) into a quagmire (or swamp), then into meadowland, silting profoundly modifies the ecosystem by reducing biodiversity: since the invertebrates that live in the mud are disappearing, migratory birds can no longer find food, because grass has replaced the tiny organisms that live in sand covered by the rising tide. The part of the bay thus covered in vegetation then loses its role as a stopover spot on the bird migratory route.

Contrary to what can happen, for example, with the *polders* in the Netherlands, it is thus not the part reclaimed from the sea that, according to the documentary, constitutes the landscape that it is important to safeguard, but the maritime landscape. An analysis of the documentary reveals the two different observer

56 "Mudflats, pebble beaches, one of the largest dune massifs in Europe make up this landscape between land and sea".

57 The presentation explains the theme in this way: "With each tide, the sea comes into the Bay of the Somme, but over time, over the years, it reaches less and less far into the estuary, which is filling up. It is undergoing a phenomenon of silting that has become so great today that it is a matter of returning to the sea the land that men have won from it since the Middle Ages. And it is this story between two waters, this fight, where this landscape's future is being played out, that I am going to tell you".

positions constructing two views of the landscape. From the first position, the three components of the place (mudflats, pebble beaches and dune massifs) are the bearers of the features that define the landscape as a representation aligned with the conceptual schemes of the "picturesqueness of the sea" (Corbin 2018 [1988], p. 158): namely the tides; the light on the water; the fauna; and flora of the foreshore (vertebrates, shellfish, migratory birds on the mudflats and seals on the sandbanks at the mouth of the estuary). This is the position of the artist or tourist. The second position is that of scientists (the geomorphologist, the biologist, the ecologist, the historian of local practices, etc.) who describe the creation and functioning of the ecosystem of which the landscape is simply a manifestation at a given time. By setting out the conditions for safeguarding the ecosystem, traces of its production by the actant "Nature" (the scientist's position), and by showing the value of its artistic representation (the tourist's position), the documentary thus contributes to presenting the landscape as heritage (therefore, de facto, to its heritagization).

The documentary on the Opal Coast, taken here in its broadest sense, which goes from the north of the Bay of the Somme up to the Belgian border, explores the relationship between touristification and maritime landscape[58]. The starting point for its topic is the way in which a privileged observation site is built facing an exceptional seascape, "sublime" in its excess, taking the observer "on vacation" from their usual obligations[59]. According to the architectural historian (Dominique Rouillard), this vacation was the reason for the creation of Le Touquet by the Paris bourgeoisie who treated themselves to living as aristocrats for a while, and, we could add, it is still this kind of vacation, albeit taking different forms and concerning a very different population, that drives the tourist influx in the seaside resorts along this coast. It is looking out over the seascape that the first actors to manifest the observer actant were to establish themselves. Another architect (Richard Klein) lists the different phases of the urbanization of Le Touquet[60]: the planting of a forest, the building of the first villas, the development of a seaside resort that presents itself as a kind of extension of Paris with a sea view. This defines what will become the emblematic figure of the seaside resort, with its two landscapes, as they appear today "[...] Violent nature with its tides, with its wind, its

58 Presentation of the documentary (Raphaël Hitier): "Wherever you are in this region, towns are never far away. The coastal towns and villages were literally pulled out of the sand in the 19th century, when the sea bathing was in fashion. Together we will see that this landscape of the Opal Coast is actually something very recent".

59 On the way in which this "vacation" characterizes vacations, I refer to Bachimon's (2013, p. 10) book (*Vacance des lieux*). One of the protagonists filmed in Wimereux, sitting facing the ocean, in a way provides the figure for this.

60 It should be noted, in addition to what this architect explains, that Le Touquet contains a great many buildings with heritage status (21 on the Mérimée database).

sun, and at the same time, an urbanity"[61]. And yet these two landscapes do not form a single entity here, as they do, on the other hand, on the less developed portions of the coastline, still dominated by "wild nature", made of the ocean, but also, on land, dunes strewn with marram grass, battered by the winds and caught up in a succession of gales, thunderstorms and sun that produces "changes of light from one moment to the next". Nature, together with the harshness of the world of small-scale fishermen, which the painters of the second half of the 19th century and the early 20th century had discovered, as explained by the curator of the Musée de France South Opal museum in Berck (Georges Dilly), commenting on the paintings and pochades[62] exhibited at the museum.

The same applies to the cliffs of Cap Blanc-Nez which, in the documentary, is essentially approached from the point of view of its geological particularities by a professor at the University of Lille (Alain Trentesaux). These pieces of non-urbanized landscape between land and sea, with dune massifs, mudflats, estuaries and chalk cliffs – and, of course, the part of the ocean that face them – have many characteristics comparable to those of the maritime landscape of the Bay of the Somme, for example, the light (hence its name, the Opal Coast), the tides or the richness of fauna and flora.

Ultimately, this documentary complements that of the Bay of the Somme in that it establishes a difference (even a conflict) between four categories of observer actants. The one who, in a "vacation" position, seeks to take advantage of it; aiming in fact at a consumption of landscape. The one who, from a position of "artialization", seeks to create a representation of it that translates the relationship established with the landscape. The one who, in a scientific position, offers an understanding of the complex functioning of the place. Finally, the one who, in a heritage position and using the previous one as a basis, devotes their actions to preserving the ecosystem of this complex entity.

61 As Dominique Rouillard explains: "We have on one side this incredible horizon – just look what we have in front of us – and we turn around slightly; and on the other side we have the town. So, two landscapes, there are always two landscapes at the seaside: violent nature with its tides, with its wind, its sun, and at the same time, an urbanity. There is no seaside resort without a very, very powerful urban fantasy. We can clearly see this architecture standing head to head with sea, head to head with nature, head to head with the violence of the elements. And it is this confrontation that I find extremely stimulating between nature and architecture, the city, urbanity".

62 Color sketches executed with a few brush strokes, with the specific aim of capturing the light. Painters in the late 19th and early 20th centuries, who were now working on motifs, now find subjects 3 hours by train from Paris.

But from the point of view of the landscape as a complex heritage entity, it is the positions of the latter two actants that are decisive: scientists who reveal the links between processes, of which the visible components of the landscape are only the manifestation, and the nature of the complex entity of the place, and the position of heritagization specialists who claim the means and procedures for safeguarding and contribute toward putting them in place. These two positions show that, in the case of natural heritage, the quasi-heritage object is a process that is revealed through studying and that it is not a material object.

5.2.3. *Landscape maintenance*

From the time they appeared, the early complex heritage objects, which could not be enclosed, because they were scattered across a territory (ecomuseums, urban complexes, national and then regional natural parks, certain dispersed mining sites, even some large archaeological sites), made it necessary for those responsible for them to find or devise the means to ensure their maintenance by protecting them, preserving them and seeking to regulate access to them. Indeed, when they do not become museum objects and continue to have a social use, these heritage objects require *maintenance* rather than conservation. Because the conservation of a monument is already, in itself, a delicate undertaking, to the point of sometimes involving either a restoration aimed at reestablishing a relative integrity of the object, or limiting visitors, the maintenance of a complex object has to reconcile the act of keeping elements alive (literally and figuratively) and perpetuating the entity in its complexity. It must therefore necessarily go beyond perpetuating a fixed state. The example of Annecy's urban heritage provided an illustration of a process combining creation and maintenance in which of course the state intervened (in particular through the administrative legal framework and labeling), but which actually fell to the municipality, who the proceedings both in terms of decision-making through its policies and in the implementation of actions through the intermediary of its departmental offices. However, the situation proves more delicate when it comes to heritage objects as complex as landscapes, insofar as, as we have just seen, an element (component, process or memory) can control, sometimes even invisibly, the heritage destiny of the landscape as a whole. If the knowledge produced by scientists, both on this element and on its impact, proves essential, it can only give encouragement to carry out the operations necessary to safeguard the landscape, but cannot implement them. Cooperation between a number of actants (therefore, a fortiori, a significant number of social actors) is therefore imperative.

In four of the seven documentaries, landscape coherence directly depends on the safeguarding of components, namely, the dry stone walls of the Cinque Terre, the canals of the Alpujarras and the water source of Agrigento, the paths and orchards of the Normandy bocage. As for the Val d'Orcia, it is almost all of the components (fields, trees and houses) that ensure this consistency. In all of these cases, the maintenance of these components is linked to the continuation of the old agricultural activity, but the documentaries, with the exception of that on the Normandy Bocage, mention the existence of a heritage institution (a park) that more or less regulates the functioning of the place. It is even this institution that takes measures to ensure the maintenance of the dry stone walls of the Cinque Terre. Finally, the two documentaries on the Bay of the Somme and the Opal Coast supply relatively little information on the maintenance of these two landscapes; complex maintenance that seems ultimately to be the responsibility of the political authority administering the territory (the General Council), when it is not up to the pressure of the dominant actors.

From the point of view of interpretive schemes more specifically involving maintenance, we can identify three types of situations: the pursuance of a type of maintenance presented in the documentary as a continuation of practice, for the Normandy bocage, Agrigento and the Alpujarras, a type of maintenance presented as being undertaken by the inhabitants, but under the supervisory framework of a park that manages it, with the Val d'Orcia and Cinque Terre and at the instigation of a political authority, for the Bay of the Somme and the Opal Coast.

5.2.3.1. *A continued traditional maintenance*

The bocage is not overtly presented in the documentary as a heritage landscape, but we have seen that a number of transitions are nevertheless interpretable according to heritage schemes. There are plenty of continuity markers of the traditional bocage and also of the continuation of ancient practices. This applies for example to observers: a photographer who tries to give an account of their "inhabited" character and who produced the book on the Champion couple, it applies to what the geographer says about the historical construction of the bocage. But it applies especially to the actors engaged in a practice of maintenance: the inhabitant who maintains the paths; the farming couple who continue to live and work as in the past; or to the two Calvados producers thinking about the planting of pear trees over several generations; it is even, more generally, the maintenance of some of divisions of the bocage, by paths and hedgerows, even though the area has been subjected to "*débocagement*" to increase the size of the plots and mechanize agriculture. The continuation of this traditional maintenance takes on its full heritage meaning in contrast to the introduction of "débocagement" or land abandonment, as in the forest below the Roche d'Oëtre.

We also find the continuation of traditional maintenance in the part of the Agrigento landscape corresponding to the citrus orchard of the Kolymbetra garden (Guiseppe Lopilato) with the citrus farmer's choice to cultivate this orchard using traditional techniques, particularly irrigation. Here again, this choice takes on its heritage meaning, in contrast to the farmer ploughing his field in order to plant cereals, who, although he says he is continuing to farm like his ancestors, holds a utilitarian discourse borrowed from agribusiness, putting forward his desire to expand, mechanize and increase production.

However, it is in the documentary on the Alpujarras that the continuation of landscape maintenance by the population is presented in the clearest and most detailed manner, putting forward the best arguments. The Sierra Nevada National Park has a role in the regulation of tourist fluxes, which contributes to the maintenance of the landscape place, but it is the need for maintenance of the network of irrigation canals (*acequias*) and the articulation of the risks that would result from the cessation of this maintenance that show the extent to which this network constitutes de facto a *common property* ensuring the maintenance of this landscape. Even though it continues to be tended by the irrigator community, that is, by the main users for agriculture, it actually benefits the whole environment. It ensures the irrigation of crops, of course (according to the rights of each farmer), but also that of the alpine pastures and the water supply to the village fountains. The survival of the whole of the ecosystem, human and nonhuman alike, depends on the upkeep of this network, which constitutes its backbone. The landscape is only the visible face of this ecosystem. However, one condition is mentioned: the maintenance of the canals entails that of the community that possesses their use and the know-how of their upkeep. It is therefore necessary that a sufficient number of inhabitants locally continue to be economically active, either living and working on the spot, or, for some, coming back at the weekend when they live and work in town[63], in order to be able to carry out the necessary maintenance work and continue to transmit the memory of know-how. The farmer (Modesto Alonso Alonso) mentions this condition when he explains that the rural exodus would have a terrible effect not only for the inhabitants but also for the landscapes and the entire ecological heritage. The documentary implies that it is the relative isolation of the site, the continuation of a traditional agricultural activity, the attachment of the inhabitants who have left to live in town and the park's tourism policy that together have made possible up to now the maintenance of practices, of the network of canals and consequently of the ecosystem as a whole. But it is with the difficulty of

63 The documentary *Cuisine des terroirs* on the Alpujarras presents an example of the continuance of ties as family and friends between those who continue to live in the country and a family who lives and works in town, returning to their village at the weekend.

managing the common character of such a heritage in the long term, in a context where human business activity is thought of in terms of instant profitability, that a paradox arises between two logics: the instrumental and managerial *mode of thinking*, and the *symbolic logic* (heritage)[64], which as we will see is decisive from the point of view of the continuance of the landscape as heritage.

5.2.3.2. *Maintenance within the framework of institutional responsibility*

The documentaries on the three "properties" (in UNESCO's language) inscribed on the World Heritage List, Val d'Orcia, the Cinque Terre and Agrigento, indirectly evoke their status as parks, but without specifying what this status consists of or, more importantly, what the mode of functioning of these parks consists of. It is true that their objective is to offer landscape discoveries, and not to explain their management or their institutional functioning as heritage. The Val d'Orcia and Agrigento parks are merely mentioned; the only one whose role is presented in a little more detail is the Cinque Terre National Park. It is therefore necessary to refer to other media texts to obtain information on this point: the most authoritative media text, and also the most easily accessible, is the property fact sheet on the UNESCO website. In the section "Elements required in terms of protection and management", at the end of the "Description" tab, we learn that the Val d'Orcia Park of Art, Nature and Culture is managed through a conglomerate of the five municipalities concerned, the Archeological and Landscape Park of the Valley of the Temples at Agrigente by an autonomous structure and that Cinque Terre National Park brings together the three municipalities covering the five villages. But, more importantly, this section specifies, for each park, the main axes of the management plan designed to ensure the continuance over time of the exceptional value of the "property", while preserving its living character (especially for the inhabitants) and organizing its development, as well as its promotion particularly through the adaptation and regulation of tourism that appears as one of the major challenges. However, this text is for guidance and its implementation is the responsibility of the park concerned, which will thus find itself confronted with the paradox of the two logics.

Let us consider the emblematic case of the Cinque Terre, insofar as the document sets out the constraints that impact on the maintenance of this landscape. The member of the tourism consortium linked to the national park (Roberto Venuti) talks about the need for a "dynamic conservation" of heritage. This involves both "agriculture, with young farmers who create wealth, good quality wines, and a

---

64 This opposition echoes the distinction drawn by Xavier Greffe (2014, p. 167) between the "existence value" and "use value" of heritage. For more critical approaches to the issue, I refer to Sgard (2010) and Tornatore (2017b).

rehabilitation of the terraces to preserve the environment"[65]. Due to the rural exodus because of the difficulty of exploiting the terraces and to the tourist influx, this kind of dynamic conservation proves to be much more delicate than it may be, apparently, in Val d'Orcia, where the maintenance issue is less likely to impact its very existence. Two other scientific media texts (Biagioli 2010 [2009]; Briffaud et al. 2010, pp. 119–131 and 227–230) help us to understand the issues specific to the Cinque Terre and the lessons that can be drawn from them. They show that the originality of the park lies in its competences, which are particularly extensive in terms of decision-making and management, because it is both a national park, the managing body of the UNESCO site and a conglomeration of the mayors of the three municipalities. These competences allow it, on the one hand, to implement a very firm mode of governance, both incentivizing and constraining, on the ownership and use of land, on the possibility of building houses in the vineyards, on the use of the ground floors of the village houses, on tourist rentals, on traffic and tourist fluxes, etc., and, on the other hand, to be in a position to seek financial resources for the replanting of the terraces and the upkeep of the landscape (European funds, creation of a tourist pass, etc.). But all of this has a flipside: the park is inevitably the object of criticism, depending on the particularities of each village or on the type of business activity of the inhabitants: some find that the park is too intrusive, while others believe that it does not go far enough (Biagioli 2010 [2009], p. 154). It must therefore work to make people understand the added value the protection of the site brings to the economy and to local society, but it must also put in place action to "promote profit redistribution, in particular profits from tourism, among all inhabitants in order to ensure and sustain the solidarity that used to be traditional here because of the isolation", Biagioli (2010, p. 155) explains. However, this solidarity is undermined by the influx of 2.5 million tourists, resulting in some getting rich while others cannot really gain any benefit from this influx.

65 Here is the full text of the response to the presenter (Raphaël Hitier) who has just asked him if the Cinque Terre has found a balance in terms of the massive influx of tourists: "Yes, you see, Raphaël, the essential point of the economic development of the Cinq Terres [*sic*], not only economic, but also cultural, is to reconcile the tourist flux with the very delicate, very precarious environmental balance. If you look at how the Cinq Terres are made, in reality, it is a natural landscape of anthropized origin, so a landscape made by man over the centuries. The populations who lived in these villages did not only have a marine economy, they experienced economic growth above all due to terrace farming. If we conserved the Cinq Terres statically, if we left the Cinq Terres territory like that, in the coming years, in 50 years, the Cinque Terres would crumble into the sea. So, we must give thought to a dynamic conservation between agriculture, with young farmers who create wealth, good quality wines, and a rehabilitation of the terraces to preserve the environment".

Some locals say that the influx generated by the tourist "success" is introducing a fracture between those tourist service providers and the farmers who undertake the safeguarding of the place. The park had to, for example, put in place assistance for the elderly and for farmers and allocate some of the income from the sale of passes to the salvaging of the abandoned terraces. These measures are all the more important for the heritage landscape since the creation of profit is not aligned with heritage maintenance, "insofar as income from tourism – and in particular from renting out rooms in residents' homes – is very much higher than income from agriculture" (Briffaud et al. 2010, p. 242)[66]. Everything suggests that this fracture in the cohesion of the collective is part of the paradox between the instrumental and managerial mode of thinking and the heritage logic, mentioned above in connection with the Alpujarras. Two actants seem to be directly affected: the park and the winegrowers. We have in fact just seen how the park was called upon to resolve in the short term, through management and financial procedures, the fracture between service providers and farmers, in order to sustain solidarity within the community, both the old community (intra-village) and the new community (inter-village), just like the integrity of the place. The farmers, especially the winegrowers, are expected to live from the exploitation of the place, while undertaking in the long term the upkeep of the place and its landscape coherence. Both, according to different modalities, are facing the need to reconcile issues relating to short time gain with issues relating to long term sustainability and gifts from generation to generation: in order to reconcile issues relating to individual property with common property, issues relating to exploitation of resources with maintenance of the environment and issues relating to the economy with solidarity.

5.2.3.3. *The maintenance of a natural heritage by the political authority*

The maintenance of the Bay of the Somme and the Opal Coast seems to have to meet the imperative of safeguarding the "natural" functioning of the landscape, especially the landscape place of the bay on one side and, in a less assertive way, the dunes along the coast.

But we also know (Icomos-France 2010) that unlike the properties inscribed on World Heritage, which are generally managed by organizations dedicated to the entity as a whole, the organizations involved in maintenance (inventories like ZNIEFF or Natura 2000, reserves, parks, Coastal Conservatory, Grand Site, etc.) are

66 I should clarify that the study on Cinque Terre presented in this research report was also carried out by Giuliana Biagioli.

multiple, specialized, fragmentary and overlap to some extent[67]. Dedicated to one part of the territory according to the element that is to be safeguarded, none of them take responsibility for the whole of the landscape entity.

Maintenance then falls to political authorities such as the territorial collectivities (mainly the Departmental Council) or, for some aspects, the ministries. There is therefore a dissociation between the actant having the scientific or heritage competence and the actant having the power to decide on maintenance actions.

In these circumstances, a number of indices in the documentary lead to us to wonder whether the fragmentation of maintenance organizations does not come from a staggering over time of the taking into account of the heritage character of certain portions of territory or certain processes, for which there are two possible reasons. On the one hand, natural processes, in order to be understood, entail the mediation of scientific knowledge, because an experience of the landscape place is not sufficient. However, the production of scientific knowledge takes time. On the other hand, this knowledge had to compete with economic, even political interests (farmers, fishermen, tourism sector, hunters, developers, elected officials, etc.), the political authority then seeking to reconcile contradictory imperatives that more or less go back to the paradox of the two logics.

The procrastination following the Bay of the Somme's "depolderization" project presented in the documentary by the geographer (Vincent Bawedin) seems to offer an example of this. First of all, we can note the difference in status granted to the two aims cited: making the port of Hourdel accessible and restoring the biodiversity of the part that will be beneath the waves. But above all, if we believe two articles in the newspaper *Libération*, a certain amount of indecision on the technique to be used hindered its accomplishment. The technique presented in the documentary consists of opening up a breach in the embankment at the back of the bay, along which the train travels, so that the water can rise high enough to produce a flushing effect at ebb tide. But several other actions were then contemplated, such as, for example,

67 An example of multiple and fragmentary recognitions (at the time the documentaries were produced): National Nature Reserve (including the Marquenterre ornithological park); Grand Site de France (labeled in 2011 for its wetland listed as a Natura 2000 site) and it is part of the Natural Marine Park of the Picardy Estuaries and Opal Sea (2012). The same phenomenon (perhaps even more pronounced): Cap Blanc Nez and Cap Gris-Nez have been labeled Grand Sites de France (2011). Part of the Opal Coast is also included in the Natural Marine Park of the Picardy Estuaries and Opal Sea (2012). There is a series of other protection institutions: the Caps et Marais d'Opale National Regional Park, the nature reserves (Marchand dune, Platier d'Oye, Canche Bay), the ZNIEFF of Saint-Frieux, the classified site of the Pointe du Touquet, zones acquired by France's National Coastal Conservatory (the Ghyvelde dune). As in the Bay of the Somme, urbanization, agricultural activities and tourism seem to have led to a fragmentation of the natural areas to be protected.

turning some agricultural plots into maritime zones or alternatively even extracting sand from the bay[68], before, it seems, eventually returning to the first solution[69].

5.2.4. *Publicization and heritage interpretive schemes*

A study of the way in which documentaries on landscapes contribute to their heritagization reveals what characterizes the emerging mode of media heritagization, of which intangible heritage had provided an initial example. Traces of practices, interpretive schemas play a decisive role in this and constitute what Joëlle Le Marec (2002) calls "composites".

5.2.4.1. *Heritage interpretive schemes as composites?*

Exploring Rivière's three showcases, then the heritagization of Annecy's old city center and finally the documentaries on landscapes has revealed the three levels brought into play by the media heritagization mode: the *social terrain*; its *representation in media texts*; and the *circulation of the interpretations of this representation*. Rivière's three showcases, in seeking to account for the complexity of an aspect of the functioning of a society in the process of disappearance, articulated the first two. From museum objects, they built a *representation* of the terrain through the staging of an exhibition, through the choice and layout of these museum elements. Even the shepherd's hut from Aubrac was only a material trace taken from a complex *reality*. Once displayed, this sample constituted a representation of this reality, which was itself based on a scholarly representation of this place; namely, the knowledge produced by an interdisciplinary survey on the "human establishment" of Aubrac (Segalen 2010). However, what constituted heritage was the unit that was taken and musealized, the staging of the exhibition having an essentially interpretive vocation. In contrast, Annecy's old city center remains a complex reality transformed in situ into urban heritage. At an operational level, the production of memorial, historical, urbanistic and architectural knowledge has a dual function. On the one hand, it serves as a scholarly representation that hybridizes with material reality to form the heritage object, on the other hand it provides interpretation tools – *heritage interpretive schemes* – of the reality experienced and perceived by residents or tourists, which are all indices or signs inserted into the urban space. Even though this interpretation is part of the heritage

68 *Libération* (2016). The slow silting up of the Bay of Somme [Online]. Available at: https://www.liberation.fr/voyages/2016/09/02/le-lent-ensablement-de-la-baie-de-somme_1478075 [Accessed 15 January 2023].

69 *Libération* (2012). Au chevet des sites naturels français (4/8): In the Bay of the Somme, the authorities will let the Channel come back into its own [Online]. Available at: www.liberation.fr/societe/en-baie-de-somme-les-autorites-vont-laisse-la-manche-reprendre-ses-droits-20220730_7NQVG2VR5NEHPLKAAW7L6FIBGM/ [Accessed 15 January 2023].

object, the interpretive schemes remain integrated into the scholarly representation and thus specific to this particular heritage object.

It works quite differently with the landscape. The chief reason for this is that even before any heritagization, landscapes are fundamentally hybrid. They are indissociably a place, lived and perceived, and a representation based on conceptual schemes mobilized by the observer in order to interpret markers. This is why, while certain landscape places become heritage, in the form of a defined heritage object, through the institutional decision of an authorized instance (such as UNESCO), others are recognized as heritage by virtue of the application of interpretive schemes through their belonging to the cultural being, Heritage. The two procedures can moreover coexist perfectly well for the same landscape, since they are very dissimilar and performed by different social actors. Nevertheless, the second case entails the schemes, whether they are present in landscape locations or implemented by user observers, being able to circulate sufficiently, the interpretation of landscape and heritage representation not being limited to one single landscape. And this is where publicization tools, such as documentaries, play their part. Compared to the exhibition, for which such circulation and its resulting association with the schemes, is due solely to visitors' actions, the documentary has a significant advantage, because not only does it circulate conceptual and/or heritage schemes in "mainstream" media, but by preserving their traces, it thereby increases the probability that these schemes will be reused, compared, associated, recomposed, either in the production of new documentaries or by interpreters (viewers, readers, visitors, etc.) of media texts about similar or related subjects. But we must also add the fact that the documentaries in the series *Paysages d'ici et d'ailleurs* have yet another advantage: in addition to their necessary conformity with scientific knowledge on the subject (a principle they share with any form of scientific publicization, including exhibitions), they are polyphonic. Each episode explicitly associates at least three voices: that of the presenter (Raphaël Hitier), that of one or more scientists (or scientific mediators), that of one or more witnesses (usually inhabitants), and this consequently increases the likelihood of the presence of heritage schemes relating to the definition and maintenance of the patrimoniality of the landscape. Each episode thus displays scientific knowledge, institutional or ordinary recognition, bonds and commitments marking their belonging to the cultural being, Heritage. It is in this that they are "composites". Jeanneret (2008, p. 16) is saying the same when he calls cultural beings composites (the cultural being, Heritage, being a form of this):

> Cultural beings are, to use the happy term of Joëlle Le Marec (2002), "composites". These mixtures of objects, representations, and practices give to the cultural forms of a society their interpretive and value markers; they serve as operators for representations of the

> world, but also for the cultural postures and judgment categories that a culture can recognize.

Interpretive schemes thus provide the link between the three levels, allowing landscape terrains to transition from singular cultural beings to their belonging to the general cultural being, Heritage: the place (objects, processes, actors, practices), the representation by the media texts and the triviality of the schemes. In practice, each level corresponds to a form of knowledge production. The first form is that of scientific knowledge produced by the natural sciences (geomorphology, hydrology, biology, ecology, etc.) and/or the social sciences (history, geography, sociology, etc.), which give us an actual means of understanding what make landscapes complex heritage entities. The second form is that of the media texts (in this case, documentaries) that mobilize this knowledge, in order to publicize landscape places, showing de facto what makes them heritage. The third corresponds to the production, use, circulation, recognition of interpretive schemes by those who create and those who interpret the documentaries in order to build the cultural being, landscape-heritage. Using the knowledge produced by scientists, documentaries produce, mobilize, bring to mind and circulate heritage interpretive schemes that are then activated by the viewer. Heritage interpretive schemes are thus traces of practices and memories belonging to the place, the production of knowledge, the writing of documentaries, their interpretation and ultimately of the heritagization of landscapes. Therefore, when I earlier performed a socio-semiotic analysis of the documentaries, I essentially plotted the way in which they were bearers of heritage interpretive schemes that functioned as traces. I would now like to show how, through publicization, these interpretive schemes are composites based on examples of their mobilization at the time of reception of these documents.

What surely becomes clear at first glance is the stark disconnect between the epistemological, methodological approach proposed by Joëlle Le Marec and my operational, restricted approach. Le Marec's theory of composites has a quite different aim – and above all a completely different scope – from the characterization of simply the evolution of a mode of heritagization. Its objective is in fact to reconfigure the construction process of the research object. In the article giving an account of the ethno-semiotic research of uses that she conducted with Igor Babou, on the computerization of ENS's library, in which they tested – in her own words (Le Marec 2002, p. 57) – the "possibility of thinking of composites as condensations that are both material and discursive, dynamic", we read that:

> Composites constitute an observation framework constructed by research, defined in advance, which allows us to shift the notion of social representation (including uses) by expanding the dimensions taken into account, while still remaining operational through having its focus on a specific task. (Le Marec and Babou 2003, p. 247)

The composite is defined here as "a heterogeneous, dynamic configuration that is to be described" (Le Marec and Babou 2003, p. 246). It is, as I see it, thus more of the order of the "situation" of research than of the place or the subject, because it entails an approach that "suspends" the interpretation, in order to allow "place" and "concept" to interact[70]. What sense can there be in applying this research approach to the production of knowledge designed to give a heritage status to a socially constructed object?

When I regard heritage interpretive schemes as composites on the basis of their mobilization in media texts, that is, documentaries, I approach them as "dynamic, heterogeneous configurations, but which constitute units of knowledge"[71], because the issue is actually the consideration of their symbolic functioning, their operativity[72]. Consequently, the proximity between, on the one hand, my approach and, on the other hand, Le Marec's epistemological aim, relates neither to the subject, nor to the approach, nor even to the method, but rather to the posture. In my analysis of documentaries as media texts, this position is that of a *socio-semiotic reading* of the traces of uses of practices of which they are the bearers, in order to show how, through circulating, the heritage interpretive schemes, mobilized in the documentaries or called by them into the viewer's mind, contribute to constructing a new heritagization mode that could be said to be less to do with institutions than with cultural beings. This is why, once all of the differences relating to situations and objects have been taken into account, it seems to me that an analysis that aims to let the production of meaning play out by opening up the complexity, therefore

70 Joëlle Le Marec (2002, p. 56) describes her discovery of "composites" as follows: "I came to think of the objects that I sought to grasp as being necessarily composite, which led me to choose the composite as being a possible concept for thinking theoretically about *the type of object that I was seeking to construct.* The concept on the one hand, the place on the other, must set constraints on each other, they must constrain each other to mutually transform each other in order to help get closer to what we glimpse which is always an imbalance to be resolved between what we would like to think and what we grasp empirically. This concept therefore takes its place in the author's mind in a whole reflection on how to investigate the "audience", the "uses" and the "social representations" (my emphasis).

71 "*Composites* emerge from the same questions as *social representations*, *discursive formations*, texts, *media*, *dispositifs*: dynamic, heterogeneous configurations, but which constitute units of knowledge" (Le Marec 2002, p. 56).

72 Joëlle Le Marec concludes that these dynamic, heterogeneous configurations "are conceived in such a way that they cannot be thought of other than as materially incarnate. But they are not structured by this materiality, except by an "order" that transcends the division between materiality and discursivity, even if it is arbitrary for that reason. This is the only way to account, both conceptually and empirically, for the symbolic dimension specific to social facts" (Le Marec 2002, p. 56–57). This "order" is the symbolizing third party that makes possible a symbolic operativity (Quéré 1982, p. 33).

seems to be not completely foreign to the posture that Joëlle Le Marec describes as consisting of:

> Staying for as long as possible in the confrontation zone between observation's reaction to questioning and questioning's reaction to observation, without precipitating conceptualization, but without rejecting it either, with prudence, patience, avoiding the supporting points given by overly critical analytical positions or overly assertive modelizations. (Le Marec 2002, p. 57)

It seems to me that it is therefore operant to classify heritage interpretive schemes, as traces of social practices in media texts, as "dynamic, heterogeneous configurations, but which constitute units of knowledge" (Le Marec 2002, p. 56).

5.2.4.2. *The operativity of the publicization of documentaries*

The seven documentaries report, or refer to, a few in situ publicization *dispositifs*, such as installations designed to facilitate the tourism practices of the place (roads, hiking trails, signage, etc.), the most noteworthy example of which is the tourist train between Noyelles and Saint-Valery, in the Bay of the Somme, or the urban installations of the Opal Coast. But, for obvious reasons given the nature of the documentary as a media text, publicization through the *production of visitor aids*, whether in paper or electronic format (maps, guides, GPS data, etc.), which is one of the most common and probably most often used forms of in situ *dispositifs*, does not appear here. Also not mentioned are classic forms of ex situ advertising, such as press articles, promotional tools or websites[73]. For the same reasons, there is no trace of publicization of scientific knowledge by the producers, whether in the form of a scientific text or a heritagization operator text. On the other hand, this knowledge is widely used for the construction of the documentaries, and some of it is also presented directly by the scientists interviewed, or is mentioned in contributions from mediators, such as tourism officers, guides or facilitators from heritage institutions (parks, reserves, landscape houses, etc.). This autocentration and this closedness of the documentary can be explained by the fact that it is designed to be broadcast by mainstream media (in this case, on television): it must therefore be informationally self-sufficient. It must in some way deliver all of the information necessary for its topic to be understood, without having to refer viewers to other texts for them to supplement their knowledge (or challenge it) in the same way as a scientific text or an encyclopedia. Therefore, the representation of the landscape thus constructed is necessarily restricted, if only because, for example, of

73 There is however the presentation of the photograph album on the Champion family (Christian Malon) of the Normandy bocage or the book on Renaissance painting in the interview at Pienza, with the curator of the Montalcino Museum (Alessandra Dani) in the documentary on the Val d'Orcia.

the editorial format that is prescribed by its duration. But this is why, on the other hand, this media text bears a large number of traces deriving from the place, from the production of knowledge, from the definition and maintenance of the landscape, as well as from its own production.

The documentary, as a publicization of the scholarly representation, has two strengths that define its operativity: the fact that it is constructed on a topic and the fact that, being broadcast, it makes representation circulate. The first of these strengths puts the documentary in parallel with the exhibition, with which it shares the rule of truthfulness (Davallon 1999, p. 32). However, in the second, it is the exactly the opposite. In fact, while the exhibition is the medium of space and physical encounters, the documentary is that of time and distance. By constructing a representation of the landscape place in the form of a document in order to circulate it, documentaries participate not only in the formation of landscape conceptual schemes, but also in the triviality of heritage interpretive schemes. This is what I would like to illustrate now, by going back to the way these schemes function in some of the documentaries.

I have shown above how this capacity to produce interpretive schemes played a role in the definition of the landscape place and in its maintenance. One of the most illuminating examples of this production is provided by the documentary on the Val d'Orcia. Particularly autocentric and closed off, the concern paid by this documentary to informational completeness results in the mobilization of explicitly enunciated conceptual and interpretive schemes, thus delivering, from the start, the reason why the place is a heritagized landscape[74], in a manner that, to some extent, is of a landscape reading. All the viewer has to do is use them in order to recognize and interpret what they see.

But to this intratextual use of schemes, as utterances facilitating the production of meaning by the viewer, is added a second, extratextual – or rather, intermediate – use, which will contribute to the constitution of the cultural being, Heritage. Through them, the documentary becomes a stakeholder in the mass of all of the media texts (publicization, mediatized, but also scientific or operational) that mobilize the *generic heritage interpretive scheme* (feeling an obligation to keep, in

74 I am referring to the documentary's presentation sequence which explicitly enunciates the program that is subsequently implemented: "When we walk in the Val d'Orcia, we have an impression, a perception of singular harmony as if each tree, each path, each house was in exactly the right place. A feeling that we can fully understand when we know that the region as a whole was thought out, designed; a landscape policy that dates back to the 16th century, when the Republic of Siena was using its campaigns as a showcase to demonstrate its good governance to the whole world. The story I am going to tell you is the story of this ideal landscape".

order to transmit, a "thing", material object or process, which came from a past where it was endowed with values), with which it can, in theory, become interactive to thus constitute a vast archive. But in practice, for documentaries at least, from the moment when quoting another text is not the norm[75], the liaison is made through the interpretation activities of the viewer.

For obvious reasons of simplicity[76], but wishing nonetheless to give an idea of this process, I will take a few examples of the way in which the mobilization of a heritage interpretive scheme by a documentary, or the bringing of it to the viewer's mind, is the beginning of a network, here limited to a few more documentaries that I was able to identify (and what is more, on the Arte website alone), which refer to current areas of interests in the maintenance of the environment. The point here is to show how an imagined matrix is constructed from traces – not to establish the content – made from a set of heritage interpretive schemes, which makes the heritage logic function.

5.2.4.3. *Composite heritage interpretive schemes*

I will start with the example of the mobilization, through the Val d'Orcia documentary, of a heritage scheme that will invite the viewer to call to mind other documentaries that are bearers of this scheme (and, potentially, other media). When the sheep farmer explains that "the ewes play an active part in the conservation of the landscape, we can even say that they are the greatest actors in this landscape" and when he explains that, the land having been abandoned in the 1970s (with the end of sharecropping), "today, without us, there would be practically nothing left but forest on the hills", he is enunciating an actual state of affairs (the landscape is kept open by the grazing of the ewes). However, in doing so, he is using as his basis the implication that this is a method of maintaining landscapes by preventing their closure through reforestation. While the documentary says enough about this for us to understand the impact of sheep farming on this landscape, the viewer is nevertheless more or less invited to call to mind what he knows, from having read or seen it in other media texts – including, of course, in documentaries to which, for simplicity of presentation, I will refer – to clarify what this modality of maintenance is all about. The summoning to mind of this heritage interpretive scheme will help

75 Although some documentaries, such as the series *Un monde nouveau*, are an exception in that they cite their sources (under the heading "Archives"). But it is true that this documentary is made by a journalist (Cyril Dion) who is "documenting" a subject.

76 The reason for this is the dual constraint, theoretical and practical: (i) of the extent of the potential network that includes not only documentaries, but diverse media (press, television, websites, social media, etc.), as well as conversations between people (who also give these schemes their operativity by making it possible for there to be a change of opinion on the topics to which they relate), as opposed to (ii) the fact that the connection is made through the interpretative activity of the readers-viewers.

the viewer to understand all the better the issue of this type of livestock farming for the safeguarding of this landscape. But also, possibly (since the connection always remains dependent on the viewer's interest in it), to replace it in what underpins this modality of maintenance; namely, on the one side, the scientific knowledge that underpins it, and, on the other side (which is obviously, here, the essential point), the heritage logic to which it responds. For example, I know that it is developed in other documentaries that show how this modality of maintenance, which can also be carried out by sheep, goats or cows, is a means to maintain not only the openness of the landscapes, but also biodiversity[77]; a means also to develop environmentally friendly livestock farming methods; even the means to use new ways of refertilizing the soil or to experiment with new farming methods[78]. In turn, the situation described in the documentary on the Val d'Orcia contributes to extending and reinforcing the scheme by applying it to a heritage landscape, and at the same time using this scheme as a marker of patrimoniality.

Another example of the construction of a heritage interpretive scheme awaiting activation can be found in the sequence on the Kolymbreta garden, in Agrigento, based on two objects: the citrus orchard and the canals. The original connection of the orchard was instigated with a documentary from the series *Cuisine des Terroirs* about the cultivation of bergamot trees in Calabria and their use in cookery[79]. The one on the canals with the documentary on the Alpujarras. What is important, however, is not the similarity of the types of objects between two documentaries, but the heritage interpretive scheme that unites them; namely, the question of the maintenance of orchards or canals, the continuation of practices and know-how for agriculture, the preservation of species, etc.; in other words, the conservation scheme for a mode of agriculture thought of here as a heritage. Even though the actors do not necessarily express their attachment in terms of heritage, it is as a viewer establishing a link between the three situations that I recognize a heritage interpretive scheme between the three documentaries. I can then extend this link to a situation found in a fourth documentary[80], in which a farmer from Andalusia (Remedios Arres) brought back into operation the network of canals (which she too says were inherited from the Arabs) in order to irrigate her fields, although this is in the plain and not in the Alpujarras mountain. The scheme of individual attachment to a heritage to be conserved is then extended (or integrated) to that of the use of "a secular irrigation system which is a sustainable solution" (in the words of "moderator" Pierre Girard), which – at a third level of the broadening of the scheme

77 "Roots : Biodiversité, saveur et résilience". The cows that maintain the mountain pastures in *Paysages d'Europe : Montagnes, forêts et marais*.

78 "Roots : Le bétail, un bien ou un mal pour le climat?". In the documentary on *The Bavarian Alps* (cutting the grass in the alpine pastures in order to put cows there again in summer).

79 *Cuisine des terroirs : Calabre*.

80 *Roots*. Desertification: extending the battlefield.

– becomes an element in the fight against desertification and the overexploitation of soils impoverished by intensive farming, by means of soil reconstitution, the soil thus being improved in a sustainable way for biodiversity (taking into account birds, foxes, etc.). Faced with climate change, the farmer says she practices "regenerative" farming in order to leave a healthier land to future generations.

A final example, borrowed from the domain we call "natural heritage", will help to clarify a little more this interweaving at multiple levels of the schemes going from concrete situations right up to the cultural being, Heritage. In the documentary on the Bay of the Somme, the desilting project is approached from two angles: that which is essentially called to mind by the geographer, of a purely "hydraulic" operation, the function of which is to free up access to the port of Hourdel and incidentally to conserve the aesthetic aspect of the bay for tourism, and that subsequently developed by the curator of the nature reserve of the Somme Bay Nature Reserve, of the preservation of a biotope that serves as a stopover for migrating birds by a *renaturing* of part of this biotope which has been transformed into an area used for livestock. The first operation is principally economic and is part of the instrumental and managerial way of thinking, even though tourism promotion is effected through maintenance of a landscape that has become heritage due to the many creations that have celebrated it. The heritage aspect here is used in this promotion. On the other hand, the second operation is directly based on an interpretive scheme for the recognition and rehabilitation of a biotope to be safeguarded for its biodiversity. However, the latter scheme is widely present in other media texts[81], chiefly in documentaries that show the extent to which nature can reclaim its rights in the case of redevelopment. This is, for example, the experience of "revitalization" (to use the actual terms of the documentary), not of a coastline, but of a river (the Inn, which runs from Switzerland to the Danube, via Austria and Germany), where a large part of the redevelopment of the river system has been rethought. The documentary shows a number of *renaturing* operations, with the restoration of wetlands bringing back animal and plant species[82], a general example, with one of the two documentaries on the *Paysages d'Europe*, in which the final part is devoted to the coasts of northern Germany. The sequences on the coastline of the Wadden Sea, where huge tides make it "one of the richest biotopes in the world", show the resting and feeding behaviors of migratory birds, but also the return of brown seals and a gradual recolonization by otters from the Baltic[83]. This documentary sets as the backdrop to the preservation the need to limit the

81 Which relate, for example, to the management by the Baie de Somme–Grand Littoral Picard joint syndicate which manages the nature reserve and classified natural sites or to controversies about development and intervention choices. Available at: www.baiedesomme.org [Accessed 25 February 2023].

82 *Des mains vertes pour la nature : Engagés pour l'Inn.*

83 *Paysages d'Europe : Champs, landes et côtes.*

devastating effects of ammonia and nitrogen generated by road traffic, factories and the agro-industry (industrial livestock rearing and intensive cropping), thus repositioning the evolution of landscapes in an ecosystemic functioning, at least at the European scale. The patrimoniality of a landscape is then associated with the preservation of the environment, or reinterpreted from it on the basis of a common conception of temporality in particular.

5.3. In conclusion: interpretive schemes and media heritagization

If only one lesson were to be taken away from the study of the heritagization of complex entities, it would be the place that heritage interpretive schemes henceforth occupy. The comparison drawn in this chapter between the three types of media texts – the showcase displaying museum objects, the old town center exhibition and the documentarized landscape – shows first of all how these texts construct a representation that seeks to account for an absence: that of the relationship that the collective maintains with the elements, even though it is through this relationship that the elements constitute an entity that is greater than simply their juxtaposition. The thing is particularly visible with the herd showcase in which the layout and formatting of the objects are not restricted to suggesting absent living beings, but linked to the characteristics of these objects; they construct the representation of a social practice. The shepherd's hut showcase does the same, except that, even more so than that of the herd, and a fortiori than that on the mining and cutting of stone, it reveals a social group's mode of inhabiting the world. Then, a further step forward is taken, as we have seen, with the putting on exhibition of the Annecy's old city center, which is reduced neither to the presentation of an uninhabited monument nor to the creation of a stage setting, but in which the living is indeed present. The issue is then the relationship that the inhabitants uphold with the old city center turned heritage; the entity as a social reality depends on their practices, their decisions and their representations. Finally, to this complexity of the museum representation or of the ancient city due to the presence of the inhabitants, the documentarized landscape adds two more factors of complexity: the fact that the landscape is already in itself a representation taken from the continuity of a milieu, and the fact that a documentary, as a media text, not only shows but at the same time offers an interpretation, in this case a heritage interpretation, of what it shows. This is why interpretive schemes occupy an important place that invites comments.

First of all, I will focus on *the coexistence of the two modes of landscape heritagization: institutional and media*. To understand the meaning of this coexistence, it is necessary to position the heritagization of these complex entities, the landscapes, in the evolution of heritagization modes. The impression that the Australian archaeologists have of the heritage character of the engravings, the recognition of the patrimoniality of this or that building by the Inventory's specialists, the attribution of the status of monument, place of memory in Vézelay by the historian, interpret objects according to the rules defining what constitutes heritage. On the other hand, the documentary *Australie : Le grand livre des Aborigènes*, by publicizing the research that is the basis for the archaeologists' impression, through the comments of Peter Eeckhout, gives visibility to this link between impressions and the definition of heritage. It thus leads viewers to call to mind in response what they know about archaeology and the rules that govern the heritage status of an object, if they want to understand the position taken by these archaeologists and ultimately to appreciate for themselves the heritage value of the engravings. In fact, we are thus looking at a mobilization of heritage interpretive schemes, but this does not extend beyond the framework of the functioning of heritagization according to its traditional mode. It seems obvious when we compare the Inventory's approach or the unfolding through the *Lieux de mémoire* with social heritagization: these schemes are part of the knowledge that underpins the declaration, for as long as heritagization remains a matter for specialists and the administration, even though the gesture of knowledge production or of publicization gains in importance. It is when knowledge is mobilized by the members of a collective that believes, for example, that an old mine, an old factory or a district of a city[84] has a *heritage value* that justifies its recognition as heritage, so that things begin to change: knowledge then feeds into the relationship of the collective to "things" upon which it feels it depends, and therefore to which it believes itself to be attached as an heir. This change relating to the use of knowledge is concomitant both with the "change" according to which collectives are no longer regarded as social entities of less value than the body politic of the nation, and with the upsurge of social heritagization that occurred simultaneously in the 1970s and 1980s in a great many places in French society. However, it was with the arrival of the UNESCO Convention on Intangible Heritage in 2003 (UNESCO 2018) that a new change occurred, the publicization of heritagized social processes, mediated by sites and documentaries, which began to replace the exhibiting of material heritage objects. The need to "document" the process, if only in order to have it officially recognized, has made the production of knowledge, whether scholarly or endogenous, the basis of this publicization, thus paving the way for the mediatized development of interpretive schemes and the emergence of the media heritagization mode. Consequently, while heritagization modes have undergone at least two abrupt changes (two points of singularity), the schemes have steadily grown in importance as the collectives, actors or audiences of heritagization have grown in power, in correlation with the development of the capacities of the mediatization of video

84 As was the case with the *Renaissance du Vieux-Lyon* association, founded in 1946 by Annie and Régis Neyret, which saved the medieval district, at the foot of the Fourvière hill, from its scheduled demolition to make a motorway sliproad, obtaining for it the status of protected area (one of the first created) in 1964, before it was listed in 1998 as a UNESCO World Heritage Site.

recordings. The example of landscapes thus merely illustrates the importance taken on by the schemes; but their originality is to present the two modes of heritagization, institutional and media. When they are declared heritage by an authorized instance, such as the state or UNESCO, this declaration, whether it be based on cultural or natural characteristics, remains an external recognition that presupposes defining a landscape place in a relatively stable way, according to limits and characteristics fixed by decree, even if some possible internal adaptability is recognized. The knowledge produced thus primarily concerns this object. The schemes are, on the contrary, labile: they sometimes apply to the landscape in its entirety, sometimes to only one of its aspects or components (objects, practices, actors, territories, etc.), sometimes several schemes are jointly applicable to one element. However, the co-presence of the official declaration and of the schemes shows that we cannot apply any scheme to any element or to any landscape. On the one hand, the schemes belong to the cultural being, Heritage, and, on the other hand, the communicational functioning of documentaries requires the existence of a framework of interpretation (a universe of reference, intentionalities, a type of communication, etc.) shared by the protagonists in the communication: in this case, the members of the collective or the specialists who are interviewed, the creators of the documentary and the viewers. This framework defines which schemes are acceptable and which are not. By enabling "cooperation" between the "author" and the "viewer", it thus guarantees the "trust" put in the document, in the way in which the authenticity of what is shown and the truthfulness of the knowledge used underpin the trust accorded to the exhibition (Davallon 1999, pp. 31–32; Le Marec 2007).

The second comment concerns the *viewer's work on the interpretive schemes involved in the media functioning of documentaries about the landscape*. In terms of its creation, the documentary uses interpretive schemes that help with its organization in order to make the creator's message understandable. For their part, the viewer mobilizes the schemes that are part of their "encyclopedia"[85]. For example, in order to exhibit the landscape, which is already a representation, and interpret its characteristics, as so many indices of "heritage value", the documentary will incorporate fragments of knowledge and heritagization rules (both constitutive and normative), through the description of the landscape that it gives and in the interpretations that it provides. To stay with a simple example, when it gives a definition of the landscape place or presents its maintenance, not only can it show the way in which this is an institutionally recognized heritage object, but also the marks of attachment or commitment of social actors or collectives. And it is to these interpretive schemes, which function in this case as organizational schemes, that viewers will be able to react by recognizing them and mobilizing other schemes that complement them, reinforce them, modify them, or even put the emphasis on a

85 A concept that, like those of "cooperation" and "interpretant", I have borrowed from Eco (1985 [1979]), transposing it from the literary text to the media text of the documentary.

different aspect of this landscape place. A few examples are given at the end of this chapter which illustrate this process of cooperation. In fact, I mobilized interpretive schemes ("interpretants") that have to do with the heritage character of the landscape by citing documentaries where these schemes were developed, in order to make the process understandable. However, of course, these are only simple examples, since the schemes mobilized can come from different types of media texts, such as websites, books, printed media, radio, TV news, etc. Essentially, all I have done is indicated proximities in order to give visibility to a potentiality that documentary viewers will in practice implement in various ways, because they cannot be precisely described, and predicted even less. I have therefore "only illustrated" the intermedia circulation-rewriting (i.e. between media texts) belonging to triviality. Unlike an automated processing of the semantic diagram that would reveal the traces, I have exemplified its functioning. Because the main thing is what underpins these logistical proximities, namely, the links that the schemes, as indices of the constitutive rules of heritage and as traces of heritagization operations, establish between media texts, that is, in this case, between documentaries. The issue here is grasping how some of the people who are practicing, developing, contributing to producing the elements of this landscape for the purpose of *maintaining* it in order to transmit it, describe their *interest* in the landscape element to which the scheme relates, in the form of an *attachment* to this element (Tornatore 2010b, p. 20) and how they mobilize *scientific knowledge* that is not only technical, but which apprehends the entity scientifically both in its complexity and in its relationships to its environment.

The preceding comments lead us, thirdly, to spend some time looking at *the proximity of heritage schemes with environmental schemes*, and more generally on *the connection that the mobilization-convocation of the schemes establishes between a heritage maintenance of the landscape and an environmental approach*. In the documentaries, it appears that committing to the former (i.e. maintenance), with the goal of transmitting these entities to future generations, amounts to being almost obliged nowadays to call to mind schemes relating to the safeguarding of the living, or, to be more precise, of living beings and their environment. However, as Lionel Charles explains, landscape and environment correspond to two profoundly different frames of reference:

> Where the landscape appears to be part of a culturally inscribed anthropocentric construction in the order of the visible, of the representation, the environment originates from a very different dynamic, which can be described as praxeo-cognitive, from a generic observation of action and ignorance/knowledge, which is part of a strong decentering. The environment consists of both potential meaning and potential limit, as a universe open to questioning that continually expands in the face of the accumulation of means acquired by the human species both to appropriate the multifaceted resources,

> but also to better identify their implications, in a particularly significant existential context: it is the future of humanity, engaged in a process of massive expansion, that is at stake, in a random approach that has no rules to rely upon, but which depends on its ability to analyze its own implications to reorient itself to the best of its intentions, its desires and an understanding of the threats it poses to itself. (Charles 2017, p. 131)

Should we deduce that in heritagizing the landscape, the interpretive schemes called to mind by the documentaries cause it to tip from the first frame of reference to the second? A detour through the analysis conducted by Tardy and Rautenberg (2013) on the scientific literature from 2000 to 2010 on the subject of natural heritage will allow us to clarify this issue. After summarizing the procedures for the declaration of official status of cultural heritage, the two authors note that the texts examined, most of which have an economic and geographical slant, approach this heritage in terms of a "resource". But this last notion is the seat of an ambiguity:

> Nature, or rather living things in the broad sense (environment, landscapes, wild or domestic fauna, regional produce), seems to be defined ambiguously, between a stock offering a range of possibilities for exploitation and the riches necessary for human life, therefore to be protected and conserved. (Tardy and Rautenberg 2013, paragraph 22)[86]

This ambiguity, they explain, opens up "a double questioning around economic development or the survival of humanity" (Tardy and Rautenberg 2013). But in order to properly address this questioning, it must be taken into consideration that these approaches in terms of "resource" introduce a social dimension that relativizes ecological and biological studies: "the acquired resource" is thus turned round to become "a constructed resource", including know-how, such that social practices form an integral part of the biological resource (Tardy and Rautenberg 2013, paragraph 26). And to conclude:

> It is accepted that heritage as a resource is transmitted through transformation and not through immutability over time. There are no natural reserves from which we can draw indefinitely without affecting them, but neither is it possible to freeze them, in seeking to protect them. The resource is dependent on the use that is made of it, the practices that create it and mobilize it. Living heritage is above all an evolving heritage, anchored in present-day activity, looking towards future generations, yet without being cut off from a

86 On the questions raised by this opposition, I refer to Tardy (2018 [2014]).

> connection with the past. It no longer about *natural* heritage only, yet does not tend towards *cultural* heritage only. Instead, we should speak of a "complex heritage", which is more difficult to label. (Tardy and Rautenberg 2013, paragraph 27)

The economic approach and the traditional heritage approach are therefore placed back to back. On the one hand, resources are not inexhaustible, contrary to what is postulated by the model of endless economic development that legitimizes current industrialized practices, but, conversely, isolating landscape places to make them frozen entities is not a sustainable position either, and would, I should add, only leave open what is left, that will not be heritagized, to a possible exploitation without discrimination or limitation. Hence, it is necessary to redefine the resource-heritage as a "complex heritage", neither natural nor cultural – or rather part-nature and part-culture. Landscapes are at the heart of this ambiguity. By treating landscape places as complex heritage entities – a classification that seems to me relatively close to that of "heritage complexes" – documentaries nevertheless shift the economy versus heritage conflict by positing that, subject to the adaptability of landscapes (the principle of maintenance attests to this), their value is indissociably heritage and environmental.

Hence, the question of *what underpins this dual heritage and environmental value of landscapes*. What link, in fact, unites heritage and environment? The documentaries' answer is clear: the same conception of temporality unites these two cultural beings. If the environmental interpretive schemes mobilized or called to mind are operant for fresh thinking about the heritagization of landscapes, they owe this largely to the fact that they themselves respond to a temporality of the environment as a thing coming from the past that we have an obligation to keep in order to transmit it to future generations. In other words, it is a conception of temporality that starts from the present as the point where we are, as the point that organizes what we can see (i.e. what we know): as the point that is located between a past, which made the present what it is, and a future, to which the present is duty-bound to bequeath what it has found (in the different senses of the term, physical and scholarly). It should be noted that this temporality is almost exactly the same one that the constitutive rule of heritage has preserved at the heart of modernity, during a period when progress gave value to a present that disowned the past in order to move toward a supposedly ever-better future, and which was – and still is, for those who continue to profit from it – the only reference value. In this, the environmental schemes, which sprang up in the same movement as the heritage "change" of the 1970s and 1980s[87], and which are based on the observation of the limitations and negative effects of what has become an ideology that justifies the current structure of society, do in fact respond to a heritage temporality. While

87 See, for example, Davallon et al. (1997).

heritage schemes show that it is the links that landscape places sustain with the environment that give them their meaning as a thing to be kept, the question obviously arises in return as to what meaning their status as heritage objects can have... The answer belongs to the future. But for the time being, we can observe that declaring certain landscapes as "heritage complexes" constructs reference points (knowledge and experience, for example) for safeguarding the environment. In any event, with the landscape, the critical *leitmotiv* of the 1980s, according to which "everything becomes heritage", then appears in a new light. It does not mean (or it no longer means) that only the present remains, but that on the contrary the present looks at landscapes from the past not only as a simple de facto heritage, received as a legacy, but, as we have seen, as what must be *maintained* for the future, that is to say, (re)built in part, sometimes "regenerated" or "revitalized". As a result, if we consider the ambiguity of the resource, the environmental schemes, like the heritage schemes, because they relate to the maintenance of complex entities, are located, by virtue of their relationship to time, on the side of a long temporal perspective, not chronological, but primarily aspectual, which maintains the link between present, past and future. A temporality that can be said to sit in complete rupture with the "presentism" of the modernist "natural" resource, conceived outside of time (except that of approaching the present from the angle of what else it will give a moment later)[88] and all reality, because it is thought of as an infinite fact that can be exploited without worrying about the limitations linked to the past or the consequences for the future. In other words, the heritage and environmental schemes mobilized by the documentaries we studied are thus manifestly contrary to industry's productivist ideology of resources and living, foreign to complexity and knowing only what the short term holds.

In summary, if the landscapes I have studied here constitute an emblematic example of a complex heritage entity, they owe it to their publicization by documentaries. These confer upon them, by activating interpretive schemes, a patrimoniality through the recognition of their consubstantial dependence on both nature and culture. It is this attribution of patrimoniality by media texts capable of activating interpretive schemes that I call *media heritagization*. Confining ourselves to the case of landscapes, it appears that this heritagization mode results from a certain number of *conditions for success* even outside other heritagization modes (as is the case of the Normandy bocage). Some of these conditions relate more to landscape heritagization itself. This is the case for the existence of a production of scientific knowledge that makes it possible to understand what characterizes the landscape, its functioning, its history, its connections with what surrounds it; but above all, the case of the existence of an interest on the part of some of its inhabitants in its future, in order to ensure its safeguarding in the long term, which

88 I mean "presentism" in the sense given to it by Baschet (2018, p. 115), when he writes: "The present of presentism is therefore not a pure present, but a present caught up in the moment after".

leads them to devise its maintenance[89]. Other conditions relate to the media texts that carry out publicization. In the first place, there must be a sufficient production of these media texts, in this case documentaries, which meet the rules that govern heritage publicization (i.e. are to do with reality and not offering a fiction; mobilize truthful knowledge). Then, it is necessary for these media texts to circulate freely in society, the *dispositifs* for their dissemination guaranteeing compliance with these rules. Finally, the existence of interpreters (readers or viewers) who are interested in landscapes, heritage and the environment for these media texts is essential. I have already mentioned the relatively large presence of documentaries about landscapes on the Arte website, the fact that on the whole they have a lot of views increases the probability of interrelationships and therefore that of the representations being consolidated, contributing to forming what I called, in the second chapter, a "propitious representational environment", both institutional and socio-affective[90]. The existence of a propitious environment such as this is the last condition, perhaps the chief one. It has to do with an awareness of the long-term temporality and complexity of living things, understood in the broadest sense, including not only humans, but also what we include under the term "nature". And it is this point, as we have seen, that brings together the cultural being, Heritage and the cultural being, Environment, jointly opposing them to the productivist ideology characterized by a short-term temporality and a simplification, a reduction of the complex that could go as far as the pure and simple extinction of living things. Against this ideology, they assert the obligation of a maintenance of complex entities – of which the landscape is the emblematic figure – over the long term, which is based on a balance between human life and the life of the rest. It is thus no longer only the constitutive rule of heritagization that applies to the living as in official attributions of heritage status; it is the dissemination and sharing of the need for this maintenance that leads us to rethink our relationship with the past and what gives it its value.

At the end of *Le don du patrimoine* (The Gift of Heritage), I noted that the heritagization of material traces of the past, which made use of interpretation to establish a knowledge of what they consisted of and their world of origin in order to keep these heritage traces for future generations, had in fact allowed us to open up a temporal perspective that henceforth makes us look further back into the past. Interpretive schemes, which shift heritagization toward collectives and audiences by using the media, now extend this perspective toward the future. Giving a central place to publicization, media texts – documentaries – show how the production of knowledge on the functioning of things of the past, from the memory that the actors have of them as well as the traces that this functioning has left, can be used to develop maintenance practices that respect the logic of these things, while adapting them to present circumstances, so that they can still exist in the future.

89 The *Sauvons la forêt* series provides examples of how this interest can be implemented.

90 Section 2.2.1.

Cited Works

Cited works[1]

Abriat, N. and de Massary, X. (2016). Les fondamentaux de l'Inventaire général. In *Une aventure de l'esprit. L'Inventaire général du patrimoine culturel*. Éditions Lieux Dits, Riotord.

Adell, N. (2011). Le patrimoine, l'éthique, l'identité. *La Ricerca Folklorica*, 64, 81–93.

Adell, N. and Pourcher, Y. (eds) (2011). *Transmettre, quel(s) patrimoine(s) ? Autour du patrimoine culturel immatériel*. Houdiard, Paris.

Ailhaud, F. and Barbe, N. (2019). À propos d'une candidature Unesco et de l'organisation des publics. In *Le patrimoine comme expérience. Implications anthropologiques*, Tornatore, J.-L. (ed.). Éditions de la Maison des sciences de l'homme, Paris.

Akrich, M., Callon, M., Latour, B. (eds) (2006). *Sociologie de la traduction. Textes fondateurs*. Presses des Mines, Paris [Online]. Available at: http://books.openedition.org/pressesmines/1181.

Amougou, E. (2004). *La question patrimoniale. De la "patrimonialisation" à l'examen de situations concrètes*. L'Harmattan, Paris.

Andrieux, J. -Y. (1997). *Patrimoine et histoire*. Belin, Paris.

Andrieux, J. -Y. (1998). *Patrimoine & société*. Presses Universitaires de Rennes, Rennes.

Appadurai, A. (1986). Introduction: Commodities and the politics of value. In *The Social Life of Things: Commodities in Cultural Perspective*, Appadurai, A. (ed.). Cambridge University Press, Cambridge.

Augé, M. (1994). *Pour une anthropologie des mondes contemporains*. Aubier, Paris.

---

1 Taking into account a significant number of works for which there is a gap between the cited edition and the original edition, for reasons of precision of historical information, the second appears in square brackets after the first.

Aykan, B. (2014). Whose tradition, whose identity? The politics of constructing "Nevruz" as intangible heritage in Turkey. *European Journal of Turkish Studies*, 19 [Online]. Available at: http://ejts.revues.org/5000.

Babou, I. and Le Marec, J. (eds) (2017). *Paysages d'énigmes : les paysages entre actions, représentations et institutions*. Éditions des archives contemporaines, Paris. doi: 10.17184/eac.9782813002402.

Bachimon, P. (2013). *Vacances des lieux*. Belin, Paris.

Balsamo, I. (1997 [1994]). André Chastel et l'aventure de l'inventaire. In *Entretiens du patrimoine 1 : science et conscience du patrimoine*, Nora, P. (ed.). Arthème Fayard/Éditions du Patrimoine, Paris.

Balsamo, I. (2003). Les enjeux politiques de la création de l'inventaire général. In *Pour une histoire des politiques du patrimoine*, Poirrier, P. and Vadelorge, L. (eds). Comité d'histoire du ministère de la culture/La Documentation française, Paris.

Balsamo, I., Gros, C., Verdier, H. (2007). *Les publications de l'inventaire général du patrimoine culturel. Collections nationales*. Ministère de la Culture et de la Communication, Paris.

Barbe, N. (2008). La mission du patrimoine ethnologique comme instrument de gouvernementalité [Online]. Available at: https://shs.hal.science/halshs-00353808/document.

Barbe, N. and Blondeau, C. (2011). Récits de dons : les réaménagements à distance d'un château et ses donateurs. In *Les formats d'une cause patrimoniale. Agir pour le château de Lunéville*, Tornatore, J.-L. and Barbe, N. (eds). Lahic/Ministère de la Culture, Paris.

Baschet, J. (2018). *Défaire la tyrannie du présent. Temporalités émergentes et futurs inédits*. La Découverte, Paris.

Baudrillard, J. (1976). *L'échange symbolique et la mort*. Gallimard, Paris.

Becker, H. (1988 [1984]). *Les Mondes de l'art*. Flammarion, Paris.

Bednarik, G.R. (2008). The Dampier Rock Art Precinct Western Australia. In *Communication Colloque UNESCO* [Online]. Available at: https://www.acade-mia.edu/9480777/THE_DAMPIER_ROCK_ART_PRECINCT_WESTERN_AUSTRA-LIA?auto=download&email_work_card=download-paper.

Bednarik, G.R. (2017). Editorial: The World Heritage List and rock art. *Rock Art Research*, 34(2), 123–129 [Online]. Available at: https://www.academia.edu/35945588/RAR_editorial_Nov._2017.

Belaën, F. (2002). L'expérience de visite dans les expositions scientifiques et techniques à scénographie d'immersion. PhD Thesis, Université de Dijon, Dijon.

Benhamou, F. (2020 [2012]). La valorisation du patrimoine immatériel, ou le triomphe de l'économisme. In *Le patrimoine culturel immatériel au seuil des sciences sociales. Actes du colloque de Cerisy-la-Salle, septembre 2012*, Csergo, J., Hottin, C., Schmit, P. (eds). Éditions de la Maison des sciences de l'homme, Paris [Online]. Available at: http://books.openedition.org/editionsmsh/15990.

Benkass, Z. (2012). Le tri : un enjeu scientifique pour l'évaluation des objets de musée contemporains. In *Que reste-t-il du présent ? Collecter le contemporain dans les musées de société*, Battesti, J. (ed.). Musée basque et de l'histoire de Bayonne, Bayonne.

Bercé, F. (2000). *Des monuments historiques au patrimoine du XVIIIe siècle à nos jours ou "Les égarements du cœur et de l'esprit"*. Flammarion, Paris.

Berthelleu, H. and Dassié, V. (2019). Les mémoires des migrations au musée. Questions patrimoniales et dynamiques anthropologiques. In *Le patrimoine comme expérience. Implications anthropologiques*, Tornatore, J.-L. (ed.). Éditions de la Maison des sciences de l'homme, Paris.

Biagioli, G. (2010 [2009]). Le parc national des Cinque Terre (Italie). In *Valeurs universelles, valeurs locales. Pour qui, pour quoi un site est-il grand ? Actes du Colloque International ICOMOS France*. ICOMOS, Paris.

de Bidéran, J., Deramond, J., Fraysse, P. (2022). Les pratiques info-communicationnelles au cœur du patrimoine culturel immatériel. *Communication & Langages*, 211, 21–30 [Online]. Available at: https://www.cairn.info/revue-communication-et-langages-2022-1-page-21.htm.

Bortolotto, C. (2007). From objects to processes. Unesco's intangible cultural heritage. *Journal of Museum Ethnography*, 19, 21–33.

Bortolotto, C. (2011). Introduction : le trouble du patrimoine culturel immatériel. In *Le patrimoine culturel immatériel : enjeu d'une nouvelle catégorie*, Bortolotto, C. (ed.). Éditions de la Maison des sciences de l'homme, Paris.

Bortolotto, C. (2013a). Authenticity: A non-criterion for inscription on the lists of UNESCO's intangible cultural heritage convention. In *IRCI Meeting on ICH 73: Evaluating the Inscription Criteria for the Two Lists of UNESCO's Intangible Cultural Heritage Convention – The 10th Anniversary of the 2003 Convention*. IRCI, Osaka [Online]. Available at: https://www.academia.edu/en/4954394/Authenticity_A_non_criterion_for_inscription_on_the_Lists_of_UNESCO_s_Intangible_Cultural_Heritage_Convention.

Bortolotto, C. (2013b). L'Unesco comme arène de traduction. La fabrique globale du patrimoine immatériel. *Gradhiva*, 18, 50–73. doi: 10.4000/gradhiva.2708.

Bortolotto, C. (2014). Le transfert d'un standard international : le patrimoine culturel immatériel vu par la France. In *Les vocabulaires locaux du "patrimoine" : traductions, négociations et transformations*, Bondaz, J., Graezer-Bideau, F., Isnart, C., Leblon, A. (eds). Lit Verlag, Zürich.

Bortolotto, C. (2017). Patrimônio e o futuro da autenticidade. *Revista do Patrimonio historico e artistico nacional*, 36, 23–37 [Online]. Available at: http://portal.iphan.gov.br/uploads/publicacao/revpat_36.pdf.

Bortolotto, C. (2020 [2012]). Le patrimoine immatériel et le tabou de l'authenticité : de la pérennisation à la durabilité. In *Le patrimoine culturel immatériel au seuil des sciences sociales. Actes du colloque de Cerisy-la-Salle, septembre 2012*, Csergo, J., Hottin, C., Schmit, P. (eds). Éditions de la Maison des sciences de l'homme, Paris [Online]. Available at: http://books.openedition.org/edi-tionsmsh/16377.

Bourdieu, P. (1980). *Le sens pratique*. Éditions de Minuit, Paris.

Briffaud, S. (1998). De "l'invention" du paysage. Pour une lecture critique des discours contemporains sur l'émergence d'une sensibilité paysagère en Europe. *Compar(a)ison: An International Journal of Comparative Literature*, 35–56 [Online]. Available at: https://shs.hal.science/halshs-00929693/document.

Briffaud, S., Brochot, A., Biagioli, G., Candau, J., Onodi, G. (2010). Paysages d'exception, paysages au quotidien. Une analyse comparative de sites viticoles européens du patrimoine mondial. Report, Ministère de l'Écologie, de l'Énergie, du Développement Durable et de la Mer [Online]. Available at: http://www.paysage-developpement-durable.fr/IMG/pdf/rf_briffaud1.pdf.

Brito, E. (2018). Le rhizome patrimonial. Analyser un mouvement hétérogène d'acteurs et de sens. PhD Thesis, Université d'Avignon et des pays de Vaucluse/Université fédérale de Rio (Unirio), Avignon/Rio de Janeiro.

Brulon Soares, B. (2015a). L'invention et la réinvention de la Nouvelle Muséologie. *ICOFOM Study Series*, 43, 57–72 [Online]. Available at: http://network.icom.museum/fileadmin/user_upload/minisites/icofom/pdf/ISS_43a.pdf.

Brulon Soares, B. (2015b). A invenção do ecomuseu: O caso do écomusée du Creusot Montceau-les-Mines e a prática da museologia experimental. *Mana*, 21(2), 267–295. doi: 10.1590/0104-93132015v21n2p267.

Brulon Soares, B. and Desvallées, A. (2015 [2014]). André Desvallées : entre muséologies (Entretien). *ICOFOM Study Series*, 43 [Online]. Available at: http://network.icom.museum/fileadmin/user_upload/minisites/icofom/pdf/ISS_2014.pdf.

Brumann, C. (1999). Writing for culture: Why a successful concept should not be discarded. *Current Anthropology*, 40(supplement), S1–S27 [Online]. Available at: https://www.journals.uchicago.edu/doi/epdf/10.1086/200058.

Brumann, C. (2013). Comment le patrimoine mondial de l'Unesco devient immatériel. *Gradhiva*, 18–48 [Online]. Available at: https://journals.openedition.org/gradhiva/2698.

Calafat, M.-C. and Chevallier, D. (2019). Une technographie par l'objet : expériences du musée national des arts et traditions populaires. *Techniques & Culture*, 71, 92–105 [Online]. Available at: http://journals.openedition.org/tc/11378.

Cameron, D.F. (1992 [1968]). Un point de vue : le musée considéré comme système de communication et les implications de ce système dans les programmes éducatifs muséaux. In *Vagues. Une anthologie de la nouvelle muséologie*, Desvallées, A. (ed.). Éditions W/MNES, Mâcon/Savigny le Temple.

Castéret, J.-J. (2017). Le numérique comme "lieu" de la sauvegarde du patrimoine culturel immatériel. *In Situ*, 33 [Online]. Available at: http://journals.openedition.org/insitu/15478.

Cendoya-Lafleur, J., Lavorel, M., Davallon, J. (2015). Patrimonialiser la mémoire de la guerre au musée : entre histoire et témoignage. In *Mémoire et nouveaux patrimoines*, Tardy, C. and Dodebei, V. (eds). OpenEdition Press, Marseille [Online]. Available at: http://books.openedition.org/oep/411.

Charles, L. (2017). Paysage, représentation, environnement. In *Paysages d'énigmes. Les paysages entre actions, représentations et institutions*, Babou, I. and Le Marec, J. (eds). Éditions des archives contemporaines, Paris.

Chastel, A. (1964). *L'inventaire général des monuments et des richesses artistiques de la France*. Ministère des Affaires Culturelles, Paris [Online]. Available at: https://hal.archives-ouvertes.fr/hal-02390614.

Chastel, A. (1990). L'invention de l'Inventaire. *Revue de l'art*, 1, 5–11 [Online]. Available at: https://www.persee.fr/doc/rvart_0035-1326_1990_num_87_1_347815.

Chastel, A. (1997 [1986]). La notion de patrimoine. In *Les lieux de mémoire*, Nora, P. (ed.). Gallimard, Paris.

Chaumier, S. (2003). *Des musées en quête d'identité. Écomusée versus technomusée*. L'Harmattan, Paris.

Chaumier, S. (2005). Introduction. Du musée au parc d'attraction. *Culture & Musées*, 5, 13–36.

Chaumier, S. (2011). La nouvelle muséologie mène-t-elle au parc ? In *Expoland. Ce que le parc fait au musée : ambivalence des formes de l'exposition*, Chaumier, S. (ed.). Éditions Complicités, Paris.

Chaumier, S. (2013). Écomusées : structures porteuses de ressources pour l'avenir : l'exemple de l'écomusée de Saint-Nazaire. In *Les musées d'ethnologie : culture, politique et changement institutionnel*, Mazé, C., Poulard, F., Ventura, C. (eds). Comité des travaux historiques et scientifiques, Paris.

Chevallier, D. (ed.) (1991). *Savoir-faire et pouvoir transmettre. Transmission et apprentissage des savoir-faire et des techniques*. Éditions de la Maison des sciences de l'homme, Paris.

Choay, F. (1992). *L'Allégorie du patrimoine*. Le Seuil, Paris.

Choay, F. (1998). Introduction. In *L'urbanisme face aux villes anciennes*, Giovannoni, G. (ed.). Le Seuil, Paris.

Clair, J. (1992 [1974]). Du marteau-pilon à l'écomusée. *L'Art vivant*, 51, 3–6.

Cometti, J.-P. (2012). La fabrique de l'art : médiation et activation. *Aisthesis. Pratiche, linguaggi e saperi dell'estetico*, 4(2), 101–110 [Online]. Available at: https://www.yumpu.com/fr/document/view/49186909/07-cometti-def-aisthesis-ri-vista-di-estetica-online.

Cominelli, F. (2020 [2012]). Patrimoine culturel immatériel : paradigmes économiques, débats et perspectives. In *Le patrimoine culturel immatériel au seuil des sciences sociales. Actes du colloque de Cerisy-la-Salle, septembre 2012*, Csergo, J., Hottin, C., Schmit, P. (eds). Éditions de la Maison des sciences de l'homme, Paris [Online]. Available at: http://books.openedition.org/editionsmsh/15990.

Conan, M. (1994). L'invention des identités perdues. In *Cinq propositions pour une théorie du paysage*, Berque, A. (ed.). Éditions Champ Vallon, Seyssel.

Condevaux, A. (2014). Le patrimoine culturel immatériel à Tonga et ses médiations touristiques. *Culture & Musées*, 23, 43–65. doi: 10.3406/pumus.2014.1769.

Condevaux, A. (2019). "Experts" et "communauté" dans la définition du patrimoine culturel immatériel : le cas du *lakalaka* tongien. In *Le patrimoine comme expérience : implications anthropologiques*, Tornatore, J.-L. (ed.). Éditions de la Maison des sciences de l'homme, Paris.

Corbin, A. (2018 [1988]). *Le territoire du vide*. Flammarion, Paris.

Cornu, M. (2020 [2012]). À qui appartient le patrimoine culturel immatériel ? In *Le patrimoine culturel immatériel au seuil des sciences sociales. Actes du colloque de Cerisy-la-Salle, septembre 2012*, Csergo, J., Hottin, C., Schmit, P. (eds). Éditions de la Maison des sciences de l'homme, Paris [Online]. Available at: http://books.openedition.org/editionsmsh/15990.

Csergo, J., Hottin, C., Schmit, P. (2020 [2012]). Le moment Cerisy. Au tournant des nouvelles approches du patrimoine culturel immatériel. In *Le patrimoine culturel immatériel au seuil des sciences sociales. Actes du colloque de Cerisy-la-Salle, septembre 2012*, Csergo, J., Hottin, C., Schmit, P. (eds). Éditions de la Maison des sciences de l'homme, Paris [Online]. Available at: http://books.openedition.org/editionsmsh/15990.

Cuisenier, J. (1987). Le corpus de l'architecture rurale française. Esquisses pour une synthèse prochaine. *Terrain*, 9, 92–99 [Online]. Available at: http://terrain.revues.org/3189.

Dal Cin, P. (2001). Quelles politiques émergentes pour un paysage-patrimoine ? *Travaux de l'Institut Géographique de Reims*, 105–106, 127–132 [Online]. Available at: https://www.persee.fr/doc/tigr_0048-7163_2001_num_27_105_1412.

Davallon, J. (1986). Philosophie des écomusées et mise en exposition. In *Claquemurer pour ainsi dire tout l'univers. La mise en exposition*, Davallon, J. (ed.). Centre de création industrielle/Centre Georges Pompidou, Paris.

Davallon, J. (1990). L'image médiatisée de l'approche sémiotique des images à l'archéologie de l'image comme production symbolique. PhD Thesis, École des hautes études en sciences sociale, Paris.

Davallon, J. (1992). Le musée est-il vraiment un média ? *Publics & musées*, 2, 99–123 [Online]. Available at: https://www.persee.fr/doc/pumus_1164-5385_1992_num_2_1_1017.

Davallon, J. (1997). L'évolution du rôle des musées. *La lettre de l'OCIM*, 57, 4–8.

Davallon, J. (1999). *L'exposition à l'œuvre. Stratégies de communication et médiation symbolique*. L'Harmattan, Paris.

Davallon, J. (2000). Le patrimoine : une "filiation inversée" ? *EspacesTemps. Les cahiers*, 74/75, 6–16 [Online]. Available at: https://www.persee.fr/doc/espat_0339-3267_2000_num_74_1_4083.

Davallon, J. (2002). Comment se fabrique le patrimoine ? *Sciences humaines*, 36, 74–77.

Davallon, J. (2004). Objet concret, objet scientifique, objet de recherche. *Hermès*, 38, 30–37.

Davallon, J. (2006). *Le don du patrimoine. Une approche communicationnelle de la patrimonialisation*. Hermès-Lavoisier, Paris.

Davallon, J. (2010). The game of heritagization. In *Constructing Cultural and Natural Heritage: Parks, Museums and Rural Heritage*, Roigé, X. and Frigolé, J. (eds). Institut Català de Recerca en Patrimoni Culural, Girona.

Davallon, J. (2012a). Comment se fabrique le patrimoine : deux régimes de patrimonialisation. In *Le patrimoine, oui, mais quel patrimoine ?* Khaznadar, C. (ed.). Maison des Cultures du Monde, Paris.

Davallon, J. (2012b). L'objet contemporain de musée, un "objet sans qualités" ? In *Que reste-t-il du présent ? Collecter le contemporain dans les musées de société*, Battesti, J. (ed.). Musée Basque et de l'histoire de Bayonne, Bayonne.

Davallon, J. (2015). Mémoire et patrimoine. Pour une approche des régimes de patrimonialisation. In *Mémoire et nouveaux patrimoines*, Tardy, C. and Dodebei, V. (eds). OpenEdition Press, Marseille [Online]. Available at: http://books.openedition.org/oep/411.

Davallon, J. (2018 [2014]). À propos des régimes de patrimonialisation : enjeux et questions. In *Conférence d'ouverture du Colloque Patrimonialização e sustentabilidade do património: reflexão e prospectiva*. NOVA University Lisbon, IHC-NOVA FCSH, Lisbon, 13–31 [Online]. Available at: https://www.academia.edu/11290454/À_propos_des_régimes_de_patrimonialisation_enjeux_et_questions.

Davallon, J. (2019a). Les musées, au cœur de la reconfiguration des patrimoines ? In *Musées, mutations…*, Le Marec, J., Schiele, B., Luckerhoff, J. (eds). Éditions universitaires de Dijon/Office de coopération et d'information muséales, Dijon.

Davallon, J. (2019b). De l'œuvre d'art à l'objet patrimonial. *Communication & langages*, 202, 133–152 [Online]. Available at: https://www.cairn.info/revue-communication-et-langages-2019-4-page-133.htm.

Davallon, J. (2019c). Les paysages, patrimoine européen ? *Culture & musées*, 33, 19–59 [Online]. Available at: https://journals.openedition.org/culturemusees/2748.

Davallon, J. (2022). Traduire un processus social en patrimoine. *Communication & langages*, 211, 31–51 [Online]. Available at: https://www.cairn.info/revue-communication-et-langages-2022-1-page-31.htm.

Davallon, J. and Jeanneret, Y. (2004). La fausse évidence du lien hypertexte. *Communication & langages*, 140, 43–54.

Davallon, J., Micoud, A., Tardy, C. (1997). Vers une évolution de la notion de patrimoine ? Réflexions à propos du patrimoine rural. In *L'esprit des lieux. Le patrimoine et la cité*, Grange, D. and Poulot, D. (eds). Presses universitaires de Grenoble, Grenoble.

Debary, O. (2002). *La fin du Creusot ou l'art d'accommoder les restes*. Éditions du Comité des Travaux Historiques et Scientifiques, Paris.

Delarge, A. (2000). Des écomusées, retour à la définition et évolution. *Publics & musées*, 17/18, 139–155 [Online]. Available at: https://www.persee.fr/doc/pumus_1164-5385_2000_num_17_1_1159.

Deleuze, G. (1986). *Foucault*. Éditions de Minuit, Paris.

Descamps, F. (2019). *Archiver la mémoire : de l'histoire orale au patrimoine immatériel*. École des hautes études en sciences sociales, Paris.

Descola, P. (2005). *Par-delà nature et culture*. Gallimard, Paris.

Descola, P. (2017 [2014]). *La composition des mondes. Entretiens avec Pierre Charbonnier*. Flammarion, Paris.

Desvallées, A. (1998). À l'origine du mot "patrimoine". In *Patrimoine et modernité*, Poulot, D. (ed.). L'Harmattan, Paris.

Desvallées, A. (2009). L'expologie de Georges Henri Rivière : des collectes systématiques aux unités écologiques : la question du contexte. In *Du folklore à l'ethnologie*, Boëll, D.M., Christophe, J., Meyran, R. (eds). Éditions de la Maison des sciences de l'homme, Paris. doi: 10.4000/books.editionsmsh.10050.

Desvallées, A. and Mairesse, F. (2011). *Dictionnaire encyclopédique de muséologie*. Armand Colin, Paris.

Dubost, F. (1994). *Vert patrimoine. La constitution d'un nouveau domaine patrimonial*. Éditions de la Maison des sciences de l'homme, Paris.

Duclos, J.-C. (1995). Place de l'écomusée dans l'évolution de la pensée muséologique. *MuzeoForum 1995*, 283–312 [Online]. Available at: https://www.etno-muzej.si/files/etnolog/pdf/0354-0316_5_duclos_vloga.pdf.

Duclos, J.-C. (2005). Depuis Rivière... *Le Monde alpin et rhodanien. Revue régionale d'ethnologie*, 1/4, 139–150 [Online]. Available at: www.persee.fr/doc/mar_0758-4431_2005_num_33_1_1883.

Duclos, J.-C. (2019). Quelques principes clefs des expositions de Georges Henri Rivière. *La lettre de l'OCIM*, 184, 18–21 [Online]. Available at: http://journals.openedition.org/ocim/2896.

Dulong, R. (1998). *Le témoin oculaire. Les conditions sociales de l'attestation personnelle.* École des hautes études en sciences sociales, Paris.

Eco, U. (1978). Pour une reformulation du concept de signe iconique. *Communications*, 29, 141–191 [Online]. Available at: https://www.persee.fr/doc/comm_0588-8018_1978_num_29_1_1438.

Eco, U. (1985 [1979]). *Lector in fabula : ou la coopération interprétative dans les textes narratifs.* Grasset & Fasquelle, Paris.

Eco, U. (1992 [1990]). *Les limites de l'interprétation.* Grasset & Fasquelle, Paris.

Eco, U. (1993). Observations sur la notion de gisement culturel. *Traverses*, 5, 9–18.

Erlande-Brandenburg, A. (1997). Notre-Dame de Paris. In *Les lieux de mémoire*, Nora, P. (ed.). Gallimard, Paris.

Ermisse, G. (1997 [1994]). L'Inventaire, aujourd'hui et demain. In *Entretiens du patrimoine. Science et conscience du patrimoine*, Nora, P. (ed.). Éditions du Patrimoine, Paris.

Farge, A. (1989). *Le goût de l'archive.* Le Seuil, Paris.

Fèvres, J. (2012). Infographie, images de synthèse et patrimoine monumental ; espace de représentation, espace de médiation. PhD Thesis, Université Michel de Montaigne – Bordeaux III, Bordeaux [Online]. Available at: https://theses.hal.science/tel-00751452.

Filipe, G. (2000). O Ecomuseu Municipal do Seixal no movimento renovador da museologia contemporânea em Portugal (1979–1999). Master's Thesis, NOVA University Lisbon, Lisbon.

Filipe, G. and de Varine, H. (2015). Que futuro para os ecomuseus? *Revista Al-madan*, II(19), 21–36 [Online]. Available at: https://www.academia.edu/21742401/Que_futuro_para_os_ecomuseus.

Flon, É. (2005). La patrimonialisation de l'archéologie : mise en scène des vestiges dans l'exposition. PhD Thesis, Université d'Avignon et des pays de Vaucluse, Avignon.

Flon, É. (2015). Les illustrations du passé archéologique : entre interprétation scientifique, témoignage et mémoire sociale. In *Mémoire et nouveaux patrimoines*, Tardy, C. and Dodebei, V. (eds). OpenEdition Press, Marseille [Online]. Available at: http://books.openedition.org/oep/411.

Flon, É. and Jeanneret, Y. (2010). La notion de schèmes organisateur, outil d'analyse sémiopragmatique des écrits d'écran. *Revue des interactions humaines médiatisées*, 11(1), 3–33.

Fournier, L.-S. (2011). La Tarasque métamorphosée. In *Le patrimoine culturel immatériel. Enjeu d'une nouvelle catégorie*, Bortolotto, C. (ed.). Éditions de la Maison des sciences de l'homme, Paris.

Fraysse, P. (2006). Le patrimoine monumental en images : des médiations informationnelles à la conversion monumentaire des documents. PhD Thesis, Université Toulouse 2, Toulouse.

Fumaroli, M. (1991). *L'état culturel : une religion moderne*. De Fallois, Paris.

Garcia, P. (2000). Les lieux de mémoire, une poétique de la mémoire ? *EspacesTemps*, 74/75, 122–142 [Online]. Available at: https://www.persee.fr/doc/espat_0339-3267_2000_num_74_1_4094.

Gauchet, M. (1977). La dette du sens et les racines de l'état. *Libre* 2, 5–43.

Gauchet, M. (1981). Des deux corps du roi au pouvoir sans corps : christianisme et politique. *Le débat*, 14/15, 147–168.

Geary, P. (1986). Sacred commodities: The circulation of medieval relics. In *The Social Life of Things. Commodities in Cultural Perspective*, Appadurai, A. (ed.). Cambridge University Press, Cambridge.

Genette, G. (1994). *L'œuvre de l'art. Immanence et transcendance*. Le Seuil, Paris.

Genette, G. (1997). *L'œuvre de l'art. La relation esthétique*. Le Seuil, Paris.

Gerbaud, M. (2000). Aux origines des écomusées : les premiers pas de Marquèze. *Publics & musées*, 17/18, 176–180 [Online]. Available at: https://www.persee.fr/doc/pumus_1164-5385_2000_num_17_1_1324.

Giovannoni, G. (1998 [1931]). *L'urbanisme face aux villes anciennes*. Le Seuil, Paris. Original publication (1931): *Vecchie città ed edilizia nuova*. UTET, Turin.

Givre, O. (2012). Savoirs et pouvoirs, stratégies et tactiques dans "l'arène patrimoniale" du nestinarstvo (Bulgarie). *Civilisations*, 61, 103–120.

Glévarec, H. (2003). De la société d'histoire à l'association du patrimoine : changements socioculturels et nouvelles raisons de mobilisation. In *Pour une histoire des politiques du patrimoine*, Poirrier, P. and Vadelorge, L. (eds). La Documentation française, Paris.

Glévarec, H. and Saez, G. (2002). *Le patrimoine saisi par les associations*. La Documentation française, Paris.

González Zarandona, J.A. (2011). The destruction of heritage: Rock art in the Burrup Peninsula. *The International Journal of the Humanities*, 9(1), 325–341 [Online]. Available at: https://www.academia.edu/1951264/The_Destruction_of_Heritage_Rock_Art_in_the_Burrup_Peninsula.

González Zarandona, J.A. (2013). Destruction of heritage or secular iconoclasm? The case of Dampier Archipelago Rock Art. In *The Challenge of the Object. 33rd Congress of the International Committee of the History of Art*. Germanisches National Museum, Nuremberg.

Goodman, N. (1990 [1968]). *Langages de l'art*. Éditions Jacqueline Chambon, Nîmes.

Goody, J. (1979 [1977]). *La raison graphique. La domestication de la pensée sauvage*. Éditions de Minuit, Paris. Original publication (1977): *The Domestication of the Savage Mind*. Cambridge University Press, Cambridge.

Goody, J. (1994 [1977]). Mémoire et apprentissage dans les sociétés avec et sans écriture : la transmission du Bagré. In *Entre l'oralité et l'écriture*. Presses Universitaires de France, Paris.

Gorgus, N. (2003 [1999]). *Le magicien des vitrines*. Éditions de la Maison des sciences de l'homme, Paris.

Greffe, X. (1990). *La valeur économique du patrimoine. La demande et l'offre de patrimoine*. Anthropos-Economica, Paris.

Greffe, X. (2014). *La trace et le rhizome. Les mises en scène du patrimoine culturel*. Presses de l'Université du Québec, Quebec.

Gregory, J. (2009). Stand up for the Burrup: Saving the largest Aboriginal Rock Art Precinct in Australia. *Public History Review*, 16, 92–116.

Guillaume, G. (1973 [1933]). Immanence et transcendance dans la catégorie du verbe. Esquisse d'une théorie psychologique de l'aspect. *Journal de psychologie*, 30, 355–372.

Guillaume, M. (1980). *La politique du patrimoine*. Éditions Galilée, Paris.

Guillaume, G. (1984). *Temps et verbe. Théorie des aspects, des modes et des temps, suivi de l'architectonique du temps dans les langues classiques*. Honoré Champion, Paris.

Guiomar, J.-Y. (1974). *L'idéologie nationale. Nation représentation propriété*. Éditions Champ Libre, Paris.

Gunter, C. (2019). La nouvelle muséologie comme mouvement politique : sa production et sa réception en France et au Québec. *Culture et gouvernance locale*, 6(1), 50–63. doi: 10.18192/clg-cgl.v6i1.4556.

Habermas, J. (1978 [1962]). *L'espace public. Archéologie de la publicité comme dimension constitutive de la société bourgeoise*. Payot, Paris.

Hafstein, V.T. (2011). Célébrer les différences, renforcer la conformité. In *Le patrimoine culturel immatériel. Enjeu d'une nouvelle catégorie*, Bortolotto, C. (ed.). Éditions de la Maison des sciences de l'homme, Paris.

Hafstein, V.T. (2018). *Making Intangible Heritage. El Condor Pasa and Other Stories from UNESCO*. Indiana University Press, Bloomington.

Hammad, M. (2015). *Sémiotiser l'espace : décrypter architecture & archéologie. Essais sémiotiques*. Geuthner, Paris.

Hammou, K. (2012). Les Mondes de l'art comme activité collective. In *Howard Becker et les mondes de l'art*, Benghozi, J.-P. and Paris, T. (eds). Éditions de l'École Polytechnique, Paris.

Harris, R. (1993). *La sémiologie de l'écriture*. Éditions du CNRS, Paris.

Harrison, R. (2013). *Heritage: Critical Approaches*. Routledge, Abingdon-on-Thames.

Hartog, F. (2003). *Régimes d'historicité. Présentisme et expérience du temps*. Le Seuil, Paris.

Heinich, N. (2009). *La fabrique du patrimoine. "De la cathédrale à la petite cuillère"*. Éditions de la Maison des sciences de l'homme, Paris.

Heinich, N. (2012). Le patrimoine culturel immatériel. Enjeux d'une nouvelle catégorie. *Gradhiva*, 15, 227–229.

Heinich, N. (2017). *Des valeurs. Une approche sociologique*. Gallimard, Paris.

Hottin, C. (2012). À la recherche du patrimoine immatériel : tâtonnements, tactiques et stratégies pour la mise en œuvre par la France de la Convention de 2003. In *Les mesures de soutien au patrimoine immatériel*, Gauthier, A. (ed.). Conseil québécois du patrimoine vivant, Quebec [Online]. Available at: https://api.patrimoinevivant.qc.ca/content/uploads/2020/09/les-mesures-de-soutien-au-patri-moine-immateriel-actes-du-colloque-2011.pdf.

Hottin, C. (2020 [2012]). Le patrimoine culturel immatériel en France en 2015 : autour de trois questions lancinantes. In *Le patrimoine culturel immatériel au seuil des sciences sociales. Actes du colloque de Cerisy-la-Salle, septembre 2012*, Csergo, J., Hottin, C., Schmit, P. (eds). Éditions de la Maison des sciences de l'homme, Paris [Online]. Available at: http://books.openedition.org/editionsmsh/15990.

ICOMOS Australia (2013 [1979]). *Charter for the Conservation of Places of Cultural Significance (Burra Charter)* [Online]. Available at: https://australia.icomos.org/wp-content/uploads/The-Burra-Charter-2013-Adopted-31.10.2013.pdf.

ICOMOS France (2010 [2009]). Valeurs universelles, valeurs locales : pour qui, pour quoi un site est-il grand ? In *Actes colloque Icomos-France*. ICOMOS, Paris.

Idjéraoui-Ravez, L. (2004). Paroles et objets : mémoires d'immigrés. Ethnologie ou muséologie ? In *Le goût des belles choses, ethnologie de la relation esthétique*, Nahoum-Grappe, V. and Vincent, O. (eds). Éditions de la Maison des sciences de l'homme, Paris.

Idjéraoui-Ravez, L. (2012). *Le témoignage exposé : du document à l'objet médiatique*. L'Harmattan, Paris.

Inventaire général du patrimoine culturel (2012). Compte-rendu du séminaire pluridisci-plinaire. Report, Ministère de la Culture et de la communication, Paris [Online]. Available at: https://aapq.org/bibliotheque/vers-la-prise-en-compte-des-paysages-culturels-dans-le-cadre-des-operations-dinventaire.

Jeannelle, J.-L. (2014). Le Kamtchatka patrimonial. *Critique*, 805/806, 494–507.

Jeanneret, Y. (2008). *Penser la trivialité. Volume 1. La vie triviale des êtres culturels*. Hermès-Lavoisier, Paris.

Jeanneret, Y. (2014). *Critique de la trivialité. Les médiations de la communication, enjeu de pouvoir*. Éditions Non Standard, Paris.

Jeanneret, Y. (2020). *The Trace Factory*. ISTE Ltd, London, and John Wiley & Sons, New York.

Jeudy, H.-P. (1977). *La publicité et son enjeu social*. Presses Universitaires de France, Paris.

Jeudy, H.-P. (1986). *Mémoires du social*. Presses Universitaires de France, Paris.

Jeudy, H.-P. (1989). *Les ruses de la communication. L'euthanasie des sages*. Plon, Paris.

Jeudy, H.-P. (2001). *La machinerie patrimoniale*. Sens & Tonka, Paris.

Jeudy, H.-P. (2003 [2001]). Inflation patrimoniale et politiques culturelles. In *Entretiens du patrimoine : le regard de l'histoire. L'émergence et l'évolution de la notion de patrimoine au cours du XXe siècle en France*, Rousso, H. (ed.). Fayard, Paris.

Kirshenblatt-Gimblett, B. (1998). *Destination Culture. Tourism, Museums and Heritage*. University of California Press, Berkeley.

Kirshenblatt-Gimblett, B. (2004). Intangible heritage as metacultural production. *Museum International*, 221/222, 52–65.

Kirshenblatt-Gimblett, B. (2006). World heritage and cultural economics. In *Museum Frictions. Public Cultures/Global Transformations*, Karp, I., Kratz, C.A., Szwaja, L., Ybarra-Frausto, T. (eds). Duke University Press, Durham.

Kopytoff, I. (1986). The cultural biography of things: Commoditization as process. In *The Social Life of Things: Commodities in Cultural Perspective*, Appadurai, A. (ed.). Cambridge University Press, Cambridge.

Lamy, Y. (ed.) (1992). *Le pouvoir de protéger. Approches, acteurs, enjeux du patrimoine en Aquitaine*. Maison des Sciences de l'Homme d'Aquitaine, Talence.

Lamy, Y. (1993). Du monument au patrimoine. Matériaux pour l'histoire politique d'une protection. *Genèses*, 11, 50–81 [Online]. Available at: http://www.persee.fr/doc/genes_1155-3219_1993_num_11_1_1171.

Lamy, Y. (2003). Patrimoine et culture : l'institutionnalisation. In *Pour une histoire des politiques du patrimoine*, Poirrier, P. and Vadelorge, L. (eds). La Documentation française, Paris.

Larrère, C. (2018). Protection de la nature et communs : allers et retours du pays au paysage. *Les carnets du paysage*, 33, 65–77.

Laurent, X. (2016). Naissance et développement de l'Inventaire. Texte repris du hors-série no. 1 de Présence d'André Malraux, 2004 : actes de la journée d'étude consacrée à "Malraux et l'Inventaire général des monuments et des richesses artistiques de la France". *BnF* [Online]. Available at: https://malraux.org/pamhs1-2/.

Lavorel, M. (2014). Patrimonialiser les mémoires sensibles. Héritage culturel et muséologie. PhD Thesis, Université d'Avignon et des pays de Vaucluse/Université du Québec à Montréal, Avignon/Montreal [Online]. Available at: https://tel.archives-ouvertes.fr/tel-01168604.

Le Goff, J. (1997 [1986]). Reims : ville du sacre. In *Les lieux de mémoire*, Nora, P. (ed.). Gallimard, Paris.

Le Marec, J. (2002). Ce que le "terrain" fait aux concepts. Vers une théorie des composites. PhD Thesis, Université Paris 7, Paris.

Le Marec, J. (2007). *Publics et musés : la confiance éprouvée*. L'Harmattan, Paris.

Le Marec, J. and Babou, I. (2003). De l'étude des usages à une théorie des composites : objets, relations et normes en bibliothèque. In *Lire, écrire, récrire : objets, signes et pratiques des médias informatisés*, Souchier, E., Le Marec, J., Jeanneret, Y. (eds). Éditions de la BPI-Centre Pompidou, Paris.

Le Nouenne, P. (1992 [1978]). Un écomusée, ce n'est pas un musée comme les autres. In *Vagues. Une anthologie de la nouvelle muséologie*, Desvallées, A. (ed.). Éditions W/MNES, Mâcon/Savigny le Temple.

Ledoux, S. (2017). La mémoire, mauvais objet de l'historien ? *Vingtième Siècle. Revue d'histoire*, 133, 113–128 [Online]. Available at: https://www.cairn.info/revue-vingtieme-siecle-revue-d-histoire-2017-1-page-113.htm.

Ledoux, S. (2019). La genèse des lieux de mémoire et leur retraduction dans les politiques du passé en France (années 1970–1980). *Cahiers mémoire et politique*, 6, 11–33 [Online]. Available at: https://popups.uliege.be/2295-0311/index.php?id=232.

Leleu-Merviel, S. (2018). *Informational Tracking*. ISTE Ltd, London, and John Wiley & Sons, New York.

Lenclud, G. (1987). La tradition n'est plus ce qu'elle était... Sur les notions de tradition et de société traditionnelle en ethnologie. *Terrain*, 9, 110–123 [Online]. Available at: http://terrain.revues.org/3195.

Lenclud, G. (1995). L'ethnologie et le paysage : questions sans réponses. In *Paysage au pluriel. Pour une approche ethnologique des paysages*, Voisenat, C. (ed.). Éditions de la Maison des sciences de l'homme, Paris [Online]. Available at: http://books.openedition.org/editionsmsh/654.

Leniaud, J.-M. (1992). *L'utopie française. Essai sur le patrimoine*. Mengès, Paris.

Leveau, P. (2012). Épistémologie de la conservation du patrimoine. Ontologie d'un domaine, ergologie d'une discipline. PhD Thesis, Aix-Marseille Université, Marseille.

Leveau, P. (2017). *L'institution de la conservation du patrimoine culturel dans l'entredeux-guerres*. Office de coopération et d'information muséales, Dijon.

Livet, P. (2018). Patrimoine culturel immatériel et processus sociaux. *Nouvelle revue d'esthétique*, 21, 61–72 [Online]. Available at: https://www.cairn.info/revue-nouvelle-revue-d-esthetique-2018-1-page-61.htm.

Livet, P. and Nef, F. (2009). *Les êtres sociaux. Processus et virtualité*. Hermann, Paris.

Lobrichon, G. (1997 [1992]). Vezelay. In *Les lieux de mémoire III : les France*, Nora, P. (ed.). Gallimard, Paris.

Losonczi, A.-M. (2019). De l'efficacité des ruines : patrimonialisation et sanctification de Bellavista, village noir détruit (Chocó, Colombie). In *Le patrimoine comme expérience. Implications anthropologiques*, Tornatore, J.-L. (ed.). Éditions de la Maison des sciences de l'homme, Paris.

Macchiarella, I. (2011). Sauvegarder l'oralité ? Le cas du canto a tenore. In *Le patrimoine culturel immatériel. Enjeu d'une nouvelle catégorie*, Bortolotto, C. (ed.). Éditions de la Maison des sciences de l'homme, Paris.

Maguet, F. (2011). L'image des communautés dans l'espace public. In *Le patrimoine culturel immatériel. Enjeu d'une nouvelle catégorie*, Bortolotto, C. (ed.). Éditions de la Maison des sciences de l'homme, Paris.

Mairesse, F. (2000). La belle histoire, aux origines de la nouvelle muséologie. *Publics & musées*, 17/18, 33–56 [Online]. Available at: https://www.persee.fr/doc/pumus_1164-5385_2000_num_17_1_1154.

Mairesse, F. (ed.). (2022). *Dictionnaire de muséologie*. Armand Colin, Paris.

Marin, L. (1977). *Détruire la peinture*. Éditions Galilée, Paris.

Martel, M.D. (2005 [2003]). Genette ou Goodman, ou la transcendance dans l'œuvre littéraire. In *Actes du Colloque Langage, Pensée, Action de la SoPhA, septembre 2003*. Public@tions Electroniques de Philosophia Scientiae, Montreal [Online]. Available at: https://sopha.univ-paris1.fr/fichiers/pdf/2003/17_martel.pdf.

Marymor, L. (2018). Australian Rock Art bibliography extracted from the Rock Art Studies Bibliographic Database for the years 1841 to 2018 – Part 1. *Rock Art Research*, 35(2), 188–248.

Marymor, L. (2019). Australian Rock Art bibliography extracted from the Rock Art Studies Bibliographic Database for the years 1841 to 2018 – Part 2. *Rock Art Research*, 36(1), 49–93.

Mayeur, J.-M. (1997 [1986]). Une mémoire frontière : L'Alsace. In *Les lieux de mémoire*, Nora, P. (ed.). Gallimard, Paris.

Mazé, C., Poulard, F., Ventura, C. (2013). Démantèlement, reconversions, créations. In *Les musées d'ethnologie : culture, politique et changement institutionnel*, Mazé, C., Poulard, F., Ventura, C. (eds). Comité des travaux historiques et scientifiques, Paris.

McDonald, J. and Veth, P. (2009). Dampier Archipelago petroglyphs: Archaeology, scientific values and National Heritage Listing. *Archaeology in Oceania*, 44, 49–69 [Online]. Available at: https://www.academia.edu/2695178/Dampier_Ar-chipelago_petroglyphs_archaeology_scientific_values_and_National_Heritage_Listing [Accessed 21 April 2009].

McDonald, J. and Veth, P. (2011). Study of the outstanding universal values of the Dampier Archipelago Site, Western Australia. Report to the Australian Heritage Council [Online]. Available at: https://www.academia.edu/10237269/Study_of_the_Outstanding_Universal_Values_of_The_Dampier_Archipelago_Site_Western_Australia [Accessed 5 March 2022].

Mejía Ayala, W. (2014). Les problématiques et enjeux de la mise en tourisme du patrimoine culturel immatériel en Amazonie : l'exemple du peuple indigène Sápara d'Équateur. PhD Thesis, Université d'Angers, Angers [Online]. Available at: https://www.academia.edu/39757098/Les_problématiques_et_enjeux_de_la_mise_en_tourisme_du_patrimoine_culturel_immatériel_en_Amazonie_Lexemple_du_peuple_indigène_Sápara_dÉquateur.

Mejía Ayala, W. (2018 [2016]). Les problématiques et enjeux de la mise en tourisme du patrimoine culturel immatériel en Amazonie : le cas du peuple Sépara d'Équateur. In *Actes des 3e Rencontres internationales du patrimoine culturel immatériel en Bretagne*, 111–121 [Online]. Available at: https://www.academia.edu/39757787/Les_problématiques_et_enjeux_de_la_mise_en_tourisme_du_patrimoine_culturel_immatériel_en_Amazonie_le_cas_du_peuple_Sépara_dÉquateur.

Melot, M. (2005). Le grand inventaire. *In Situ*, 6 [Online]. Available at: http://insitu.revues.org/8797.

Melot, M. (2012). *Mirabilia. Essai sur l'Inventaire général du patrimoine culturel.* Gallimard, Paris.

Micoud, A. (1995 [1994]). Le Bien Commun des patrimoines. In *Actes du colloque des 12 et 13 décembre 1994*. La Documentation française/École nationale du Patrimoine, Paris.

Micoud, A. (1996). Musée et patrimoine : deux types de rapport aux choses et au temps ? *Hermès*, 20, 115–123 [Online]. Available at: https://www.cairn.info/revue-hermes-la-revue-1996-2-page-115.htm.

Micoud, A. (2005). La patrimonialisation ou comment dire ce qui nous relie (un point de vue sociologique). In *Réinventer le patrimoine : de la culture à l'économie, une nouvelle pensée du patrimoine ?* Barrère, C., Barthélémy, D., Nieddu, M., Vivien, F.-D. (eds). L'Harmattan, Paris.

Montpetit, R. (1996). Une logique d'exposition populaire : les images de la muséographie analogique. *Publics & musées*, 9, 55–103. doi: 10.3406/pumus.1996.1071.

Morsel, J. (2016). Traces ? Quelles traces ? Réflexions pour une histoire non passéiste. *Revue historique*, 680, 813–868 [Online]. Available at: https://www.cairn.info/revue-historique-2016-4-page-813.htm.

Nas, P.J.M. (2002). Masterpieces of oral and intangible culture. Reflections on the UNESCO World Heritage List. *Current Anthropology*, 43(1), 139–148.

Navarro, N. (2015). Le patrimoine métamorphe. Circulation et médiation du patrimoine urbain dans les Villes et pays d'art et d'histoire. PhD Thesis, Université d'Avignon et des pays de Vaucluse/Université du Québec à Montréal/École du Louvre, Avignon/Montreal/Paris.

Nikočević, L., Gavrilovic, L., Hrovatin, M., Jelincic, D.A. (2012). Discussion "culture or heritage? The problem of intangibility". *Etnološka tribina*, 35(42), 57–112 [Online]. Available at: https://www.academia.edu/56233358/Culture_or_Heritage_The_Problem_of_Intangibility.

Nora, P. (ed.) (1997). Préface à l'édition "Quarto". In *Les lieux de mémoire 1*. Gallimard, Paris.

Nora, P. (ed.) (1997 [1984]a). Présentation. In *Les lieux de mémoire*. Gallimard, Paris.

Nora, P. (ed.) (1997 [1984]b). Entre mémoire et histoire. La problématique des lieux. In *Les lieux de mémoire*. Gallimard, Paris.

Nora, P. (ed.) (1997 [1984]c). *Les lieux de mémoire I : la République*. Gallimard, Paris.

Nora, P. (ed.) (1997 [1986]a). *Les lieux de mémoire II : la Nation*, volumes I & II. Gallimard, Paris.

Nora, P. (ed.) (1997 [1986]b). *Les lieux de mémoire II : la Nation*, volume III. Gallimard, Paris.

Nora, P. (ed.) (1997 [1986]c). Le patrimoine [présentation]. In *Les lieux de mémoire*. Gallimard, Paris.

Nora, P. (ed.) (1997 [1992]a). Comment écrire l'histoire de France ? In *Les lieux de mémoire*. Gallimard, Paris.

Nora, P. (ed.) (1997 [1992]b). L'ère de la commémoration. In *Les lieux de mémoire*. Gallimard, Paris.

Nora, P. (ed.) (1997 [1992]c). *Les lieux de mémoire III : les France*, volume I. Gallimard, Paris.

Nora, P. (ed.) (1997 [1992]d). *Les lieux de mémoire III : les France*, volumes II & III. Gallimard, Paris.

Nora, P. (ed.) (1997 [1994]). Introduction des entretiens. In *Entretiens du patrimoine : science et conscience du patrimoine*. Arthème Fayard/Éditions du Patrimoine, Paris.

Nora, P. (2011 [1996]). Les lieux de mémoire, mode d'emploi. In *Présent, nation, mémoire*. Gallimard, Paris.

Nora, P. (2011 [2006]). L'explosion du patrimoine. In *Présent, nation, mémoire*. Gallimard, Paris.

Nora, P. (2011 [2008]). Les lieux de mémoire, ou comment ils m'ont échappé. In *Présent, nation, mémoire*. Gallimard, Paris.

Noyes, D. (2006). The judgment of Solomon. Global protections for tradition and the problem of community ownership. *Cultural analysis*, 5, 27–55 [Online]. Available at: http://socrates.berkeley.edu/≈caforum/volume5/volume5_article2.html.

Noyes, D. (2011). La fête ou le fétiche, le geste ou la gestion. In *Le patrimoine culturel immatériel. Enjeu d'une nouvelle catégorie*, Bortolotto, C. (ed.). Éditions de la Maison des sciences de l'homme, Paris.

Olivier, L. (2008). *Le sombre abîme du temps. Mémoire et archéologie*. Le Seuil, Paris.

Paquot, T. (2016). *Le paysage*. La Découverte, Paris.

Passeron, R. (2006). *Le raisonnement sociologique : l'espace non poppérien de l'argumentation*. Albin Michel, Paris.

Pavis, P. (1983). Production et réception au théâtre : la concrétisation du texte dramatique et spectaculaire. *Revue des sciences humaines*, 189, 51–88.

Pianezza, N. (2017). La patrimonialisation selon l'immatériel ou la mémoire agissante : circulations des savoirs en contexte partenarial de production audiovisuelle. PhD Thesis, Université d'Avignon et des Pays de Vaucluse/Unirio, Avignon/Rio de Janeiro [Online]. Available at: https://tel.archives-ouvertes.fr/tel-01774377.

Poirrier, P. (2003). Politique affichée/politique discutée. Patrimoine versus "tout culturel" 1981–1993. In *Pour une histoire des politiques du patrimoine*, Poirrier, P. and Vadelorge, L. (eds). La Documentation française, Paris.

Poirrier, P. (2003 [2001]). L'évolution de la notion de patrimoine dans les politiques culturelles menées en France sous la Ve République. In *Le regard de l'histoire. L'émergence et l'évolution de la notion de patrimoine au cours du XXe siècle en France*, Rousso, H. (ed.). Fayard, Paris.

Pomian, K. (1981). Musée archéologique : art, nature, histoire. *Le débat*, 12, 57–68.

Pomian, K. (1987 [1978]). Entre l'invisible et le visible : la collection. In *Collectionneurs, amateurs et curieux. Paris, Venise : XVIe–XVIIIe siècles*. Gallimard, Paris.

Pomian, K. (1999 [1996]). Histoire culturelle, histoire des sémiophores. In *Sur l'histoire*, Pomian, K. (ed.). Gallimard, Paris.

Pomian, K. (2003). *Des saintes reliques à l'art moderne. Venise-Chicago : XIIIe–XXe siècle*. Gallimard, Paris.

Pouivet, R. (2012). L'ontologie de régime de Gérard Genette. *Aisthesis. Pratiche, linguaggi e Saperi dell'estetico*, 4(1), 39–49. doi: 10.13128/Aisthesis-10981.

Poulot, D. (1988). The birth of heritage: "le moment Guizot". *The Oxford Art Journal*, 11(2), 40–56.

Poulot, D. (1992). Le patrimoine universel : modèle culturel français. *Revue d'histoire moderne et contemporaine*, 39(1), 29–55.

Poulot, D. (1997). *Musée, nation, patrimoine : 1789–1815*. Gallimard, Paris.

Poulot, D. (2006). *Une histoire du patrimoine en Occident, XVIIIe–XXIe siècle : du monument aux valeurs*. Presses Universitaires de France, Paris.

Poulot, D. (2009). Le patrimoine immatériel en France entre renouveau muséographique et "territoire de projet". *Ethnologies*, 31(1), 165–200.

Quéré, L. (1982). *Des miroirs équivoques. Aux origines de la communication moderne*. Aubier, Paris.

Raulin, H. (1977). *L'architecture rurale française : corpus des genres, des types et des variantes. Savoie*. Berger-Levrault, Paris.

Rautenberg, M. (1997 [1994]). La connaissance du patrimoine : le métier de conseiller à l'ethnologie dans une direction régionale des affaires culturelles. In *Entretiens du patrimoine : science et conscience du patrimoine*, Nora, P. (ed.). Fayard, Paris.

Rautenberg, M. (1998). L'émergence patrimoniale de l'ethnologie : entre mémoire et politiques publiques. In *Patrimoine et modernité*, Poulot, D. (ed.). L'Harmattan, Paris.

Rautenberg, M. (2003a). *La rupture patrimoniale*. Éditions La Croisée, Paris.

Rautenberg, M. (2003b). L'intervention ethnologique. Témoignage et éléments de réflexion sur les relations entre recherche et action culturelle dans une direction régionale des affaires culturelles. In *Pour une histoire des politiques du patrimoine*, Poirrier, P. and Vadelorge, L. (eds). La Documentation française, Paris.

Rautenberg, M. (2004). La patrimonialisation, entre appropriation sociale et désignation institutionnelle. In *L'effet géographique. Construction sociale, appréhension cognitive et configuration matérielle des objets géographiques*, Debarbieux, B. and Fourny, M.-C. (eds). CNRS-MSH-Alpes, Lyon.

Renfrew, C. (1986). Varna and the emergence of wealth in prehistoric Europe. In *The Social Life of Things: Commodities in Cultural Perspective*, Appadurai, A. (ed.). Cambridge University Press, Cambridge.

Ricœur, P. (1983). *Temps et récit. L'intrigue et le récit historique*. Le Seuil, Paris.

Ricœur, P. (2000). *La mémoire, l'histoire, l'oubli*. Le Seuil, Paris.

Riegl, A. (1984 [1903]). *Le culte moderne des monuments. Son essence et sa genèse*. Le Seuil, Paris.

Rivière, G.H. (ed.) (1989). *La Muséologie selon Georges Henri Rivière. Cours de muséologie, textes et témoignages*. Dunod, Paris.

Rivière, G.H. (1992 [1980]). L'écomusée, un modèle évolutif (1971–1980). In *Vagues. Une anthologie de la nouvelle muséologie*, Desvallées, A. (ed.). Éditions W/MNES, Mâcon/Savigny le Temple.

Roger, A. (1978). *Nus et paysages. Essai sur la fonction de l'art*. Aubier, Paris.

Roger, A. (1997). *Court traité du paysage*. Gallimard, Paris.

Rojon, S. (2019). Les productions numériques du patrimoine par des amateurs en réseaux : le cas des pratiques d'exploration et d'exposition de friches industrielles. In *Le patrimoine comme expérience : implications anthropologiques*, Tornatore, J.-L. (ed.). Éditions de la Maison des sciences de l'homme, Paris.

Rolland-Villemot, B. (2016). Unités écologiques, "period rooms" : des ensembles d'objets mobiliers, de la collecte à la valorisation muséographique. *In Situ*, 29 [Online]. Available at: http://journals.openedition.org/insitu/13373.

Rosanvallon, P. (1985). *Le moment Guizot*. Gallimard, Paris.

Sagne, S. (2019). Au mirorir de la médiation : le présent du patrimoine. In *Le patrimoine comme expérience : implications anthropologiques*, Tornatore, J.-L. (ed.). Éditions de la Maison des sciences de l'homme, Paris.

Sandroni, C. (2011). L'ethnomusicologue en médiateur du processus patrimonial : le cas de la samba de roda. In *Le patrimoine culturel immatériel. Enjeu d'une nouvelle catégorie*, Bortolotto, C. (ed.). Éditions de la Maison des sciences de l'homme, Paris.

Santos, S. (2017). Ecomuseus e museus comunitários no Brasil: estudo exploratório de possibilidades museológicas. Master's Thesis, Universidade de São Paulo, São Paulo.

Schaeffer, J.-M. (1987). *L'image précaire. Du dispositif photographique*. Le Seuil, Paris.

Schaeffer, J.-M. (1999). *Pourquoi la fiction ?* Le Seuil, Paris.

Schall, C. (2010). La médiation de l'archéologie à la télévision. La construction d'une relation au passé. PhD Thesis, Université d'Avignon et des pays de Vaucluse/Université du Québec à Montréal, Avignon/Montreal [Online]. Available at: https://www.academia.edu/65745087/La_médiation_de_l_archéologie_à_la_télévision_la_construction_d_une_relation_au_passé.

Schall, C. and Lemaître, S. (2010). Filmer l'archéologie : documentaire versus docufiction. *Les nouvelles de l'archéologie*, 122, 24–29.

Schapiro, M. (1982 [1969]). Sur quelques problèmes de sémiotique de l'art visuel : champ et véhicule dans les signes iconiques. In *Style, artiste et société*. Gallimard, Paris.

Searle, J.R. (1995). *The Construction of Social Reality*. Free Press, New York.

Segalen, M. (2005). *Vie d'un musée 1937–2005*. Éditions Stock, Paris.

Segalen, M. (2008). Des ATP au Mucem : exposer le social. *Ethnologie française*, 38, 639–644.

Segalen, M. (2010). L'enquête de la RCP Aubrac (1963–1966) : une stratégie intellectuelle, un enjeu institutionnel. In *En France rurale*, Simon, J.-F., Paillard, B., Le Gall, F. (eds). Presses Universitaires de Rennes, Rennes [Online]. Available at: https://books.openedition.org/pur/102770.

Severo, M. (2022). Le patrimoine culturel immatériel entre écritures amateure et institutionnelle : le cas de l'inventaire français. *Communication & langages*, 211, 53–70 [Online]. Available at: https://www.cairn.info/revue-communication-et-langa-ges-2022-1-page-53.htm.

Severo, M. and Cachat, S. (eds) (2016). *Patrimoine culturel immatériel et numérique. Transmission, participation, enjeux*. L'Harmattan, Paris.

Sgard, A. (2008). Quels enjeux pour une action publique au nom du paysage : du patrimoine au bien commun. In *Colloque international de l'ASRDLF*, August, Rimouski [Online]. Available at: https://shs.hal.science/halshs-00326373/document.

Sgard, A. (2010). Le paysage dans l'action publique : du patrimoine au bien commun. *Développement durable et territoires*, 1(2) [Online]. Available at: http://developpementdurable.revues.org/8565.

Sitte, C. (1996 [1889]). *L'art de bâtir les villes. L'urbanisme selon ses fondements artistiques*. Le Seuil, Paris.

Skounti, A. (2012). Réflexions sur les critères d'inscription sur les listes de la Convention du patrimoine culturel immatériel. In *The First ICH Researchers Forum: The Implementation of UNESCO's 2003 Convention*. IRCI, Osaka.

Skounti, A. and Tebbaa, O. (eds) (2011). *De l'immatérialité du patrimoine culturel*. Bureau régional de l'UNESCO/Université Cadi Ayyad, Rabat/Marrakech [Online]. Available at: https://www.academia.edu/4058994/De_limmaterialite_du_patri-moine_culturel.

Smith. L. (1999). The last archaeologist? Contested identities. *Australian Aboriginal Studies*, 2, 25–34.

Smith, L. (2006). *Uses of Heritage*. Routledge, Abingdon-on-Thames.

Smith, L. (2011). All heritage is intangible: Critical heritage studies and museums. In *Reinwardt Memorial Lecture* [Online]. Available at: https://www.ahk.nl/fileadmin/download/reinwardt/lectoraat/All_heritage_is_intangible.pdf.

Smith, L. (2014). Patrimoni immaterial: un repte per al discurs de patrimoni autoritzat? *Revista d'Etnologia de Catalunya*, 39, 12–22.

Smith, L. and Akagawa, N. (2009). *Intangible Heritage*. Routledge, Abingdon-on-Thames.

Smith, L. and Campbell, G. (2017). The tautology of "Intangible values" and the misrecognition of intangible cultural heritage. *Heritage & Society* 10(1), 26–44 [Online]. Available at: https://www.academia.edu/35563575/The_tautology_of_Intan-gible_values_and_the_misrecognition_of_intangible_cultural_heritage.

Smith, L. and Watterton, E. (2009). Introduction: Heritage and archaeology. In *Taking Archaeology Out of Heritage*, Watterton, E. and Smith, L. (eds). Cambridge Scholars Publishing, Newcastle.

Soulages, J.-C. (2005). *Les mises en scène visuelles de l'information. Étude comparée France, Espagne, États-Unis*. Armand Colin, Paris.

Soulier, V. and Roigé, X. (2022). Comment valoriser le patrimoine culturel immatériel via un musée numérique ? Le projet Prometheus.museum. *Communication & langages*, 211, 87–109 [Online]. Available at: https://www.cairn.info/revue-communica-tion-et-langages-2022-1-page-87.htm.

Tardy, C. (2009). Introduction. *Culture & musées*, 14, 13–18 [Online]. Available at: https://www.persee.fr/doc/pumus_1766-2923_2009_num_14_1_1504.

Tardy, C. (2018 [2014]). La patrimonialisation du point de vue du vivant. In *Patrimonialização e sustentabilidade do património: reflexão e prospectiva*, Filipe, G., Vale, J., Castaño, I. (eds). IHC-NOVA FCSH, Lisbon [Online]. Available at: https://www.academia.edu/11290454/À_propos_des_régimes_de_patri-monialisation_enjeux_et_questions.

Tardy, C. and Rautenberg, M. (2013). Patrimoines culturel et naturel : analyse des patrimonialisations. *Culture & musées*, 115–138. doi: 10.4000/culturemusees.734.

Tauschek, M. (2012). Filmer le patrimoine, produire de la valeur. *Ethnographiques.org*, 24 [Online]. Available at: https://www.ethnographiques.org/2012/Tauschek.

Tobelem, J.-M. (2005). *Le nouvel âge des musées. Les institutions culturelles au défi de la gestion*. Armand Colin, Paris.

Tolentino, Á.B. (2016). Museologia social: apontamentos históricos e conceituais. *Cadernos de Sociomuseologia*, 52(8), 21–44.

Tomas, F. (1994). Du paysage aux paysages, pour une autre approche paysagère. *Revue de géographie de Lyon*, 69(4), 277–286 [Online]. Available at: https://www.persee.fr/doc/geoca_0035-113x_1994_num_69_4_4270.

Tomas, F. (2004). Les temporalités du patrimoine et de l'aménagement urbain. *Géocarrefour*, 79(3), 197–212 [Online]. Available at: http://journals.openedi-tion.org/geocarrefour/722.

Tornatore, J.-L. (2010a). Les banquets de la mémoire à l'auberge du patrimoine. In *L'invention de la Lorraine industrielle. Quêtes de reconnaissance, politiques de la mémoire*, Tornatore, J.-L. (ed.). Riveneuve, Paris [Online]. Available at: https://www.academia.edu/12910414/Les_banquets_de_la_mémoire_à_lauberge_du_patrimoine.

Tornatore, J.-L. (2010b). L'esprit de patrimoine. *Terrain*, 55 [Online]. Available at: http://terrain.revues.org/index14084.html.

Tornatore, J.-L. (2011). Du patrimoine ethnologique au patrimoine culturel immatériel : suivre la voie politique de l'immatérialité culturelle. In *Le patrimoine culturel immatériel. Enjeu d'une nouvelle catégorie*, Bortolotto, C. (ed.). Éditions de la Maison des sciences de l'homme, Paris.

Tornatore, J.-L. (2012). Retour d'anthropologie : "le repas gastronomique des Français". Éléments d'ethnographie d'une distinction patrimoniale. *Ethnographiques.org*, 24, [Online]. Available at: http://www.ethnographiques.org/2012/.

Tornatore, J.-L. (2014). Des noms pour dire ce à quoi nous tenons : continuité et exté-riorité de l'expérience patrimoniale. In *Les vocabulaires locaux du "patrimoine". Traductions, négociations et transformations*, Bondaz, J., Graezer-Bideau, F., Isnart, C., Leblon, A. (eds). Lit Verlag, Berlin/Münster/Vienna/Zurich/London [Online]. Available at: https://www.academia.edu/10837705/Des_noms_pour_dire_ce_que_à_quoi_nous_tenons_Continuité_et_extériorité_de_lexpérience_patrimoniale.

Tornatore, J.-L. (2017a). Questions de démocratie patrimoniale. In *De la participation à la co-construction des patrimoines urbains. L'invention du commun ?* Auclair, E., Herzog, A., Poulot, M.-L. (eds). Éditions le Manuscrit, Paris.

Tornatore, J.-L. (2017b). Patrimoine vivant et contributions citoyennes. Penser le patrimoine "devant" l'Anthropocène. *In Situ*, 33 [Online]. Available at: http://insitu.revues.org/15606.

Tornatore, J.-L. (2019a). Expériencer le patrimoine. In *Le patrimoine comme expérience. Implications anthropologiques*, Tornatore, J.-L. (ed.). Éditions de la Maison des sciences de l'homme, Paris.

Tornatore, J.-L. (2019b). Pour une anthropologie pragmatiste et plébéienne du patrimoine : un scénario contre-hégémonique. *In Situ*, 1 [Online]. Available at: https://journals.openedition.org/insituarss/449.

Tornatore, J.-L. and Sébastien, P. (2003). Publics ou populations ? La démocratie culturelle en question de l'utopie écomuséale aux "espaces intermédiaires". In *Le(s) public(s) de la culture : politiques publiques et équipements culturels*, Donnat, O. and Tolila, P. (eds). Presses de Sciences Po, Paris.

Toulier, B. (1997 [1994]). Les arpenteurs du patrimoine. In *Entretiens du patrimoine : science et conscience du patrimoine*, Nora, P. (ed.). Arthème Fayard/CNMHS–Éditions du Patrimoine, Paris.

Toulier, B. (2016). De la naissance à la refondation de l'Inventaire général du patrimoine culturel. In *Une aventure de l'esprit. L'Inventaire général du patrimoine culturel.* Éditions Lieux Dits, Riotord.

Truscott, M. (2003). Valeurs immatérielles – patrimoine d'Australie. Speech. In *14e Assemblée générale et Symposium scientifique d'Icomos*, Victoria Falls [Online]. Available at: https://www.icomos.org/victoriafalls2003/truscott_fre.htm.

UNESCO (2018). *Textes fondamentaux de la Convention de 2003 pour la sauvegarde du patrimoine culturel immatériel.* UNESCO, Paris.

UNESCO (2021). *Textes fondamentaux de la Convention du Patrimoine mondial de 1972.* UNESCO, Paris.

Vadelorge, L. (2003). Le patrimoine comme objet politique. In *Pour une histoire des politiques du patrimoine*, Poirrier, P. and Vadelorge, L. (eds). La Documentation française, Paris.

Valensi, L. (1995). Histoire nationale, histoire monumentale : les lieux de mémoire (note critique). *Annales : histoire, sciences sociales*, 50(6), 1271–1277 [Online]. Available at: https://www.persee.fr/doc/ahess_0395-2649_1995_num_50_6_279431.

de Varine, H. (1992 [1978]). L'écomusée. In *Vagues. Une anthologie de la nouvelle muséologie*, Desvallées, A. (ed.). Éditions W/MNES, Mâcon/Savigny le Temple.

de Varine, H. (2013 [1991]). *L'initiative communautaire.* Éditions W/MNES, Mâcon/Savigny-le Temple [Online]. Available at: http://www.hugues-devarine.eu/book/view/25.

de Varine, H. (2017). *L'écomusée singulier et pluriel : un témoignage sur cinquante ans de muséologie communautaire dans le monde.* L'Harmattan, Paris.

Verdier, H., de Massary, X., Coste, G. (2007). *Principes, méthode et conduite de l'Inventaire général du patrimoine culturel*, 2nd edition. Ministère de la Culture et de la Communication, Paris [Online]. Available at: http://www.culture.gouv.fr/culture/dp/inventaire/extranetIGPC/normes/livretPMC/livretPMC_2007.pdf.

Verón, E. (1983). Il est là, je le vois, il me parle. *Communications*, 38, 98–120 [Online]. Available at: https://www.persee.fr/doc/comm_0588-8018_1983_num_38_1_1570.

Veyne, P. (2008). *Foucault : sa pensée, sa personne.* Albin Michel, Paris.

Vivien, F.-D. (2005). Et la nature devint patrimoine… In *Réinventer le patrimoine. De la culture à l'économie, une nouvelle pensée du patrimoine ?* Barrère, C., Barthélémy, D., Nieddu, M., Vivien, F.-D. (eds). L'Harmattan, Paris.

Voisenat, C. (ed.). (1995). *Paysage au pluriel. Pour une approche ethnologique des paysages.* Éditions de la Maison des sciences de l'homme, Paris.

Waterton, E. and Smith, L. (2009). There is no such thing as heritage. In *Taking Archaeology Out of Heritage*, Waterton, E. and Smith, L. (eds). Cambridge Scholars Publishing, Newcastle.

Waterton, E., Smith, L., Campbell, G. (2006). The utility of discourse analysis to heritage studies: The Burra Charter and social inclusion. *International Journal of Heritage Studies*, 12(4), 339–355. doi: 10.1080/13527250600727000.

Winkin, Y. (1996). *Anthropologie de la communication. De la théorie au terrain.* De Boeck, Brussels.

Cited documentaries

Agrigente (2012). Episode of the series *Paysages d'ici et d'ailleurs*. 26 minutes. Presenter: Raphaël Hitier. Director: François Chayé. Arte France, System TV.

Alpes bavaroises (Les) : entre lacs et collines (2021). 52 minutes. Director: Joachim Haupt. ©ZDF/Parnass Film.

Alpujarras (2013). Episode of the series *Paysages d'ici et d'ailleurs*. 26 minutes. Presenter: Raphaël Hitier. Director: Emmanuel Descombes. Arte France, System TV.

Alpujarras (Les) (2021). Episode of the series *Cuisine des terroirs*. 27 minutes. Directors: Alba Vivancos and Marie Viletelle. ©ZDF.

Australie : l'aventure des premiers hommes (2013). 1st episode: *Les grands nomades*, 52 minutes. 2nd episode: *Les grandes inondations*, 52 minutes. Writers and directors: Bentley Dean and Martin Butler (National Documentary Programme). Screen Australia.

Australie, le grand livre des Aborigène (2016). Episode of the series *Enquêtes archéologiques*. 26 minutes. Presenter: Peter Eeckhout. Directors: Agnès Molia and Clémence Lutz. Arte France and Tournez S'il Vous Plaît.

Baie de somme (La) (2014). Episode of the series *Paysages d'ici et d'ailleurs*. 26 minutes. Presenter: Raphaël Hitier. Director: François Chayé. Arte France, System TV.

Bassin minier (2014). Episode of the series *Paysages d'ici et d'ailleurs*. 26 minutes. Presenter: Raphaël Hitier. Director: Jean Michel Vennemani. Arte France, System TV.

Bocage normand (Le) (2012). Episode of the series *Paysages d'ici et d'ailleurs*. 26 minutes. Presenter: Raphaël Hitier. Director: Jean-Michel Vennemani. Arte France, System TV.

Bulgarie, danser sur les braises ardentes (2019). 43 minutes. Director: Sibylle Smolka. MDR.

Calabre (La) (2022). Episode of the series *Cuisine des terroirs*. 27 minutes. Director: Caroline Haertel. ©ZDF.

Cantu in paghjella profane et liturgique de Corse de tradition orale (Le) (2009). 10 minutes. Director: Michele Guelfucci-Glinatsis. Association Cantu in paghjella [Online]. Available at: https://www.youtube.com/watch?v=FU4_MCEjUlM.

Carnaval de Schignano, étrange et archaïque (Le) (2020). Episode of the series *360° Geo Reportage*. 43 minutes. Director: Alessandro Cassigoli. Arte France, System TV.

Cinque terre (2014). Episode of the series *Paysages d'ici et d'ailleurs*. 26 minutes. Presenter: Raphaël Hitier. Director: Jean-Michel Vennemani. Arte France, System TV.

Côte d'opale (La) (2012). Episode of the series *Paysages d'ici et d'ailleurs*. 26 minutes. Presenter: Raphaël Hitier. Director: Emmanuel Descombes. Arte France, System TV.

Engagés pour l'Inn, une rivière alpine revitalisée (2021). Episode of the series *Des mains vertes pour la nature*. 52 minutes. Director: Michaël R. Gärtner. ZDF-Längengrad Filmproduktion GmbH.

Frioul, une vallée tout en musique (2017). Episode of the series *360° Geo Reportage*. 53 minutes. Director: Gordian Arneth. Arte France, Medien Kontor.

Île de Chios, en Grèce (L') (2018). Episode of the series *Cuisine des terroirs*. 27 minutes. Directors: Elke Sasse and Chrysanthi Goula. ©ZDF.

Japon, aux racines du soleil (2018). 1 hour 31 minutes. Writer: Thierry Aguila. Director: Xavier Lefebvre. Gedeon Programmes/NHK/ZDF.

Kihnou : Une île d'un autre temps (2018). 44 minutes. Director: Julia Finkernagel. Kelvinfilm/MDR.

Moutons blancs, nez noirs : les chouchous du Valais (2022). 43 minutes. Director: Antje Christ. ©ZDF.

Notre-Dame de Paris, le chantier du siècle (2022). 1/3: *La quête de la hauteur*, 53 minutes. 2/3: *L'Harmonie des forces*, 53 minutes. 3/3: *La fabrique du sacré*, 53 minutes. Director: Vincent Amouroux. Arte France, ZED, CNRS Image, EPND, INRAP.

Paghjella (A) (1994). Émission FR3 Corse. Série Territoires, 51 minutes. Presenter: Pierre-Jean Luccioni. FR3 [Online]. Available at: https://www.youtube.com/watch?v=GjdHqnaxPKg.

Patrimoine immatériel, La peau rouge. (2010). Director: Charlie Buffet. ZED (Available on SLICE FR).

Paysages d'Europe (2020). 1/2: *Montagnes, forets et marais*, 43 minutes. 2/2: *Champs, landes et côtes*, 43 minutes. Director: Jan Haft. Nautilus film.

Péloponnèse, la pasionaria de l'agroforesterie (2022). Episode of the series *360° Geo Reportage*. 32 minutes. Director: Marie-Laure Daguenel. Arte France, Medien Kontor.

Pont d'herbes tressées de Q'eswachaka (Le) (2022). 43 minutes. Director: Michael Schumacher. ©ZDF.

Rituels du monde (2019). Documentary series with 15 26 minute episodes. Presenter: Anne-Sylvie Malbrancke. Directors: Agnès Molia with Marie-France Barrier, Nathalie Laville, Mikaël Lefrançois, Thibaud Marchand, Hélène Maucourant, Olivier Mirguet, Jacques Plaisant, Camille Robert and Amira Souilem. Arte France and Tournez S'il Vous Plaît.

Roots (2022). Series with five 33 minute episodes. 1/5: *Désertification : extension du domaine de la lutte.* Directors: Stenka Quillet and Marc Meillassoux. 2/5: *De la Terre mère au sol vivant.* Director: Anaïs van Ditzhuysezen. 3/5: *Le bétail, un bien ou un mal pour le climat ?* Directors: Frédérique Veith and Marc Meillassoux. 4/5: *Biodiversité, saveur et résilience.* Director: Frédérique Veith. 5/5: *Carbon farming la culture du carbone.* Director: Stenka Quillet. Moderation and screenplay: Pierre Girard. Labo M/SR.

Saisir un patrimoine immatériel (2011). 53 minutes. Director: Laurence Scarbonchi, in partnership with the University of Lyon. Cap Canal/C Productions Chromatiques.

Sauvons nos forêts. Les héros des bois (2019). Documentary series with five 26 minute episodes. 1/5: *Préserver le climat.* 2/5: *Lutter contre l'industrie.* 3/5: *Cultiver la biodiversité.* 4/5: *Combattre la déforestation illégale.* 5/5: *Laisser faire la nature.* Directors: Vivien Pieper and Johannes Bünger. Doc|Station, WDR, Arte France.

Secrets de l'indigo (Les) (2019). 43 minutes. Director: Janina Möbius. ©ZDF/Medea Film.

Secrets du Parthénon (Les) (2007). 79 minutes. Directors: Gary Glassman and Christine Le Goff, with the assistance of Jean-Luc Delbat. Arte France, Studio International, Providence Pictures.

Sicile (La) (2021). Episode of the series *Cuisine des terroirs.* 27 minutes. Director: Elke Sasse. ©ZDF.

Trésors du patrimoine équestre (Les) (2019). 1 hour 29 minutes. Directed by Céline Gualde. Martange Production, FR5.

Un monde nouveau (2022). Series with five 53 minute episodes. 1/3: *Résister.* 2/3: *S'adapter.* 3/3: *Régénérer.* Writer, director and presenter: Cyril Dion. Co-writer: Valérie Rossellini. Arte France, Galaxie, Le cinquième rêve.

Val d'Orcia (2014). Episode of the series *Paysages d'ici et d'ailleurs*, 26 minutes. Presenter: Raphaël Hitier. Director: François Chayé. Arte France, System TV.

Glossary

Concepts marked with an asterisk correspond to a glossary entry.

Actant: An operator that performs or is the subject of an operation (an act or an action), and that can be activated by an individual or collective actor, human or nonhuman.

Actant subject: An actant* established in order to perform an action, undergo tests and ultimately be "sanctioned" (i.e. recognized as having passed or not passed the tests).

Actantial modality: See actantial role.

Actantial role: The place of actors endowed with competences* (e.g. the modalities of wanting to do, being able to do and knowing how to do in the case of activation of the actant subject*) that enables them to activate the operations specific to an actant*.

Actantial status: See actantial role*.

Adjuvant (actant-): An actant* facilitating an action that helps an actant-subject* to perform an action, in particular by conferring a competence upon it.

Administrative legal (heritagization): A heritagization mode* characterized by the implementation of national or international protection measures, chiefly concerning material objects, buildings or territories (e.g. museum objects, monuments, archaeological sites, natural sites or urban complexes).

Allographic immanence: Immanence* occurring in an ideal object* that must be manifested in the form of a document, recitation, printed material, performance, ritual, etc., in order to possess social instantiation*.

Audience: An ephemeral collective composed of individuals occupying the position of actant receiver* of a *dispositif** of communication.

Authenticity: In the heritagization* of material objects, the numerical identity* of the heritage object* over time which makes it an index* of its relationship to its world of origin. A relationship that is established and guaranteed by knowledge about its production and its future.

Autographic immanence: Immanence* based on knowledge about the production of the object that thus makes it possible to establish that the object has a numerical identity*.

Competence: A capacity given to an actant-subject* either to perform an action or to communicate, conferred by the sender or by the adjuvant, especially in the form of modalities (wanting to do, knowing how to do, being able to do).

Complex heritage entity: An entity that has been heritagized either by synecdoche (one heritage object conferring patrimoniality upon the entity as a whole), or by declaration* from a legitimate national or international authority, or by media heritagization*; these three modalities can coexist.

Conducive representational environment: The institutional, social, cognitive and affective environment that is the condition of possibility for the manifestation of a social interest (gesture of rediscovery*) in heritagizable traces, in particular through the existence and circulation of heritage interpretive schemes*.

Culturalization: The transforming into a cultural offering of diverse productions (such as exhibitions) in heritage institutions as part of the development of publicization* for one or more "audiences"*, without turning it into (or before turning it into) a tool at the service of a group or a community.

Declaration (gesture): The operation by which a legitimate authority confers upon a thing its status as a heritage object*.

Depiction: A process of representation* in a media text* by inscription.

Dispositif: A technical or organizational object (device) which, technically and symbolically, contributes to linking "discourses, institutions, architectural constructions, regulatory decisions, laws, administrative measures, scientific utterances, and philosophical, moral or philanthropic proposals" (Foucault) with a view to turning them into a social element exerting a power[1].

---

1 Since English has no word to satisfactorily render the link between the two dimensions (technical and societal), I opted to retain the French term.

Enunciative event: The enunciation of a media text* that introduces a change in how a social phenomenon is conceived.

Experience (actors): The lived relationship of actors to an object or situation that brings knowledge or a competence*.

Experiment (institutions): The modification of a state brought about by the actions of a heritage institution, introducing a new way of approaching phenomena.

Faithfulness: In the heritagization of a social process*, the maintenance* of both the principles and the spirit of the culture of origin (in the anthropological sense), both by the mobilization of memory (knowledge and transmitted competences) and by appropriate adjustments relative to the circumstances and constraints of a new implementation.

Figure: The recognition of an object as a sign, resulting from a process of representation*.

Generic identity: The fact that heritage objects* belonging to the same genre retain common characteristics.

Gestures (of heritagization): The operations composing the scenario of the transformation of a thing into a heritage object*: the gestures of rediscovery*, production of knowledge*, declaration* or recognition*, maintenance* and publicization*.

Heritage: Unless otherwise specified, this category does not refer here to the movable or immovable property transmitted within the family or in settlement, but to what is usually called "cultural heritage" or "natural heritage", that is, resulting from the process of heritagization*.

Heritage interpretive scheme: Traits of scholarly knowledge or circulating representations, drawn from the triviality of media texts*, more or less seen as contributing to the ensuing representation, which function, for the receivers of these texts, as indices that a thing (material object, process or complex entity) has a heritage value* justifying its declaration or simply its recognition* as a heritage object*. They thus serve as a basis for heritagization*, especially media heritagization*.

Heritage object: The social and symbolic status of a thing at the culmination of its heritagization*.

Heritage value: The quality of being bearer of a relationship to the past that is recognized in the heritage object* or object to be heritagized, taking into account its characteristics or those of the social circumstances of its production and its future, based on attachment, knowledge or memory.

Heritagization: The process, formalizable in a series of operations (or "gestures"), namely rediscovery*, production of knowledge*, recognition* or declaration*, maintenance* and publicization*, by which a thing acquires the social and symbolic status of heritage, thus becoming a heritage object*.

Heritagization mode: A way of producing heritage objects* from a certain type of thing and dependent on a given conducive representational environment*; the whole relative to the state of society at the time in question. Four heritagization modes are easily identifiable today: administrative legal heritagization* (what we could call "institutional" or "traditional"), social heritagization*, heritagization of social processes* (what we could call "intangible heritage") and media heritagization* (easily identifiable in the case of social processes or complex entities, though the procedure used by this mode is by no means limited to the latter two cases). These four modes are not exclusive; although they arrived successively in time, they henceforth overlap and combine.

Ideal object: In allographic immanence*, the ideal object is the set of constituent properties that is established on the basis of its manifestations (e.g. implementations of the social process*) and its denotations (cultural knowledge through the production of scientific knowledge or sourced from collective memory).

Immanence: Every heritage object possesses, conjointly, two modes of immanence, autographic* and allographic*. Unlike a work of art, which is governed by one or the other.

Index: Recognition through interpretation of the characteristics of a thing (object or process) that establishes a trace* of the past as a heritage object*.

Inscription: Formatting a medium, making it interpretable as writing.

Instance: The institutional operator of a given operation.

Interpretant: A "sign"[2] that gives a meaning for the sign that is being considered by the interpreter*, a meaning that results from the fact that this interpretant sign can itself be interpreted by another sign, and so on.

2 The inverted commas indicate that this term, reductive in view of the complexity of the semiotic process involved, is used here for convenience.

Interpreter: An actor who undertakes to find out the meaning of a thing by mobilizing an interpretant*. The "model interpreter" is the position and function of the interpreter provided by a media text*.

Line of force: See tipping point*.

Logic: The rationale governing the organization or functioning of a phenomenon (object, process, relationship, etc.).

Maintenance (gesture of): An operation that consists of safeguarding (i.e. keeping alive) a heritage object (or entity), either by conservation (e.g. a museum object), protection (e.g. a monument), reiteration (e.g. a process*), or by maintaining a continuity by means of diverse simultaneous operations (e.g. a complex heritage entity*).

Media heritagization: A heritagization mode* characterized by the circulation, interweaving and interpretation of media texts* conveying knowledge produced by scientists, by actors and (or) by heritage administrations that will generate heritage interpretive schemes* enabling the recognition of heritage characters in things either already heritagized according to another mode or that will henceforth be recognized as heritage.

Media text: A semiotic process that circulates in a media *dispositif** in documentary form.

Mediation: A communicational act through which an object is socially identified, interpreted and shared.

Mediatized mediation: An act of mediation* effected by a material and technical *dispositif**.

Modalization: The phrasing of an enunciation intended to signify the enunciator's position with regard to what they are communicating.

Mode: The state at a given moment of a social and semiotic functioning resulting from diverse transformations: interpretation modes*, heritagization modes*, etc.

Model viewer: The position and function of the viewer targeted by the media text*.

Modelization: A descriptive formalization that is used in sociosemiotic analysis to reveal the logic of a process through a model (e.g. according to the actantial scenario of the transformation of the socio-symbolic status of a thing).

Numerical identity: The fact that a heritage object* remains “one”, irrespective of the circumstances or its own transformations.

Object (actant-): The actant* on which an action is performed.

Observer (actant-): An actant* in a disengaged position in relation to the phenomenon in which they are participating (e.g. an inhabitant) or which they approach as a tourist for example (process, complex object, media text, etc.) in order to perform a cognitive act on it.

Operativity: The social and symbolic effectiveness of a *dispositif**.

Patrimoniality: The set of the characteristics of an object that make it socially recognized as being heritage.

Performance: The execution of an operation by an actant*.

Point of singularity: A trace of the event on the basis of which a tipping point* occurs.

Production of knowledge gesture: The creation of knowledge and/or the recording of memory about the thing to be heritagized, about its production and about its temporal future, and also about the context of this production and this future.

Publicization gesture: An in situ and ex situ operation through which a heritage object*, together with knowledge, relating to it, to its heritagization*, to heritagization* in general or to heritagization modes*, is disseminated to the group of origin or to the audience*.

Receiver (actant-): In communication, the actant* being addressed by the actant-sender* that operates the interpretation in reception.

Recognition gesture: The operation by which a collective or individual authority, on the basis of knowledge (knowledge produced, observation, memory, interpretive schemes), considers that an object, because of its characteristics and context, has a heritage value* justifying the status of heritage object*.

Rediscovery gesture: The manifestation of an interest (through attachment, curiosity and/or knowledge) in a thing regarded as a trace* of the past potentially endowed with heritage value*.

Representation: A basic process of the semiotic relationship by which a thing that is present is in the place of (or makes reference to) another thing that is absent, according to very different and more or less complex forms and functions, both from a semiotic or media point of view and also from a social point of view. “In representation” means that a thing or an actor is present not physically but in the form of a figure*.

Scholarly representation: Knowledge of a memorial or scientific origin which as an ideal object* is an integral part of the heritage object* and turns it into a media text*, and which can also exist in diverse autonomous documentary forms.

Sender (actant-): An actant* that either establishes a subject with a view to accomplishing an action, or that, in communication, aims to produce a communicational effect (cognitive, affective, social, etc.) on the actant-receiver*.

Social heritagization: A mode of heritagization* characterized by the fact that heritagizable or heritagized things are attached to a social group. Either the group considers them part of its culture, or its members have the memory of them.

Social instantiation: The social categorization of a heritage object according to a reference mode of immanence*, variable according to heritagization mode and society, for example, only autographic immanence of the monument (administrative legal heritagization), or, conversely, only allographic immanence of a heritagized element of culture (heritagization of a social process) category of intangible heritage).

Social process (heritagization of a): A heritagization mode* characterized by the transformation of an element of culture into an element of heritage that presupposes both its continuing implementation and the production of a scholarly representation* leading either to a declaration* as “intangible heritage”, or to recognition* arising from the application of heritage interpretive schemes*.

Sociosemiotic analysis: Description and modeling* of circulating narratives (ordinary, scholarly or media) and situations relating to the acquisition, by things (material or intangible), of the status of heritage object*, and also the various modes* of this acquisition.

Specific identity: This includes the fact that heritage objects* that belong to the same type retain common characteristics (e.g. a similar type of ideal object* in the case of allographic immanence*).

Temps impliqué/temps expliqué[3]: An opposition borrowed from Gustave Guillaume's theory on the "operative time" of the verb and on the genesis of the image of time in discourse (chronogenesis) applied to the "aspect" in the heritagization of social processes*. Here, "*temps impliqué*" corresponds to the idea of the process as it is suggested (for example) by the name of the element of culture: in this, all time is thus condensed. "*Temps expliqué*" is the time of the action separated into present, past or future according to the various moments of implementation of the heritagized process, as they appear in the enunciation of the media texts* that carry out their publicization*.

Tipping point: The moment when a new line of force opens, for example, the appearance of a new heritagization mode* or a variant of an already existing mode, which involves discourses, practices and institutions.

Trace: A material object (or simply material marks) from a more or less recent or distant past, which, because it seems to be endowed with values* and the bearer of indices* of its origin, is likely to be interpreted as constituting heritage. Alongside these heritage traces that are candidates for heritagization*, there are also traces* of the practice of heritagization* itself.

3 Since the opposition between *impliqué/expliqué* here combines two semantic oppositions (one being of the internal/external type, one being of the condensed/deployed type), I have opted to retain the original French terminology.

Index of Author Names

E

F

G

H

I

J, K

T, U

V, W

Other titles from

iSTE

in

Science, Society and New Technologies

2024

BARTHES Angela, CIBIEN Catherine, ROMAGNY Bruno
Biosphere Reserves and Sustainable Development Goals 1: Scientific and Practical Educational Issues in the Mediterranean
(Territory Development Set – Volume 3)
Biosphere Reserves and Sustainable Development Goals 2: Issues, Tensions, Processes and Governance in the Mediterranean
(Territory Development Set – Volume 4)

KOHLMANN Émilie
Linking with Nature in the Digital Age
(Communication, Environment, Science and Society Set – Volume 1)

2023

APCHAIN Thomas
Favela Tours: Building Otherness in the Favelas of Rio de Janeiro
(Tourism and Mobility Systems Set – Volume 6)

CARDON Alain, EL HAMI Abdelkhalak
Fundamental Generation Systems: Computer Science and Artificial Consciousness, the Informational Field of Generation of the Universe, the Sixth Sense of Living Beings
(Digital Sciences Set – Volume 4)

CURULLI Irene G., IKIZ KAYA Deniz, KHAEFI Arghavan
Heritage Education for Climate Action
(Research in Architectural Education Set – Volume 2)

ELAMÉ Esoh
The Sustainable City in Africa Facing the Challenge of Liquid Sanitation
(Territory Development Set – Volume 2)

JURCZENKO Emmanuel
Climate Investing: New Strategies and Implementation Challenges

LYUBAREVA Inna, WALDECK Roger
Diversity of Methodological Approaches in Social Sciences: Example of the Analysis of Media and Online Information
(Modeling Methodologies in Social Sciences Set – Volume 3)

PARET Dominique, CRÉGO Pierre, SOLÈRE Pauline
Smart Patches: Biosensors, Graphene and Intra-Body Communications

VENTRE Daniel, GUILLOT Philippe
Electronic Communication Interception Technologies and Issues of Power

VENTRE Daniel, LOISEAU Hugo
Cybercrime during the SARS-CoV-2 Pandemic (2019-2022): Evolutions, Adaptations, Consequences
(Cybersecurity Set – Volume 3)

2022

AIT HADDOU Hassan, TOUBANOS Dimitri, VILLIEN Philippe
Ecological Transition in Education and Research

CARDON Alain
Information Organization of The Universe and Living Things: Generation of Space, Quantum and Molecular Elements, Coactive Generation of Living Organisms and Multiagent Model
(Digital Sciences Set – Volume 3)

CAULI Marie, FAVIER Laurence, JEANNAS Jean-Yves
Digital Dictionary

DAVERNE-BAILLY Carole, WITTORSKI Richard
Research Methodology in Education and Training: Postures, Practices and Forms
(Education Set – Volume 12)

ELAMÉ Esoh
Sustainable Intercultural Urbanism at the Service of the African City of Tomorrow
(Territory Development Set – Volume 1)

FLEURET Sébastien
A Back and Forth Between Tourism and Health: From Medical Tourism to Global Health
(Tourism and Mobility Systems Set – Volume 5)

KAMPELIS Nikos, KOLOKOTSA Denia
Smart Zero-energy Buildings and Communities for Smart Grids
(Engineering, Energy and Architecture Set – Volume 9)

2021

BARDIOT Clarisse
Performing Arts and Digital Humanities: From Traces to Data
(Traces Set – Volume 5)

BENSRHAIR Abdelaziz, BAPIN Thierry
From AI to Autonomous and Connected Vehicles: Advanced Driver-Assistance Systems (ADAS)
(Digital Sciences Set – Volume 2)

BÉRANGER Jérôme
Societal Responsibility of Artificial Intelligence: Towards an Ethical and Eco-responsible AI
(Technological Prospects and Social Applications Set – Volume 4)

CORDELIER Benoit, GALIBERT Oliver
Digital Health Communications
(Technological Prospective and Social Applications Set – Volume 5)

DOUAY Nicolas, MINJA Michael
Urban Planning for Transitions

GALINON-MÉLÉNEC Béatrice
The Trace Odyssey 1: A Journey Beyond Appearances
(Traces Set – Volume 4)

HENRY Antoine
Platform and Collective Intelligence: Digital Ecosystem of Organizations

KOLOKOTSA Denia, KAMPELIS Nikos
Smart Buildings, Smart Communities and Demand Response
(Engineering, Energy and Architecture Set – Volume 8)

LE LAY Stéphane, SAVIGNAC Emmanuelle, LÉNEL Pierre, FRANCES Jean
The Gamification of Society
(Research, Innovative Theories and Methods in SSH Set – Volume 2)

RADI Bouchaïb, EL HAMI Abdelkhalak
Optimizations and Programming: Linear, Non-linear, Dynamic, Stochastic and Applications with Matlab
(Digital Sciences Set – Volume 1)

2020

ALAKTIF Jamila, CALLENS Stéphane
Migration and Climate Change: From the Emergence of Human Cultures to Contemporary Management in Organizations

BARNOUIN Jacques
The World's Construction Mechanism: Trajectories, Imbalances and the Future of Societies
(Interdisciplinarity Between Biological Sciences and Social Sciences Set – Volume 4)

ÇAĞLAR Nur, CURULLI Irene G., SIPAHIOĞLU Işıl Ruhi, MAVROMATIDIS Lazaros E.
Thresholds in Architectural Education
(Engineering, Energy and Architecture Set – Volume 7)

DUBOIS Michel J.F.
Humans in the Making: In the Beginning was Technique
(Social Interdisciplinarity Set – Volume 4)

ETCHEVERRIA Olivier
The Restaurant, A Geographical Approach: From Invention to Gourmet Tourist Destinations
(Tourism and Mobility Systems Set – Volume 3)

GREFE Gwenaëlle, PEYRAT-GUILLARD Dominique
Shapes of Tourism Employment: HRM in the Worlds of Hotels and Air Transport
(Tourism and Mobility Systems Set – Volume 4)

JEANNERET Yves
The Trace Factory
(Traces Set – Volume 3)

KATSAFADOS Petros, MAVROMATIDIS Elias, SPYROU Christos
Numerical Weather Prediction and Data Assimilation
(Engineering, Energy and Architecture Set – Volume 6)

MARTI Caroline
Cultural Mediations of Brands: Unadvertization and Quest for Authority
(Communication Approaches to Commercial Mediation Set – Volume 1)

MAVROMATIDIS Lazaros
Climatic Heterotopias as Spaces of Inclusion: Sew Up the Urban Fabric
(Research in Architectural Education Set – Volume 1)

MOURATIDOU Eleni
Re-presentation Policies of the Fashion Industry: Discourse, Apparatus and Power
(Communication Approaches to Commercial Mediation Set – Volume 2)

SCHMITT Daniel, THÉBAULT Marine, BURCZYKOWSKI Ludovic
Image Beyond the Screen: Projection Mapping

VIOLIER Philippe, with the collaboration of TAUNAY Benjamin
The Tourist Places of the World
(Tourism and Mobility Systems Set – Volume 2)

2019

BRIANÇON Muriel
The Meaning of Otherness in Education: Stakes, Forms, Process, Thoughts and Transfers
(Education Set – Volume 3)

DESCHAMPS Jacqueline
Mediation: A Concept for Information and Communication Sciences
(Concepts to Conceive 21st Century Society Set – Volume 1)

DOUSSET Laurent, PARK Sejin, GUILLE-ESCURET Georges
Kinship, Ecology and History: Renewal of Conjectures
(Interdisciplinarity Between Biological Sciences and Social Sciences Set – Volume 3)

DUPONT Olivier
Power: A Concept for Information and Communication Sciences
(Concepts to Conceive 21st Century Society Set – Volume 2)

FERRARATO Coline
Prospective Philosophy of Software: A Simondonian Study
(Technological Prospects and Social Applications Set – Volume 3)

GUAAYBESS Tourya
The Media in Arab Countries: From Development Theories to Cooperation Policies

HAGÈGE Hélène
Education for Responsibility
(Education Set – Volume 4)

LARDELLIER Pascal
The Ritual Institution of Society
(Traces Set – Volume 2)

LARROCHE Valérie
The Dispositif: A Concept for Information and Communication Sciences
(Concepts to Conceive 21st Century Society Set – Volume 3)

LATERRASSE Jean
Transport and Town Planning: The City in Search of Sustainable Development

LELEU-MERVIEL Sylvie, SCHMITT Daniel, USEILLE Philippe
From UXD to LivXD: Living eXperience Design

LENOIR Virgil Cristian
Ethically Structured Processes
(Innovation and Responsibility Set – Volume 4)

LOPEZ Fanny, PELLEGRINO Margot, COUTARD Olivier
Local Energy Autonomy: Spaces, Scales, Politics
(Urban Engineering Set – Volume 1)

METZGER Jean-Paul
Discourse: A Concept for Information and Communication Sciences
(Concepts to Conceive 21st Century Society Set – Volume 4)

MICHA Irini, VAIOU Dina
Alternative Takes to the City
(Engineering, Energy and Architecture Set – Volume 5)

PÉLISSIER Chrysta
Learner Support in Online Learning Environments

PIETTE Albert
Theoretical Anthropology or How to Observe a Human Being
(Research, Innovative Theories and Methods in SSH Set – Volume 1)

PIRIOU Jérôme
The Tourist Region: A Co-Construction of Tourism Stakeholders
(Tourism and Mobility Systems Set – Volume 1)

PUMAIN Denise
Geographical Modeling: Cities and Territories
(Modeling Methodologies in Social Sciences Set – Volume 2)

WALDECK Roger
Methods and Interdisciplinarity
(Modeling Methodologies in Social Sciences Set – Volume 1)

2018

BARTHES Angela, CHAMPOLLION Pierre, ALPE Yves
Evolutions of the Complex Relationship Between Education and Territories
(Education Set – Volume 1)

BÉRANGER Jérôme
The Algorithmic Code of Ethics: Ethics at the Bedside of the Digital Revolution
(Technological Prospects and Social Applications Set – Volume 2)

DUGUÉ Bernard
Time, Emergences and Communications
(Engineering, Energy and Architecture Set – Volume 4)

GEORGANTOPOULOU Christina G., GEORGANTOPOULOS George A.
Fluid Mechanics in Channel, Pipe and Aerodynamic Design Geometries 1
(Engineering, Energy and Architecture Set – Volume 2)
Fluid Mechanics in Channel, Pipe and Aerodynamic Design Geometries 2
(Engineering, Energy and Architecture Set – Volume 3)

GUILLE-ESCURET Georges
Social Structures and Natural Systems: Is a Scientific Assemblage Workable?
(Social Interdisciplinarity Set – Volume 2)

LARINI Michel, BARTHES Angela
Quantitative and Statistical Data in Education: From Data Collection to Data Processing
(Education Set – Volume 2)

LELEU-MERVIEL Sylvie
Informational Tracking
(Traces Set – Volume 1)

SALGUES Bruno
Society 5.0: Industry of the Future, Technologies, Methods and Tools
(Technological Prospects and Social Applications Set – Volume 1)

TRESTINI Marc
Modeling of Next Generation Digital Learning Environments: Complex Systems Theory

2017

ANICHINI Giulia, CARRARO Flavia, GESLIN Philippe, GUILLE-ESCURET Georges
Technicity vs Scientificity: Complementarities and Rivalries
(Interdisciplinarity Between Biological Sciences and Social Sciences Set – Volume 2)

DUGUÉ Bernard
Information and the World Stage: From Philosophy to Science, the World of Forms and Communications
(Engineering, Energy and Architecture Set – Volume 1)

GESLIN Philippe
Inside Anthropotechnology: User and Culture Centered Experience
(Social Interdisciplinarity Set – Volume 1)

GORIA Stéphane
Methods and Tools for Creative Competitive Intelligence

KEMBELLEC Gérald, BROUDOUX Evelyne
Reading and Writing Knowledge in Scientific Communities: Digital Humanities and Knowledge Construction

MAESSCHALCK Marc
Reflexive Governance for Research and Innovative Knowledge
(Responsible Research and Innovation Set - Volume 6)

PARK Sejin, GUILLE-ESCURET Georges
Sociobiology vs Socioecology: Consequences of an Unraveling Debate
(Interdisciplinarity Between Biological Sciences and Social Sciences Set – Volume 1)

PELLÉ Sophie
Business, Innovation and Responsibility
(Responsible Research and Innovation Set – Volume 7)

2016

ANDRÉ Michel, SAMARAS Zissis
Energy and Environment
(Research for Innovative Transports Set – Volume 1)

BLANQUART Corinne, CLAUSEN Uwe, JACOB Bernard
Towards Innovative Freight and Logistics
(Research for Innovative Transports Set – Volume 2)

COHEN Simon, YANNIS George
Traffic Management
(Research for Innovative Transports Set – Volume 3)

EL FALLAH SEGHROUCHNI Amal, ISHIKAWA Fuyuki, HÉRAULT Laurent, TOKUDA Hideyuki
Enablers for Smart Cities

GIANNI Robert
Responsibility and Freedom: The Ethical Realm of RRI
(Responsible Research and Innovation Set – Volume 2)

GRUNWALD Armin
The Hermeneutic Side of Responsible Research and Innovation
(Responsible Research and Innovation Set – Volume 5)

LENOIR Virgil Cristian
Ethical Efficiency: Responsibility and Contingency
(Responsible Research and Innovation Set – Volume 1)

PELLÉ Sophie, REBER Bernard
From Ethical Review to Responsible Research and Innovation
(Responsible Research and Innovation Set – Volume 3)

REBER Bernard
Precautionary Principle, Pluralism and Deliberation: Science and Ethics
(Responsible Research and Innovation Set – Volume 4)

TORRENTI Jean-Michel, LA TORRE Francesca
Materials and Infrastructures 1
(Research for Innovative Transports Set – Volume 5A
Materials and Infrastructures 2
(Research for Innovative Transports Set – Volume 5B)

VENTRE Daniel
Information Warfare – 2nd edition

YANNIS George, COHEN Simon
Traffic Safety
(Research for Innovative Transports Set – Volume 4)

2015

BRONNER Gérald
Belief and Misbelief Asymmetry on the Internet

2009

VENTRE Daniel
Information Warfare

Printed and bound by CPI Group (UK) Ltd, Croydon, CR0 4YY
13/09/2026
14961048-0007